AF335116

INTEGRATING SCIENTIFIC DISCIPLINES

SCIENCE AND PHILOSOPHY

This series has been established as a forum for contemporary analysis of philosophical problems which arise in connection with the construction of theories in the physical and the biological sciences. Contributions will not place particular emphasis on any one school of philosophical thought. However, they will reflect the belief that the philosophy of science must be firmly rooted in an examination of actual scientific practice. Thus, the volumes in this series will include or depend significantly upon an analysis of the history of science, recent or past. The Editors welcome contributions from scientists as well as from philosophers and historians of science.

WILLIAM BECHTEL

editor

Integrating Scientific Disciplines

1986 **MARTINUS NIJHOFF PUBLISHERS**
a member of the KLUWER ACADEMIC PUBLISHERS GROUP
DORDRECHT / BOSTON / LANCASTER

Distributors

for the United States and Canada: Kluwer Academic Publishers, 190 Old Derby
Street, Hingham, MA 02043, USA
for the UK and Ireland: Kluwer Academic Publishers, MTP Press Limited,
Falcon House, Queen Square, Lancaster LA1 1RN, UK
for all other countries: Kluwer Academic Publishers Group, Distribution Center,
P.O. Box 322, 3300 AH Dordrecht, The Netherlands

Library of Congress Cataloging in Publication Data

```
Main entry under title:

Integrating scientific disciplines.

   (Science and philosophy)
   Revisions of papers presented at a conference held at
Georgia State University, May 1984, sponsored by the US
National Endowment for the Humanities.
   Includes index.
   1. Biology--Research--Congresses.  I. Bechtel,
William.  II. National Endowment for the Humanities.
III. Series.
QH315.I645  1986        507.2           85-21740
```

ISBN 90-247-3242-5 (this volume)

This volume is dedicated to the memory of
WILLIAM PEARL FRENCH

Table of Contents

VIII

Preface

Interdisciplinary research has been a popular idea with many people in the last 20 years. Academic administrators have admonished their faculty to become more interdisciplinary. Students often request the chance to pursue an interdisciplinary degree. While the issue of managing interdisciplinary projects has received a fair amount of attention by those interested in science management, interdisciplinary research has received little attention from historians, philosophers or sociologists of science or from scientists themselves. Yet, there are a number of cases within the life sciences where researchers have been actively engaged in endeavors that take them across disciplinary boundaries. These are ripe for investigation by those interested in the process of science.

To provide an in-depth study of some historical or contemporary cases of cross-disciplinary research activity in the life sciences, a conference was held at Georgia State University in May, 1984. This conference was supported by the National Endowment for the Humanities (U.S.A.) through their research conference program. Over a three-day period historians, philosophers, and researchers who were actively engaged in various of the life sciences discussed specific examples of interdisciplinary research and tried to analyze what was needed for successful crossing of disciplinary boundaries. After the conference, each of the participants revised their original presentations, partly in light of the discussion at the conference. The papers in this volume are the fruits of that endeavor. These papers offer sifnigicant insight into a number of cases of cross-disciplinary research in the life sciences. Although there are many more cases that remain to be considered, and a variety of other perspectives that can be taken with respect to these cases, they offer an initial data-base that can help us better appreciate the factors that motivate cross-disciplinary research, the problems faced in pursuing such research, and the outcomes that can be expected.

To provide a framework for these analyses, I have included an introductory essay

Bechtel, W (ed), Integrating Scientific Disciplines. ISBN 90-247-3242-5.
© *1986, Martinus Nijhoff Publishers, Dordrecht. Printed in The Netherlands.*

in which I explore a number of basic conceptual and social issues that are relevant for understanding cross-disciplinary research. This introduction draws from much of the recent scholarship in the history, philosophy, and sociology of science. Rather than setting forth a definitive view of my own, I have adopted an eclectic perspective and have tried to show how considerations drawn from a number of orientations can be useful in understanding the character of cross-disciplinary research.

Many people and organizations have made major contributions to the development of this volume. The National Endowment for the Humanities and Georgia State University contributed each major financial support that made the initial conference possible. The authors whose papers are included in this volume have all exhibited their interest and commitment to the study of the nature of cross-disciplinary research. Robert Arrington and Adele Abrahamsen, both of Georgia State University, were of tremendous assistance in developing the conference and preparing this volume. I am grateful to these organizations and individuals, as well as the many others who made contributions to this project.

INTRODUCTION

The Nature of Scientific Integration

WILLIAM BECHTEL
*Department of Philosophy, Georgia State University, Atlanta, Georgia
30303–3083, U.S.A.*

We have heard frequent calls in recent years from academic administrators for more interdisciplinary research.[1] Such calls are often accompanied by criticisms of the narrowness and isolation of established academic disciplines. Behind these calls seems to lie the uncritical assumption that being interdisciplinary is itself a virtue. This attitude is well described by Gusdorf (1977):

> "Interdisciplinarity is a very topical subject, as can be seen from the frequency with which the word appears in philosophical debate and academic discussion. Everyone invokes interdisciplinarity; no one dares say a word against it. Its success is all the more remarkable in that even those who advocate this new image of knowledge would find it hard to define. The appeal to interdiscplinarity is seen as a kind of epistemological panacea, designed to cure all the ills the scientific consciousness of our age is heir to" (580).

The positive attitude towards interdisciplinary work is not shared by many scientists, who tend to be more dubious about its virtues:

> "Interdisciplinarity is an ambivalent term in science. Usually, it is discussed in the context of urgent practical problems which manifestly need a team of various specialists to be dealt with more or less effectively. ... For practical problems it is considered valid and unavoidable but for theoretical purposes in science, interdisciplinarity is handled with great caution and even with suspicion. While they pay lip service to the principle, most scientists loop upon their own discipline as either too incomplete or too immature to be coupled with another one. The prevailing attitude seems to be: first disciplinarity before engaging in interdisciplinarity" (De Mey, 1982, p. 140).

Yet, as Chubin (1976) and other sociologists of science have argued, there is a

---

[1]   Appeals to be interdisciplinary are not a totally new phenomenon; they were heard even in earlier centuries. The endeavors of the French Encyclopediaists and the Logical Positivists' proposal for an *International Encyclopedia of Unified Science* are two prominent appeals to weave an integrated picture of the domains of learning. I will discuss the mode of integration proposed for the *Encyclopedia* below.

Bechtel, W (ed), Integrating Scientific Disciplines. ISBN 90-247-3242-5.
© 1986, Martinus Nijhoff Publishers, Dordrecht. Printed in The Netherlands.

4

significant amount of cross-pollination of ideas between scientific disciplines. This
suggests that there may in fact be a great deal of activity that crosses disciplinary
boundaries which does not fit the preconception academic and science
administrators hold when they bemoan the lack of interdisciplinary activity or of
actual scientists when they caution against greater interdisciplinarity.

This circumstance points to the need for further inquiry as to what cross-
disciplinary research should involve.[2] This volume is directed toward satisfying that
need. There have been recent studies directed at the management of interdisciplinary
research projects (see, for example, Barth and Steck, 1979, Epton, Payne, and
Pearson, 1983). The studies in this volume are directed at a different level–at
identifying (1) problems within the life sciences that have lent themselves or seem
ripe for interdisciplinary endeavors, (2) the cognitive frameworks that have guided
or are guiding cross-disciplinary endeavors, and (3) the results that have come or
can be expected from such endeavors.

Within the basic research areas of the life sciences there are numerous active areas
where cross-disciplinary research is now being pursued, three of which will be
examined in subsequent parts of this volume. To provide some perspective on these
current pursuits, though, the volume will begin with two historical cases. As a
reference point for the discussions that follow, I offer below a very brief overview
of the five areas in which these cross-disciplinary endeavors have developed and
indicate the kind of cross-disciplinary work in these areas that is discussed in later
papers.

1. *The Coming Together of Biochemistry*. Between 1900 and 1940 investigators
involved in a number of endeavors in chemistry and the bio-medical sciences worked
out the sequence of chemical reactions which constituted the pathways of
intermediary metabolism, of which the citric acid cycle (Krebs' Cycle) is perhaps the
best known exemplar. These pathways provided part of the subject matter of what
became increasingly a separate discipline. However, the researchers who initially
developed these pathways came from a variety of disciplines at a time when it was

---

[2]   Sometimes a distinction is made between interdisciplinary work, which is taken to involve the
direct interaction of representatives of different disciplines, and multi-disciplinary work, which lacks
such interaction (see Birnbaum, 1983). While that may be an important consideration in determining the
mode of management of practical projects, it will not figure centrally in this study which is directed more
at the cognitive aspects of crossing discipilinary boundaries. I will follow Epton, Payne, and Pearson
(1983) in using the term ''cross-disciplinary'' for endeavors that require contributions from more than
one discipline, without specifying the kind of research unit carrying out these endeavors.

still uncertain biochemistry would become an autonomous discipline. At that period, biochemistry was still a cross-disciplinary endeavor.

2. *Dobzhansky's Contribution to the Evolutionary Synthesis.* What became known as the synthetic theory of evolution integrated the endeavors of various disciplines within biology, including genetics, systematics, paleontology, and botany. Part of the task in achieving this synthesis was to bring together analyses at different levels (those of genes, organisms, and species) to provide a coherent account of the origin of species. Additionally, experimental and field studies at these levels had to be integrated with abstract mathematical accounts offered by theoreticians. Dobzhansky played a central role in both of these endeavord that resulted in a fertile cross-disciplinary research enterprise.

3. *Incorporating Developmental Biology into the Evolutionary Synthesis.* As broad as the synthetic theory of evolution was, it did not assign a prominent role to developmental biology. However, recent theorizing about the mechanisms of development, especially those having to do with the control of development, suggest perspectives that have consequences for other disciplines included in the synthesis. Hence, there are arguments for expanding the synthesis still further.

4. *Extending Cognitive Science.* During the 1960s and 1970s cognitive science emerged as an active area of cross-disciplinary research, incorporating artificial intelligence and cognitive psychology most centrally, and linguistics, neuroscience, anthropology, and philosophy more peripherally. Researchers from these disciplines have collaborated in an endeavor to explain higher processes such as perception, cognition and language. However, the focus has been almost exclusively on the information processing capacities of human adults, raising the question as to whether additional insight can be gained by studying the neurological substrates of cognition or by studying other species.

5. *Infusing Cognitive Approaches into Animal Ethology.* Although interest in the mental processes of nonhuman animals was widespread at the turn of the century, the excesses of an uncritical anthropomorphic stance and the rise of radical behaviorism led to the virtual elimination of research on the cognitive processes of animals. More recently, a number of comparative psychologists and ethologists have adopted the cognitive perspective which was first developed in human psychology, and so have begun investigating the mental processes of various non-human animals. Some researchers have adopted a cognitive approach to animal ethology in hopes of illuminating the animal studies, while others see opportunities to gain further insight into human psychology.

The main endeavor of this volume is to offer an empirical inquiry into actual cases

6

of cross-disciplinary work. These areas present a variety of patterns of cross-disciplinary research upon which to build an analysis of what cross-disciplinary research involves and can produce in the life sciences. A data-base of five cases, each approached only from a limited number of perspectives, however, is not adequate to draw definitive conclusions about the character of cross-disciplinary research. The hope is that this analysis can provide the basis for further investigations.

This introduction is intended to provide a framework for these case studies. As indicated above, the studies in this volume are aimed at the cognitive endeavors of those engaged in cross-disciplinary research–the problems that lend themselves to cross-disciplinary investigation, the frameworks that have or are being pursued in these investigations, and the products that have resulted. However, as numerous social studies of the sciences have made clear, science does not work in a social vacuum, and even these cognitive concerns are affected by the social character of the scientific investigation and the social context in which science is pursued. Thus, in this introduction I will indicate both cognitive and social factors that are critical to understanding the development of cross-disciplinary research. I will focus on five questions:

A. What are various factors that characterize the basic units of science for which I have adopted the term "discipline"?
B. What factors have led to the isolation and "ethnocentrism" of disciplines?
C. What kinds of concerns have led scientists to cross disciplinary boundaries and seek assistance in other disciplines?
D. What kinds of institutional products emerge from research that crosses disciplinary boundaries, and what are the social/institutional factors affecting the formation of these institutional arrangements?
E. What kinds of conceptual products are developed by scientists when they cross disciplinary boundaries?

Although I will indicate the general conclusions from some of the case analyses that appear in the papers that follow, I will not develop them in detail at this point. Following each set of papers I have included an Editor's Commentary in which I comment specifically on the features of cross-disciplinary research that are revealed by those papers and draw some comparisons with papers included in other parts of this volume.

1. The Units of Science

In order to discuss cross-disciplinary endeavors in science, we need first to clarify what disciplines are and how they figure in the activities of science. The term "discipline" comes from ordinary parlance and is used to characterize the basic branches of science. However, when one goes beyond ordinary discourse and tries to develop a careful analysis of the functioning of science, one discovers the need to break up the scientific enterprise in a variety of different ways, all of which capture aspects of what we initially construed as disciplines. Thus, historians, philosophers, and sociologists of science have introduced a variety of terms for what they see as the basic units or branches in the scientific enterprise (e.g., disciplines, fields, domains, paradigms, research programmes, scientific specialties, research groups, research networks). I will continue to use the term "discipline" for the basic units that are to be bridged in cross-disciplinary endeavors, but I will not try to settle what is the proper analysis of these basic units. Rather, I lay out the space of possibilities for characterizing them. This is a necessary prolegomena to identifying the factors that contribute to the segregation of science into isolated units and to analyzing what is involved in successful integration of these units.

Our discussion of disciplines is made more complex by the fact that even in ordinary parlance we differentiate disciplines at a variety of levels of specificity. We identify biology itself as a discipline, distinguishing it from chemistry and psychology, for example. But we also treat some of the divisions of biology, such as physiology, as disciplines. Physiology can itself be broken down into subunits, such as cardiovascular physiology, that are also counted as disciplines; within cardiovascular physiology there are different groups of scientists working on different research problems. For different types of analyses, units at different levels in this hierarchy become relevant. For example, the least specialized unit, biology, is relevant to issues concerning the inculcation of values in undergraduate education, whereas it would be the unit focused on the specific problem area that would be most germane to understanding daily research activities. The level at which we wish to identify units will affect which criteria will be most relevant to defining them. Similarly, the level at which one is trying to bridge or integrate disciplines will partly determine the kinds of problems cross-disciplinary work will present.

Since cross-disciplinary endeavors can occur between units at any of these levels of specificity, I will not restrict myself to any one level. Rather, I will look at a variety of charcteristics of disciplines, recognizing that some of these will only be applicable at particular levels of specificity. The criteria that scholars have proposed in their attempts to define the basic units of science fall on three major dimensions:

8

(1) the objects studied; (2) the cognitive activities involved; and (3) the social and institutional organization. I will briefly explore how criteria from each of these dimension have been used to characterize the basic units of science. To facilitate this discussion, I will initially ignore the historical character of these units of science, but will return to an historical perspective below.

Objects of Study

Perhaps the most common way in which people describe disciplines is in terms of the things they study. Thus, at the global level, astronomy is construed as studying suns, planets, and the like, while biology studies living things and psychology studies mental activities or behaviors. Shapere captures this feature of our ordinary conception of disciplines when he introduces the term "domain" for "the set of things studied in an investigation" (Shapere, 1984a, p. 320; for Shapere's classic treatment of domains, see Shapere, 1974). Shapere's conception of a domain, however, is much more sophisticated than the common sense idea, for he argues that domains are not simply presented to us. Rather, scientists must decide what items (Shapere's term for the constituents of domains) to group together in a domain. Thus, he shows how during the 19th century chemists made basic elements a domain for study because they recognized these elements as the constituent parts of ordinary substances (see Shapere, 1984b).

Shapere resists the temptation to set forth once and for all general standards of how objects must be related in order to belong to the domain of a common discipline. He maintains that the kinds of reasons scientists may offer for grouping objects together in a common inquiry change as science evolves. Without denying this claim of Shapere's, though, we can still point to some of the common kinds of relations that serve to group items into a common domain. One prominent relationship that might bind objects into a domain would be a causal relationship, which has been demonstrated by previous investigation, or is presupposed in the specification of a research problem. The causal relationship may be interaction between domain objects (e.g., atoms causally interacting in a molecular unit), or it may be a common kind of interaction with other objects (e.g., codons that can substitute for each other in DNA chains).

In addition to looking at relationships that group items into a common domain, one can focus on the kinds of relationships that tend to separate items into different domains. Parts of entities, for example, often fall into a different domain than the whole they constitute. While the parts of a whole tend to interact mostly with one another, the whole itself tends to interact with other wholes, thus allowing us to distinguish a hierarchy of organized systems. An incomplete example of a part-whole hierarchy for living organisms would be: atom, molecule, cell, organ, organism, social group, species. Wimsatt (1976) argues that such hierarchies are not just defined within local systems, but can be generalized across nature so as to yield a generalized hierarchy of levels of organization. There are problems with completely explicating this notion of levels in nature and at present it remains largely an intuitive one.[3] Even left at this intuitive level, though, the notion of levels in nature offers one suggestive way of cutting up the domains of disciplines. As the papers of Darden and Bechtel in this volume suggest, an important kind of cross-disciplinary work involves relating objects at different levels of organization in nature.

The notion of levels is not enough to divide up the domains of science. There are generally multiple domains at a given level, for the following reasons among others. First, there are particular local networks of interaction at any given level, which generate different domains. For example, at the level of bodily organs, the heart and the processes in which it participates constitute the domain studied by cardiovascular physiology, whereas other organs (at the same level) fall within the domains of such other disciplines as neurophysiology, nephrology, and the like. Second, even the same entities may participate in different phenomena, and hence be involved in more than one domain. Toulmin makes this point in commenting on Shapere's notion of domains:

> "If we mark sciences off from one another (using Shapere's term) by their respective 'domains', even these 'domains' have to be identified, not by the types of objects with which they deal, but rather by the questions which arise about them. Any particular type of object will fall in the domain of (say) 'biochemistry' only in so far as it is a topic for correspondingly 'biochemical' questions; and the same type of object will fall within the domains of several difference sciences, depending on what questions are raised about it. The behavior of a muscle fibre, for instance, can fall within the domains of biochemistry, electrphysiology, pathology, and thermodynamics, since questions can be asked about it from all four points of view..." (149).[4]

---

[3]   Wimsatt proposes to differentiate these levels of organization in nature in terms of a network of objects that causally interact with each other in direct ways. Note that in order to define levels in terms of causal interaction, one needs some way of distinguishing direct causal interaction from cases where objects interact only because they are parts of two higher level objects which are interacting. Salmon's (1971, 1984) notion of screening off may provide the best way of doing this. Brandon (1982) has proposed this as a way for differentiating levels at which selection as a causal process occurs.

[4]   This comment should not taken as an objection to Shapere, since Shapere does not require an exclusive division of items into domains.

Third, even where the same entities and same phenomena are involved, several disciplines may address the common domain differently. For example, the disciplines of mainstream social psychology and ethology take very different approaches to studying the phenomena exhibited by social groups. Less dramatically, cardiovascular medicine and cardiovascular physiology take different approaches to the domain defined above.

The objects of study constitute an important part of what characterizes disciplines, but as Shapere's sophisticated analysis of domains already indicated, we cannot identify such sets of objects without considering the factors that lead scientists to group them together. Some of these factors have a basis in nature, but they are also influenced by both the cognitive and social features of disciplines.

Cognitive Activities of a Discipline

This second dimension focuses on the cognitive tools a discipline uses to describe and study the items in its domain. It corresponds with what Whitley (1980) distinguishes as the "intellectual context of research," which consists in

> "that abstracted set of norms and procedures which both govern and constitute what is done to what phenomena, in which cognitive setting, and how it is understood. It consists of the cognitive structures which, on the one hand, represent what is known and, on the other hand, constitute the resources with which to change and develop what is known" (302).

It will be noted that this characterization of the cognitive features of a discipline includes much more than facts and theories, which provided the main focus of attention for more traditional historical and philosophical studies of science. The impetus for recognizing some of these additional features of the cognitive endeavors of science came from Kuhn's (1970) characterization of a paradigm or disciplinary matrix. In this section I will discuss how the following cognitive factors figure significantly in the identity of disciplines: the laws and theories of the discipline; the central problems addressed by the discipline; and the methods by which the problems are addressed.

The notion that disciplines are partly defined by their laws and theories is an old one. For example, physics might be defined as the discipline that includes Newton's laws of motion, genetics as the discipline that includes Mendel's law of segregation, and so forth. This approach of defining disciplines in terms of their theories might have worked fairly well if science had turned out to be cumulative. Then one could define a discipline in terms of its basic laws and view it as evolving through a series of amplifications and applications of these laws. But the theories and laws of a discipline are not permanent. Although only rarely are the fundamental theories and laws of a discipline replaced, they are subject to revision and modification. These changes can alter the work done within the discipline (they may, in fact, make some practitioners obsolete), but one can easily make out a continuity of the discipline through the theory change.[5]

While laws and theories are important charcteristics of disciplines, there are problems that arise if too much emphasis is placed on them. First of all, not all units that we might want to call disciplines are clearly identified with such laws. If one were to take the global unit of biology as the discipline, one would be hard pressed to identify any defining laws. Second, even when one can identify a set of laws or theories with a discipline, such a focus produces a very static conception of disciplines, providing little understanding of the dynamics of disciplines. To capture the dynamics of disciplines at a cognitive level, one might instead focus on the problems that investigators are trying to solve and the ways they go about trying to solve them. It is to capture these aspect of disciplines that Darden and Maull (1977) introduce the notion of a field. They characterize a field in terms of

> "a central problem, a domain consisting of items taken to be facts related to that problem, general explanatory facts and goals providing expectations as to how the problem is to be solved, techniques and methods, and, sometimes, but not always, concepts, laws and theories which are related to the problem and which attempt to realize the explanatory goals" (1977, p. 144).

Shapere (1984a) largely endorses this conception of fields, but cautions that one must be sensitive to the fluidity of fields and to the fact that often different practitioners within a field will not share exactly the same methods. We shall see some of the difficulties to which Shapere is alluding in the discussion that follows.

Focusing on the problems motivating inquiry is particularly appropriate when dealing with the most specific conception of discipline, that concerned with individual research enterprises. These research groups are often directed at specific problems. For example, Morgan's research group was directed towards determining the location of genes on chromosomes while Delbruck's phage group was directed at the question of the molecular unit of heredity. As we move to somewhat larger units, though, it is not clear that there is one defining central problem guiding research or that researchers will share a common conception of the problem.

---

[5] See Lakatos, 1970, for an attempt to account for the continuity of theoretical statements in a discipline through a process of change in terms of a distinction between the hard core and periphery of research programmes. Many scholars remain dubious of the viability of such a distinction.

12

Furthermore, over time researchers solve problems and go on to new ones, but that does not seem to change their disciplinary affilitation.

Toulmin (1973) tries to overcome this difficulty by introducing the notion of a genealogy of problems. To do this one needs some sense of what makes for a continuity of problems, which Toulmin tries to give by considering how problems arise for scientists: they arise as a difference between scientists' aspirations or scientific ideals and their current capacities. For this ultimately to work, one needs an account of how aspirations become focused for a discipline. Determining what are the ideals guiding any particular investigation is extremely difficult, especially with historical cases (see Darden's paper in this volume). Researchers do not publish their goals while conducting their investigations; reports of scientific investigations are written *ex post facto* to reflect what goals were achieved. They may not be the ones with which they initiated the work. With more recent science, where grant support depends on formal proposals, we may be able to turn to these proposals for an indication of the goals researchers hold in undertaking particular investigations. These expressions of goals, though, may be tailored toward the funding agency. (For an intriguing study of how research goals were tailord toward funding potentials from the Rockefeller Foundation, see Kohler, 1977.)

The proposal that one may be able to identify a lineage of problems, nonetheless, is certainly a promising one. At least working backwards through the historical record, one can identify such lineages of research problems. A far more difficult task is to determine what factors determined the succession of problems investigated within a discipline. One factor that makes this task difficult is that there are often a plethora of problems available for investigation at any given time. What needs to be explained is why some problems are found to be more urgent or more promising, and so come to direct the activity of investigators. Several philosophers have tried to provide criteria by which scientific problems may be assessed in terms of the importance to the science itself and the availability of techniques to solve them (see Nickles, 1980, Laudan, 1978, and Shapere, 1984c.) Sociologists have pointed to additional factors that affect how scientists weigh problems, such as availability of support and importance to career development (see Shapin, 1982).

In defining fields, Darden and Maull refer not only to domains and problems, but also to the methods, techniques, and theoretical tools available for use in solving these problems. While these have not played a major role in philosophical analyses in the past, they were given a central role by Kuhn (1970) in his discussion of paradigms or disciplinary matrices. Kuhn draws attention to the fact that science instruction does not consist primarily in the learning of abstract laws and theories, but in learning how to apply these laws and theories to a variety of cases. One becomes "disciplined" so as to perform experiments with a particular design, to use particular kinds of instruments, and to interpret results in a standardized way. These tools commonly differ between disciplines so that the kinds of experiments that one does are influenced by the discipline one was brought up in. Thus, even when two disciplines work on a common problem area, they may have different ways of investigating that domain and interprating their results. Shapin (1982), for example, traces a number of conflicts between disciplines, such as the conflict between Mendelians and biometricians, to differences in methodologies. The way in which researchers in a discipline design studies and interpret their results has been most attended to by sociologists of science who have engaged in "laboratory studies," where they have followed the daily activities in various laboratories. (For examples, see Latour and Woolgar, 1979, and Knorr, 1981.)

In addition to methodologies for investigating nature, disciplines may also be characterized by the "argumentation format"[6] that members of a discipline accept for presentation of their claims in, for example, journal articles. The fact that there are significant differences here can be readily attested to by anyone who has looked into the literature of another discipline and found not just the content to be foreign, but also the structure of the reports. An example of an aspect of methodology that belongs to the argumentation format would be the requirement that tests of statistical significance must be reported in professional psychology journals, with only results surpassing the arbitrarily set .95 level of significance counting as adequate support for a claim. This contrasts with the argumentation format in physiology, where one can report measurements made on a small sample (sometimes one organism) and not carry out statistical evaluations. The argumentation format includes the claims that are assumed within a discipline and need not be stated explicitly. An example would be the use of a technique with a standardized procedure, where one would not report the procedure unless the standard were violated.

There are thus a variety of cognitive factors that are important to the identity of a discipline. Theories and laws are clearly a central factor, but as well there are the problems addressed, the tools for investigating these problems, and the procedures for presenting solutions. While much can be learned about these features of disciplines by a purely cognitive and logical investigation, it is widely recognized that scientists identify problems, apply their research methodology, and present their findings and theories in a social milieu, which needs to be considered as well.

[6] The term as well as the importance of this feature of methodology has been suggested to me by Steve Fuller in personal communication.

14

The Social and Institutional Features of a Discipline

The claim that scientific activity is embedded and influenced by institutional and social factors, and so does not occur purely in Popper's (1972) world 3, has been argued most vigorously by sociologists of science. Sometimes the discussion of social and cognitive factors is framed in terms of exclusive alternatives–was the advance the result of cognitive factors or was it due to social factors? In some cases, philosophers and historians of ideas have been cast as defending the claim that cognitive factors were the determining factors while sociologists and social historians have defended the claim that social and institutional factors were what mattered. This view was nurtured by the controversies surrounding the "strong programme in the sociology of knowledge" (Barnes, 1977, and Bloor, 1976). However, a much less confrontational position is available, one that recognizes the role of social factors and cognitive factors as interacting variables.[7] This potential for interaction is partly created by the fact that much recent sociology of science takes as its conceptual starting point the theses of the underdetermination of scientific theories and the theory ladenness of observations. Sociologists see these theses as giving a place for social negotiation in determining the conceptual direction of science, since according to these theses rational factors do not totally settle the issue (see Shapin, 1982, and Knorr-Cetina and Mulkay, 1983). But the very way in which these theses are advanced points to an important role for rational evaluation as well. Underdetermined theories are ones that meet all the rational considerations for being acceptable scientific theories.

The relationship between social and cognitive factors is many faceted. One aspect of it stems from the fact that only through the social structure of disciplines are the endeavors of individual scientists regulated to insure compliance with accepted principles. Thus, Polanyi introduces "the principle of mutual control," which

---

[7]   The potential for raprochment is found both in the works of many recent sociologists of knowledge, who explicitly recognize a role for cognitive factors, and in recent philosophy of science. In different ways, Toulmin (1972), Shapere (1984), Lakatos (1970) and Laudan (1977), as philosophers of science, have all argued that the rationality of scientific decision making is dependent on features of the scientific community making the decisions. At the same time, sociologists have recognized that they cannot ignore cognitive factors in developing accounts of the social factors influencing the development of science (see Krohn, 1980). Shapin (1982), while arguing that science is developed in a context of "collectively sustained goals", insists that this does not deny the rationality of science: "That rationality is expressed in the instrumental character of their behavior. Their calculations may, as a matter of fact, take into consideration goals pertaining to the wider society or they may not. Actors' judgments which are informed by wider social interests seem no less intelligible and competent than those which do not (180)."

"consists ... of the simple fact that scientists keep watch over each other. Each scientist is both subject to criticism by all others and encouraged by their appreciation of him. This is how *scientific opinion* is formed, which enforces scientific standards and regulates the distribution of professional opportunities" (1966, p. 72).

This aspect of the social structure of science was much emphasized by the Mertonian tradition in sociology of science (See Merton, 1973).

Recent sociology and social history of science has pushed beyond the Mertonian tradition to show how the institutional structure of disciplines influences the content of scientific research by, for example, setting the structure of a particular scientist's endeavors. Thus, Rosenberg (1979) comments:

"It is the discipline that ultimately shapes the scholar's vocational identity. The confraternity of his acknowledged peers defines the scholar's aspirations, sets appropriate problems, and provides the intellectual tools with which to address them; finally, it is the discipline that rewards intellectual achievement" (p. 444).

These institutional forces not only impact on individual scientists but also serve to "mediate between science and the political, cultural, and economic institutions on which science depends for material and support" (Kohler, 1982, p. 2). Thus, the discipline becomes not only a unit of social organization but also a mediator between individual scientists and other social units.

When one thinks of the social and institutional aspects of scientific work, one is likely early on to focus on academic departments. Departments in colleges and universities, though, are in many respects problematic features of disciplines. On the one hand, the decision as to which faculty belong to a particular department may often be determined by administrative convenience. There are a plethora of ways in which universities divide faculty into departments, with the result that disciplines identified in terms of departmental structures may not resemble disciplines identified in terms of other social and cognitive factors. On the other hand, departments play critical roles in determining the historical continuity of disciplines and in securing the cognitive and social allegiances of members of a discipline.[8]

One of the respects in which university departments play a critical role as the inheritance mechanism of disciplines is through awarding degrees, which serve the function of certifying the competence of its holder to work in the profession. This

---

[8]   Whitley (1980) comments: "Educational institutions ... form the basic commitments of scientists in nearly all fields, and constitute the fundamental unit of social and cognitive identity in the sciences, which is one reason why the term 'discipline' is usually understood to refer to units of organisation in universities" (310).

is most true of the Ph.D., "The trade union ticket for obtaining academic posts and access to research positions and facilities" (Whitley, 1980, p. 310). (One should note, though, that although departments award degrees, candidates typically work with one professor. How professors direct the activities of those in their research group can critically affect how effectively their research tradition is carried on. See Krantz and Wiggins, 1973.) Directly related to certifying competence is the role of departments in determining ability to continue functioning within the discipline. For many areas of science, the most common mode of employment is through appointment to university faculties. (The complexion of the discipline may be quite different when opportunities for careers are prevalent outside of academic departments. See Whitley, 1982 and 1984, who sees in these other opportunities for employment a factor that is lessening the power of academic departments.)

In conjunction with controlling these gateways, departments also provide the framework in which one acquires the cognitive tools of the discipline–the theories and information accumulated in the discipline, the ability to identify central problems, and the tools to work on these problems. There is a feature of this education in graduate departments that requires special mention. Often the theories, information, and tools acquired in the course of graduate education will be broader than those which one is likely to actually in later research. A student takes courses in a variety of topics within the general discipline and interacts with fellow students who will pursue different specializations. Further, one is frequently required to take preliminary examinations that encompass the areas incorporated within the department. Never again is one typically exposed to the discipline construed in a broad manner. Hence, life in a graduate department may determine a scientist's broadest commitments. However, not all departments maintain equal breadth in their focus and so the nature of the graduate department in which one is educated may shape one's conception of the overall discipline.

University departments thus provide a critical aspect of disciplinary identity. However, there are some serious problems with using departmental structures alone to characterize disciplines. First, there is considerable variability in the ways in which universities divide faculty into departments (a few schools in fact do not have a departmental structure at all). At smaller colleges, departments tend to correspond to disciplines as defined at the most global level (e.g., chemistry, biology, and psychology). At larger universities, finer distinctions are often made, introducing considerable potential for variation. Biochemistry, for example, is sometimes found as an independent department, sometimes joined into a department of biochemistry and biophysics, sometimes joined into a department of physiology and chemistry, and at other times served by an interdisciplinary committee incorporating members of departments of biology and chemistry. Moreover, there are individuals whose departmental assignments do not at all correspond to the kind of work they do (Joel Cracraft, a taxonomist and evolutionary theorist, is located in the anatomy department at the University of Illinois; Stuart Kauffman, a theorist of evolutionary and developmental biology, is housed in a department of biochemistry and biophysics at the University of Pennsylvania). Moreover, there are departments that are shifting their emphasis away from their own discipline. Rossini, Porter, Chubin, and Connolly (1983), in a study of placement and hiring patterns amongst anatomy departments, identified a trend towards hiring people with a background outside of anatomy, often in neuroscience. Thus, while departments may play critical roles, especially in the inheritance of disciplinary identity, they may not be coextensive with disciplinary identity.

Professional organizations and journals are two other key institutional ingredients of disciplinary identity. Very often one finds professional organizations and journals corresponding to disciplines. Moreover, organizations and societies can be found at a variety of levels of specificity. At the broadest levels one finds national organizations of science and general science journals (e.g. *Naturwissenschaften, Nature* and *Science*. Major disciplines have their own professional societies and journals (e.g., the American Chemical Society and its associated journal); as do major subfields within these organizations (e.g., the Biochemical Society and the *Biochemical Journal*). More specialized research areas have their own groups and often journals.

One way in which these have an impact on the activities of a discipline is by providing outlets for the dissemination of information. However, given the fact that apparently few people read a great deal of the journal literature or attend significant numbers of papers at professional conferences, we should probably look further to identify the importance of professional organizations and journals to disciplinary identity. Two of the most significant reasons to present material at professional meetings and to publish in professional journals may be to establish claims to certain ideas and to gain professional recognition. Given the importance of giving or publishing papers, professional organizations and journals gain a powerful position by setting rules that affect advancement within disciplines. Thus, Whitley (1980) comments: "the existing set of journals in a science constrain and direct research topics and ways of working on them" (316). It is not surprising that an important aspect of creating a new discipline is the establishment of new professional societies and new journals. (Such decisions are often controversial, for in creating new institutions one is to some degree cutting one's affiliations with other organizations and journals. See Cambrosio and Keating, 1983, for a discussion of the controversy between Colin Pittendrigh and Franz Halberg over the creation of new journals and institutions for chronobiology.)

While departments, major disciplinary organizations, and journals do tend to provide organizational and social structuring to scientific activities, they tend not to be the units most important for communication and interaction in research. These tend to employ more specialized structures. Some large research areas may have their own conferences and journals; smaller areas may employ more informal channels of communication.[9] It was for the smallest of these groups that Hagstrom (1965) introduced the name speciality. Specialities can be characterized as small groups of researchers who are continually aware of each other's work and are able to comprehend it well enough to criticize and develop it. It has been these more specialized units that have been the major focus of recent sociological inquiry.

One of the first proposals for identifying these basic working groups in science was Price's notion of a "invisible college." (See Price, 1961; for further development of the notion, see Crane, 1972 and Chubin, 1982). This term is usually applied to a group of researchers who are in regular communication, usually by word of mouth. They typically share a common conceptual framework, problem focus, and set of techniques for dealing with the problem, although they may disagree on many particular theories or empirical claims. These groups sometimes develop around a particular intellectual leader, who provides much of the direction for the group (Delbruck's role in the development of the phage group is frequently pointed to as an example – see Mullins, 1972). Informal communication is very hard to identify and so quite naturally sociologists have sought other means of identifying scientific research communities. They have therefore developed a variety of bibliometric techniques for identifying working clusters in science on the basis of citations in journal articles (Garfield, 1979; see Edge, 1979, for a critique). At best, however, these quantitative measures can provide a guide to the degree of contact between different investigators, not the qualitative character of their interactions. For a more thorough assessment, one must carry out a much more detailed and qualitative study of the record of interactions between various investigators (for examples, see Mullins, 1972, and Law, 1973).

---

[9]   The more global disciplines may differ greatly in how much they promote these more specialized structures. Whitley (1976) distinguishes two types of disciplines – umbrella and polytheistic disciplines – which differ in their sponsorship of specialties. Umbrella disciplines promote specialty and research areas which end up being quite autonomous from the master discipline, while in polytheistic disciplines research in more specialized areas is continually being related to questions about the overall approach of the discipline. He attributes this difference to the degree of development and articulation of what he calls the discipline's "ordering principle" – which sets ways of conceiving objects and ordering them in some general framework. The more developed and articulate this principle is, the more likely the discipline is to have specialities and research areas that function autonomously.

On the social and institutional dimension, there are thus a wide variety of ways to characterize the units of science. Some of these ways of characterizing disciplines may correlate quite well with those identified on the domain and cognitive dimensions, others may diverge considerably. As the conflicts between cognitive and social approaches to the study of science diminish, one of the useful tasks will be to examine the degree to which social and cognitive factors parallel each other, and identify the points where they diverge and the significance of those points of divergence.

Disciplines as Historical Entities

So far I have looked at the characteristics of disciplines largely as if they were non-temporal entities, although the discussion of both the cognitive and social features of disciplines makes it clear that disciplines, at whatever level of specificity, are changing entities. This might provide a serious difficulty if one hoped to define particular disciplines. However, that is almost certainly a misguided idea, as it seems very unlikely that one would ever find necessary and sufficient conditions for identifying, for example, biochemistry, molecular biology, or genetics as discrete disciplines. A far better approach is to view disciplines as historical individuals or lineages, characterized not by a defining property but by their historical continuity. This suggestion comes from Hull (1982), who previously made a similar suggestion that evolutionary biology view species as individuals with a lineage, rather than as natural kinds (see Hull, 1978).

The characteristic aspects of an individual is that it is born, maintains itself as a coherent entity through space and time, gives birth to new individuals, and dies. One could view disciplines at various levels of inclusiveness in such a way. They begin by splitting off from a parent disciplines (or disciplines), maintain themselves through a life history (often including significant transformations), sometimes spawn new disciplines, and eventually die out. Adopting this framework, the kinds of questions one will ask will be of the order: out of what discipline(s) did the first practitioners of the new discipline come? How did they establish themselves as an independent discipline? What led them to eventually fracture into new disciplines with the death of the original discipline? To some degree the analogy drawn from looking at individual organisms as entities with life histories, may have to be modified. For example, it may sometimes happen that specializations will not be born from predecessor specializations but from the more general discipline in which it is a part (generalists in physiology may give rise to a speciality that looks only at one organ system), Nonetheless, viewing disciplines as individuals provides a perspective within which important questions about the life history of disciplines can be addressed. The various characteristics of disciplines I have considered (domain, cognitive character, and social structure) can then all be viewed as features of these individuals. Some of the difficulties of applying the criteria discussed earlier will not arise within in this perspective. For example, it will not count as a difficulty that the central problems under investigation changed as long as the historical continuity can be traced. (There will be a question of how much change in a historical entity will be consistent with the continuity of the same historical lineage, and when one should mark the death of the old and the birth of a new lineage. Pragmatic criteria, though, may suffice for answering these questions when they arise.)

Viewing disciplines as historical entities provides an important perspective on how we construe disciplines at various levels of specificity. At the broadest level, numerous scholars have argued that through the Middle Ages and Renaissance, Arts and Sciences constituted one faculty in universities, demarcated from professional faculties of law, medicine, and theology (Machlup, 1982). Gusdorf (1977) traces the fractionation of knowledge into different disciplines largely to Napoleon's division of the Imperial University into distinct faculties of letters and of science. This division was carried further within German Universities, where departmental structures were developed and education began to be linked with doing research within these departments. Whitley (1982) views this department structure that developed in the German universities as largely responsible for the development of disciplines as "reputational structures" that have subsequently played a central role in governing the life of science.[10]

Once the division into departmental faculties began, it created a pattern where new disciplines (even specific ones like molecular biology) sought their own disciplinary status. Whether this was possible, however, often depended on external factors, such as whether universities were expanding so that new departments could be justified (in European univerities, this involved the establishment of academic chairs). Without room for expansion, however, it is difficult to create new departments. Then new specializations must be handled within existing departments. Kohler (1982) discusses the problems this posed for physiological chemistry in its attempt to become an autonomous discipline. According to his analysis, it was the opportunities provided by the expansion and reorganization of

[10]   Whitley comments further: "the specific identification of intellectual goals and procedures with units of education and training probably increased the degree of intellectual and social closure to a greater extent than if professionalization had taken a different form" (1982, p. 318).

universities in the United States that provided the opportunity for developing new departments of biochemistry.

When one moves down to more specific diciplinary levels, the historical perspective can be even more fruitful. Studies of specialty groups reveals how they change dramatically in structure through time. Mullens (1972) analyzed the history of research on phage until it became part of the general specialty of molecular biology into four eras, each with different conceptual orientations and social structures. De Mey (1982) has tried to generalize from this and proposes an account of the life cycle of a scientific specialty in terms of which he tries to correlate institutional and social structures, modes of disseminating ideas, and stages in the development of methodological orientations and cognitive content. In what he calls "the pioneering stage" there is a highly informal structure, in which ideas are disseminated informally (e.g., through photocopied manuscripts and preprints). The goal is to formulate the cognitive framework, and the emphasis is on originality and programmatic work. The building stage witnesses the beginning of social structures (symposia and invisible colleges), a switch to journal articles as the primary mode of disseminating ideas, and an emphasis on developing empirical support for key ideas. Next comes the stage of internal criticism; de Mey claims that here one finds an increase in formal structure (formal societies and formal meetings) and formal modes of dissemination (including the development of textbooks), but also a new element of attention to anomalies and a decrease in the volume of well-confirmed empirical advances. In the stage of external criticism, the specialty achieves formal recognition by, for example, having independent departments in universities, but ceases to expand its empirical cognitive content and degenerates into philosophical controversies.

While this model has some appeal, it seems unlikely that all cases will adhere to such a simple linear progression. For example, it seems improbable that invisible colleges always disappear when more formal institutions are developed or that empirical advances cease with the formation of academic departments. Despite such problems as may arise for the specifics of de Mey's model, it does call attention to the fact that the various features of disciplines discussed above change in their manifestations over time. One must thus identify the time stages in the development of a discipline when discussing its various domain, cognitive, and social features.

2. The Ethnocentrism of Disciplines

The frequent calls for greater interdisciplinarity is, to a great degree, the result of a perceived isolation of disciplines or what Campbell (1969) called the "ethnocentrism of disciplines." Campbell characterizes this ethnocentrism as "the symptoms of tribalism or nationalism or ingroup partisanship in the internal and external relationships of university departments, national scientific organizations, and academic disciplines" (p. 328). This gives rise, according to Campbell, is "a redundant piling up of highly similar specialties, leaving interdisciplinary gaps." Since such ethnocentrism, to the degree that it exists, constitutes a serious obtacle for successful cross-disciplinary work, it is worth considering how such ethnocentrism results from various of the charcteristics of disciplines that I discussed in the previous section.

Let us begin with some of the factors driving ethnocentrism that Campbell identified in his analysis: institutional factors like departmental structures and professional orgnizations. Taking departments first, Campbell argues that departments do not simply divide up the conceptual trritory benignly but alter the terrain to make certain parts of it more important. He hypothesizes that those topics which fall within the central area of the department's domain generally attain greater value as their practitioners control such matters as departmental chairpersonships, core curriculum and degree requirements, promotions and tenure. (For Campbell, "central" is identified topologically in his multi-dimension topology of disciplines. One can equally interpret these points by identifying centrality in terms of other factors like historical precedence.) For example, Campbell suggests that unless there is an unresolvable conflict between two figures in central areas in a department, the chairpersonship will rarely go to someone working in a peripheral specialty. Campbell also considers mechanisms which force peripheral specialties to become more directed toward the central focus of the department and not allied with related specialties in other departments. For example, to insure that their students survive, faculty at the periphery must require that their students become well prepared in the central specialties and encourage pursuit of these over cross-disciplinary work. This turn away from cross-disciplinary work with other departments is further exacerbated by competition between departments for resources, which may encourage the deprecation of other disciplines.

Saxberg and Newell (1983) interviewed a large number of individuals involved in or responsible for administrating interdisciplinary work and compiled a catalogue of objections that departmental members are prone to raise to interdisciplinary endeavors, of which the following were the most prominent:

"(1) interdisciplinary research by its very nature is shallow, (2) the researcher's skills and competence are underutilized and so lose their edge, (3) joining an interdisciplinary programme is an escape from the rigour of the discipline and thus an indication of failure, (4) the quality of interdisciplinary research is suspect because of the absence of applicable standards of evaluation'' (p. 208).

Without at this point evaluating the legitimacy of these complaints, we can treat them as evidence that across disciplines there is a significant amount of prejudice against department members crossing beyond the domains of their particular department. Saxberg and Newell also recorded complaints about how interdisciplinary endeavors harmed individuals' careers because senior faculty within a department would be less familiar with the cross-disciplinary projects and would tend to discredit publications co-authored with members of other disciplines. As a result, they came to the conclusion ''that a non-tenured faculty member should not become involved in interdisciplinary research'' (208).

Training within a department can be a factor in developing ethnocentric attitudes in students, since such training serves not only to inculcate into students the skills and values of a particular discipline, but also a sense of group loyalty and distrust of other disciplines. Gold and Gold (1983) describe the extent of this experience of being socialized into a discipline:

> ''The training and socialization that the student in a discipline undergoes lead to an identification with the disciplinary community that comes prior to, and is generally regarded as more persistent than, identification with a specific employer or particular task. In combination with normative differences these feelings of identification and loyalty to the disciplinary community can, and often do, develop to the point of professional chauvinism, providing barriers severe enough to defeat interdisciplinary collaboration at the earliest phases. A milder manifestation of community identification is that individuals from different disciplines tend at first to regard each other as exchangeable representative specimens of their respective disciplines'' (94).

Campbell also explores the ethnocentric aspects of professional organizations and journals. He argues that the way professional organizations and journals are organized provides additional pressures to keep researchers' attention directed inwards to the core of their discipline rather than to related specialties in other disciplines. For example, referees for journals require one to cite appropriate work in the discipline represented by the journal and to use the research tools approved by it, giving less credence to citations and research tools of other disciplines. To gain a hearing by the discipline with which one wants to collaborate one must become almost as competent in that discipline as (or perhaps more competent than) its own members, clearly a foreboding task.

Campbell focused predominantly on the organizational and institutional factors contributing to the ethnocentrism of disciplines. He does indicate, as well, factors

24

impinging on the cognitive activities of individual researchers. One of the most significant is the constraint of time which limits not only one's ability to keep up with journal literature, but also one's ability to maintain contacts with others in professions of interest. These are important in terms of learning of recent findings that may affect one's work as well as learning newly developed techniques for research. Pursuing each of these requires a commitment and investment, an investment that may be taken away from one's core area.

Other cognitive aspects of disciplines also help enforce ethnocentrism. Not only can the research techniques, theories, and background knowledge of a discipline be a barrier against those from other disciplines, they also represent an investment made by members of the discipline itself. Like all investments, researchers may not want to relinquish them. Shapin (1982) comments:

"The analysis in terms of socially acquired technical competences may even be extended to encompass scientists' investments in the practical or interpretive line of their previous work. If a group of scientists have accomplished a body of publicly available research in which they argue for a given point of view, theory or interpretation, they may well wish to defend that position from attack and display its value and scope over other positions-even if they are technically able to work from another cognitive or practical orientation. What is involved is a strategy for defending and furthering interests, based on complex calculations about the consequences of various courses of action." (165).

Shapin analyses a number of historical conflicts, including the conflict over taxonomical approaches between morphologically trained botanists and those trained in cytology or genetics, in these terms.

These differences can arise even when one is not trying to push one's investment in a particular methodology but is trying to cooperate on an interdisciplinary team. Thus, Barmark and Wallen (1980), in describing an interdisciplinary forest ecology project, note intense conflict between those who were conversant and comfortable with the mathematics and systems models being used in the project, and those more comfortable with traditional biological techniques. The results were the development of incommensurable pieces of data that could not be readily assimilated to the theoretical model being developed.[11] Gold and Gold (1983) present these differences in conceptual frameworks as more general problems for bridging disciplinary boundaries, making it difficult for researchers from different disciplines to agree on a common project.

---

[11]   Barmark and Wallen analyze this inability and reluctance to work within new conceptual frameworks in terms of costs to the researchers: "The cost of learning a new competence was too high and was worthless as a skill when they had to return to their original disciplines after the project" (226).

"As a result of cognitive, subject matter, and normative differences, scientists of different disciplines may have difficulty agreeing upon appropriate sets of goals, an appropriate framework for pursuing those goals, and an appropriate evaluative framework" (93).

Disciplines are also divided by standards for reviewing research. As I noted above, such standards are critical both to maintaining quality within the discipline and to giving direction to individual researchers. But this can often be an obstacle to developing cross-disciplinary connections and so foster ethnocentrism (Russell, 1983). An example will dramatize this effect. In American experimental psychology experimenters tend to utilize a particular conception of data, experimental design, and statistical analysis that emphasizes control over some factors and systematic manipulation of others, rendering all subjects as comparable "units." Piaget and Inhelder's seminal work in developmental psychology generally did not fit this conception; much of their data was generated using the "revised clinical method," by which different children are presented with somewhat different questions or situations depending on the responses they are giving. For Piaget and Inhelder's contributions to become more widely accepted, it was necessary for their major findings to be replicated by others in the American manner–that is, presenting children in rigidly-defined age ranges with standardized questions in standardized or counterbalanced order, and submitting the results to inferential statistical analysis using conventional levels of significance.[12]

So far I have approached ethnocentrism as largely a secondary consequence of the social and institutional aspects of disciplines. But at the stage of discipline founding, there may often be a deliberate attempt to foster an ethnocentric perspective and to set one's own discipline off from any others. Gaining the recognition and status of a discipline can often be very important, for it is as a discipline that one can have academic appointments, award degrees, gain grant support, and the like. Several sociologists have thus characterized the process of discipline creation in conflict terms. Bourdieu, for example, introduces the concept

[12] During the conference at which the papers in this volume were originally presented, Mason noted how a number of these factors impeded his attempt to pursue cross-disciplinary work. In his work as a comparative psychology he arrived at a physiological proposal to account for the differences between two species of monkeys. However, he faced a number of difficulties in getting a physiologist to work with him. First, he had to deal with significant methodological differences. For example, whereas psychologists require experiments on multiple subjects with results submitted to statistical analysis, physiologists view a single organism as representative of a species. Moreover, the particular physiological idea he wished to pursue was now considered passe by most physiologists. Finally, neither journals in physiology nor psychology were interested in the particular problem.

of a "field" (using the term in a very different way than Darden and Maull) in which disciplines struggle to establish their stake:

"As a system of objective relations between positions already won (in previous struggles), the scientific field is the locus of a competitive struggle, in which the *specific* issue at stake is the monopoly of *scientific authority*, defined inseparably as technical capacity and social power, or, to put it another way, the monopoly of *scientific competence*, in the sense of a particular agent's socially recognized capacity to speak and act legitimately (i.e., in an authorised and authoritative way in scientific matters" (Bourdieu, 1975, p. 19).

When one focuses on the activities of creating disciplines, one can see at the outset a powerful force for making disciplines ethnocentric. Setting up such a structure is not a neutral activity, for the way it is done and who does it may determine what status particular individuals will hold in the science. Creating a discipline is a means of gaining credibility for oneself as a scientist. Thus, Cambrosio and Keating (1983), in chronicling the development of chronobiology as a separate discipline, offer a reason why it was so important for some practitioners to gain disciplinary status:

"To constitute oneself in a disciplinary form, which defines the rules of the rules of the game, is to release oneself from the domination of one's competitors and to dictate, in relative autonomy, one's own rules of the game: the disciplinary form thus functions as a machine which (re)-produces its own rules of legitimation" (328).

This characterization of the scientific activities in economic terms may seem to constitute a denial of the cognitive pursuits of disciplines. However, it need not. A significant part of carving out a disciplinary stake (Cambrosio and Keating's term) is to establish a distinctive intellectual context of research. Thus, in describing the endeavors of Franz Halberg in establishing chronobiology, Cambrosio and Keating describe how Halberg developed the notion of rhythms in direct contrast with the notion of homeostasis, which he viewed as the central notion in the competing discipline of physiology. Coleman (1985), who does not work within this conception of science as economic conflict, nonetheless, in discussing Bernard, describes how he was developing a cognitive framework to distinguish experimental physiology for both vitalism and experimental chemistry. Thus, even the cognitive endeavor of creating a new discipline often involves drawing boundaries between oneself and others which can be the fuel for ethnocentrism within the discipline.

In this section I have discussed a number of factors that contribute to the differentiation of disciplines and their ethnocentrism. Recognizing why disciplines have an ethnocentric character is important as we now turn to considering the process of crossing disciplinary lines to integrate various endeavors in science.

3. Reasons for Crossing Disciplinary Boundaries

It may seem almost sacrilegious to some people to seek reasons for crossing disciplinary lines and becoming interdisciplinary. Many have operated as if being interdisciplinary or participating in interdisciplinary endeavors is a good in itself that does not require a justification. However, there may be something pathological, as Gusdorf (1977) suggests, in many calls for interdisciplinarity. They may simply represent an inability to accept the complexity and accompanying fractionation of modern science and a desire for "a reaffirmation of a lost wholeness" (p. 581). There are, however, somewhat better reasons that can be adduced for participating in cross-disciplinary endeavors in science. Campbell's discussion of ethnocentrism in science which I discussed in the preceeding section was part of a call for a more interdisciplinary approach to science. While he would disown the "Leonardesque aspiration" of individual omniscience, he endorses interdisciplinarity as a vehicle for omnisence within the species. Adopting for a moment this *a priori* perspective on human knowledge, I can suggest two additional benefits which are complementary to the one noted by Campbell. First, problems that receive insufficient attention primarily because of disciplinary enthnocentrism could get the attention due them in a fully interdisciplinary environment. Second, if ethnocentrism were to wane, scholars would find it easier to locate other scholars who could aid their particular inquiry.

Although it sounds like a worthy goal, one may question how important the kind of omniscience Campbell advocates really is. One may also question my extension of his view by asking whether there really are significant problem areas untouched by current disciplines. As it stands, scientists and their disciplines tend to be politically opportunistic, looking for problem areas into which they can expand. As a result, most problems ripe for solution are likely to be attacked by disciplines capable of solving them. Moreover, interdisciplinarity is not without costs.[13] One is that the heritage of what a science has accomplished is only preserved through the disciplinary structure. If there were no disciplinary cohesion, the result might not be greater progress, but a loss of past wisdom (necessitating retreading old ground and re-committing old mistakes). Second, if there were not a cohesive discipline, there would not be the kind of peer review process that, as discussed above, has been instrumental in guiding research and ensuring that mistaken ideas are rejected. Third, disciplines tend to develop specialized discourse in order to develop their insights into a particular domain. While these specialized modes of

[13] Some of these costs were pointed out by Donald Campbell himself in his introductory remarks to the conference from which this volume is drawn.

discourse sometimes constitute an obstacle to interdisciplinary discourse, sacrificing them might also prove counterproductive.

Rather than continuing in this *a priori* vein, it will be more productive to look at the actual motivations scientists have for crossing disciplinary boundaries. (Scientists do not always reveal their motives for pursuing a particular line of inquiry; hence, some of what follows will involve attempts to interpolate the motivation from the inquiry pursued.) The kinds of factors that motivate crossing disciplinary boundaries can be both sociological and cognitive. Since most of the emphasis in this volume will be on cognitive motivations for cross-disciplinary boundaries, let me briefly mention some of the more social factors that sociologists and social historians have identified as motivations for engaging in cross-disciplinary work. This list of motivating factors is not intended to be exhaustive, but rather to provide a starting perspective that can be extended through further investigations.

An important consideration suggested by social studies of scientific communities is that while there is often a strong bias towards ethnocentrism of disciplines, there is also a significant amount of openness in the scientific community. While the notion of an "invisible college" suggests a closed unit, and much of the analysis has focused on the importance of the central figures in such "invisible colleges," Crane (1969) herself has also noted the importance of openness in such units:

> "Most problem areas are open to influences from other fields. The desire for originality motivates scientists to maintain contacts with scientists and scientific work in areas different from their own in order to enhance their ability to develop new ideas in their own areas" (p. 349).

Hagstrom (1970) claims that many researchers end up working in more than one specialty at a time and do change specialties over time, albeit the changes may be between closely related specialties. While recognizing that for most individual scientists, the discipline in which one is trained constrains the potential for later changes, Chubin (1976) proposes that the boundaries of scientific specialties are often fluid intellectually and socially. This fluidity may involve members actually working in multiple specialties or switching between them as well as the inclusion of outliers in the specialty who serve as conduits between specialty groups.

Sociologists have given the label "migration" to the phenomenon of scientists changing from one area of science to another. The most commonly adduced reason for such migration is opportunity for advancement. For example, Ben-David and Collins (1966) trace Wundt's move into psychology to the lack of opportunities for career development in physiology. According to their analysis, Wundt left a high prestige field where opportunities for advancement were very limited, physiology, for a less prestigious discipline, philosophy, where he nonetheless recognized an opportunity to develop a new research domain. In Mullins' (1972) characterization of Delbruck's migration into biology, the factors were not so much personal ambition as a sense that major new developments were not in the offing in physics, the recognition of problems that could be addressed using the methods of physics on biological problems, and finally the hope that these biological pursuits might end up forcing a revision in the understnding of basic principles in physics. Lemain, MacLeod, Mulkay, and Wingart (1976), in introducing a collection of papers on the emergence of new scientific disciplines, identify migration as a factor in developing new disciplines and indicate a variety of social factors that can lead to migration:

> "Scientific migration is not a random process, for the scientists moving into a new field tend to come from other areas with specifiable characteristics. In particular, they come from research areas which have experienced a pronounced decline in the significance of current results; from areas where there are few or no avenues or research easily available; from areas whose members have special competence in or knowledge of techniques which appear to have wider application; and from areas which have been disrupted, often by events originating outside the research community, and whose members have consequently no firm commitment to an established field. They tend to move into areas which appear to offer special opportunity for productive research, for the utilization of their particular skills, and, consequently, for career development" (5).[14]

While migration may be a particularly startling way of crossing disciplinary boundaries, the notion of interdisciplinary research is more commonly applied to researchers who fundamentally retain their disciplinary identity but pursue work that involves utilizing and interacting with another discipline. Here too there can be social factors motivating the activities. From the same interviews where they derived a number of reasons for not engaging in cross-disciplinary endeavors, especially if one lacks tenure, Saxberg and Newell (1983) also elicited a number of positive benefits that were found in cross-disciplinary activities:

> "(1) funding ... may be available for time released for research, (2) an approved budget may include allowance for travel to professional meetings, printing costs of journal articles, supplies, computer time, etc., hard to come by within a university's regular operating budget, (3) the opportunity exists to generate papers and articles for publication,

---

[14] Lemain, MacLeod, Mulkay, and Wingart see in these factors a possible explanation for the phenomenon of multiple discoveries in science. They propose that due to the kind of factors they itemize, numerous researchers may migrate at the same time without communication. Coming from similar backgrounds, they will then direct themselves at the same obvious and general problems, resulting in similar simultaneous solutions.

(4) tenure, promotion, and merit may be favourably effected by extensive and successful research, (5) new avenues of teaching and research may open up as a result of interdisciplinary involvement, (6) interactions and collaboration with colleagues may prove stimulating and ensure research productivity, and (7) interested graduate students with a commitment may be available and supported to assist in the research project'' (209).

While the negative aspects they identified raised the prospect that cross-disciplinary endeavors will not be valued, these positive features all point to the fact that opportunities may be found in cross-disciplinary endeavors where there may be limited prospects in the heartland of the discipline. In choosing to work intra- or interdisciplinarily, then, one may be making a calculated gamble as to whether to pursue low probability but highly valued activities within the mainstream of the discipline or higher probability but low valued activities at the fringes where contact with other disciplines is possible.[15]

Besides the social factor of perceived opportunity in cross-disciplinary work, there is almost always also a strong cognitive factor, at least amongst those designing the project. One very common cognitive factor influencing a decision to cross disciplinary lines is a recognition that the problem one is encountering cannot be adequately dealt with within one's own discipline. Darden and Maull (1977) raise this as a common reason for developing what they call interfield theories (see section 5 below). One of the things investigators might seek in another discipline is a mechanism that can explain the phenomenon which has been identified within one discipline but cannot be explained within it. Both biochemistry and the synthetic theory of evolution can be seen as stemming from the inability to explain phenomena within one discipline alone. In both these cases, the mechanism behind the phenomena resided at different levels. Thus, the mechanism responsible for physiological processes like oxidation and fermentation required appeal to the chemical level while, according to Darden's analysis (this volume) the problem of speciation could not be addressed at either the level of natural selection or genetic replication and required appeal to a new level at which isolating mechanisms could divide populations.

A second reason is found in cases where the problem was not one of finding a mechanism, but rather of acquiring guidance in developing a theoretical explanation. The exchanges between psychology and linguistics both in the late 19th century (Blumenthal, forthcoming) and in the 1960s (Reber, forthcoming) are partly due to the fact that first linguists and later psychologists saw in the other discipline

---

[15] Barmark and Wallen (1980), in discussing why some researchers joined an interdisciplinary forest project, noted that for some it was the only available opportunity to find work related to their field, supporting the idea that engaging in interdisciplinary work may appeal most strongly to those without opportunity within the mainstream of their discipline.

models that might be fruitfully used to solve problems within their own discipline. These were not mere analogies in the sense in which nuclear physicists drew upon analogies from astronomy. The relationship was thought to be much more integrative, since the generative models for one phenomenon (e.g., language) were thought to be readily extended to provide models for the other (cognitive activity). Several of the papers below show a similar reason for crossing disciplinary boundaries. For example, Mason and Anderson are crossing the disciplinary lines between animal ethology and human cognitive psychology in the hope of gaining a theoretical perspective for guiding their inquiry in their own discipline: Mason appeals to cognitive frameworks that have been established first to understand human behavior so as to understand animals, while Anderson is turning to animal ethology for help in finding ways to incorporate ecological perspectives into human cognitive psychology. Kauffman presents a slightly more complex case of the same thing. While trying to bridge the gaps between evolutionary and developmental biology, he appeals to yet a third area, mathematical thermodynamics. Using models drawn from that field, he hopes to represent the regulatory system in the genome. This provides a basis for discovering implications for evolutionary theory.

A third reason for working across disciplinary lines is the awareness of incommensurabilities in the approaches to the same domain by different disciplines. One response to such incommensurabilities would be to simply affirm the rightness of one of the approaches. Another would be to press each against the other, in the hope that the dialectical interaction might advance the understanding in both enterprises. This is exemplified in recent work on language aphasias discussed by Richardson in this volume. This work began with the recognition that the kinds of functions aphasia researchers considered important in accounting for language disruption (comprehension and production) were orthogonal to the set of distinctions which linguists take as central (phonology, syntax, semantics, pragmatics, and lexical structure). Initially, the endeavor was to recast the aphasia research in terms of the linguistics distinctions (so as to treat Broca's aphasia as involving a syntactic deficit – see Bradley, Garrett, and Zurif, 1980). Now the aphasia work is suggesting that a further modification may be required in the linguistics categories so as to focus on a distinction between automated and non-automated functions (Grodzinsky, Swinney, and Zurif, 1983). Thus, a dialectic is occurring between different approaches to distinguishing cognitive function, with ideas from each discipline being modified to fit the other.

A fourth cognitive motivation for crossing disciplinary boundaries is found in the paper of Rumbaugh and Sterritt in this volume. In their work they are not seeking a mechanism at another level to explain a phenomenon observed first at one level. Rather, they are trying to explain the origin of a phenomenon that has traditionally fallen within the domain of one discipline, but where the origin requires entering the domain of another discipline. Thus, intelligence is an attribute much discussed in human psychology. To explain its origin, however, requires a phylogentic inquiry. An attempt to trace its origins, however, requires one to enter the domains of disciplines where the concept of intelleigence has not played a central role. As in the interchange between aphasia work and linguistics, a dialectical process results from this inquiry. In turning to the study of other animals to try to explain the origin of intelligence, Rumbaugh and Sterritt discover the need to develop a new conception of intelligence, since its expression in other animals cannot be detected by the same measures as are employed in linguistic species.

Sometimes the basis for building a bridge to another discipline may not originate with any special interest in building such a bridge. This is found in the case of Savage–Rumbaugh and Hopkins. While the proposal they have developed is here considered from the perspective of how it might enhance cognitive science, which has already developed as an interdisciplinary endeavor focused on human cognition, that is not the motivation for developing their proposal. Rather, they recognized that the set of research techniques they were using in teaching forms of language to primates yielded results that did not fit either the models of animal learning or animal ethology. To handle these results, Savage–Rumbaugh and Hopkins propose a theoretical framework with a central emphasis on intentional communication. This framework then offers a suggestive basis for extending the cross-disciplinary work already occurring in cognitive science.

The reasons noted here are ones that have provided motivations for scientists to cross-disciplinary boundaries; they do not depend on any idealized conception of knowledge but on problems that arise in the actual conduct of research. Drawing attention to these motivations is important for two reasons: (1) as these factors are ones that have drawn scientists to cross-disciplinary endeavors in the past, they are ones likely to do so again; and (2) the motivations scientists have for pursuing cross-disciplinary work may well affect the kinds of products that result from that work. As the early parts of this section acknowledge, these motivations may be both social and cognitive. We must bear these different motivations for crossing disciplinary boundaries in mind as we turn now to the institutional and cognitive products of interdisciplinary work.

4. The Institutional Products of Cross-Disciplinary Research

Just as I have adopted the term "discipline" to refer to the units of science in as theoretically neutral a manner as possible, I intend for the term "integrating" to be understood in a neutral manner, covering various possible modes of relationship that can be developed between disciplines. When attempts to cross-disciplinary boundaries are at least partly successful, both social and theoretical arrangements tend to emerge. In this section I will focus on the social and institutional structures that result from cross-disciplinary endeavors and reserve discussion of their theoretical and empirical products to the next section.

In the first section I noted some of the organizational and institutional characteristics of disciplines, such as having academic departments, professional organizations, and journals. I also explored some of the reasons why they are important to the survival of the discipline. Similar organizational and institutional structures must be developed by cross-disciplinary researchers if they are to flourish and maintain themselves. For example, researchers must find ways of training new researchers if the cross-disciplinary project is to continue. As I indicated earlier, one vital function of academic departments is to educate new students to accept the goals and employ the techniques and theories produced by the research endeavor. Journals and professional organizaticns provide channels for the communication and transmission of ideas and for establishing reputations. Hence, it is unsurprising that interdisciplinary endeavors often seek to establish comparable structures.

The importance of such institutional and organization mechanisms can be recognized by considering the numerous handicaps their absence places on researchers. Building a community of co-workers can be very difficult without such mehcanisms. When Delbruck made his first forays into biology, he was still in a department of physics, where few graduate students were interested in joining his endeavor. Many of the early converts to the phage program came through Delbruck finding an alternative institutional arrangement–running a summer phage course at Cold Springs Harbor. Through this course other researchers were educated and recruited into the program (see Mullins, 1972). Job opportunities in the academic world follow departmental structure, making it difficult for those working outside of established departmental units to find employment. Professional organizations are often the ones responsible for putting on conferences and publishing journals, so without such organizations, scholars encounter difficulties in getting their ideas and results published. Finally, the struture of grant-giving agencies often mirrors the organizational structure of academia, so that those whose work falls outside of the established boundaries often have difficulty getting financial support.

A factor that needs to be considered is whether establishing such social and institutional arrangements has the effect of creating new disciplines out of cross-disciplinary endeavors. This seems to be what happened in the case of biochemistry, which began with researchers drawn from a number of different research areas with focuses on different levels of organization, but resulted in a new discipline with not only its own conceptual unity (which is discussed in Holmes' and Bechtel's papers) but also its own social and institutional unity (see Kohler, 1982). But this is not the only possible outcome. One can maintain a much looser organizational and institutional structure for the cross-disciplinary endeavor. Its institutions can coexist with the basic departmental institutions and not replace them.

There have been a number of experiments in recent decades whereby institutions have tried to alter departmental structures to combat their role in furthering ethnocentrism (some of these are discussed in Campbell, 1969; see also Malchup, 1982). The simplest approach has been to develop inter-departmental programs or committees in which faculty may hold secondary appointments and in which students can receive degrees. (The University of Chicago is well known for such committees as the Committee on Evolutionary Biology.) In many universities, such committees are authorized degree-granting units, whereas in others the degrees are still governed by one of the participating departments. To be sure, there are a variety of tensions in such arrangements, both for students and faculty. If faculty hold their primary appointment within a disciplinary department, evaluations of them for tenure and salary will likely follow the policies of the discipline, which may or may not be compatible with significant investment in cross-disciplinary activities. For students, a major challenge is often finding employment. Since the structure of cross-disciplinary units varies quite widely between institutions, students may find difficulty matching their training to the requirements for departmental appointments. For example, they may not be judged qualified to teach the core courses of the home disciplines. A further problem with such structures is that they are viewed as a threat to existing departments, especially when there is a limit to resources. (See Campbell, 1969, for a discussion of the dissolution of Yale University's Institute for Human Relations as a result of pressure from other departments. See also Saxberg and Newell, 1983, for accounts of the difficulties such arrangements face.)

Another approach is to employ an administrative division orthogonal to that which differentiates departments. Such an approach might divide faculty into colleges independently of their department affiliation. I am aware of one case in which college membership rather than departmental membership determined one's office location (University of California, Santa Cruz) and one in which each college had its own departments, resulting in considerable duplication of courses (Rutgers University). In both of these cases, changes since the mid-70s have brought these schools to structures more like that of the University of California, San Diego, in which college affiliations exist but do not usually determine faculty location or result in duplication of departments. Another approach, which seems fairly stable where it has been tried, is to do away with departmental divisions and simply have very broadly defined units, such as the School of Social Sciences at the University of California, Irvine. These have been meccas for those academics with a strong interdisciplinary bent but may be viewed less favorably by those with more traditional disciplinary interests.

Another vehicle for institutionalizing cross-disciplinary endeavors within acadmia is through research institutes. (In discussing research institutes, one should bear in mind that probably the majority of such institutes are undisciplinary structures, not vehicles of interdisciplinary activities.) Insofar as these institutes are organized primarily for research, they avoid some of the conflicts that are encountered by developing educational programs orthogonal to the departmental structure of the university. (An educational function can still occur in such institutes, but it usually involves students working on specific projects with faculty from their own departments. Thus, their status is that of a research assistant within the institute; they fulfill their degree requirements in a standard department.) However, there are still a host of difficulties facing such institutes. Externally, they are in competition with other components of the university for resources. Internally, there is often competition between researchers from different disciplinary orientations as to how to define the research objectives of the institute. One might expect that once established, such institutes would further cross-disciplinary endeavors, since many of the difficulties that might imped individual projects could be worked out at the initial establishment of the institute. A study by Birnbaum (1978) covering several interdisciplinary projects in the United States, however, does not support that assumption:

> "Permanent institutes with full managerial hierarchies, permanent staffs, expensive equipment and permanent space were found to provide significantly more integrating devices but were not found associated with higher performing projects, to provide significantly greater interdisciplinary activity, or to reduce the time spent by principal investigators in planning and assembling resources when compared to projects operating independently of institutes" (95).

(For further evaluation of cross-disciplinary research in specially designated institutes and a discussion of the problems encountered by such institutes see Saxberg, Newell, and Mar, 1979; Teich, 1979; and Cutler, 1979.)

It seems clear that there are significant problems in arriving at a suitable organizational relationship for cross-disciplinary research in academic institutions. Some of these difficulties are overcome when the place of employment is outside the university and part of the corporate world. Here, project goals can provide focus for research endeavors. However, what is often lost in the corporate world is the commitment to pursuing basic science. One of the rare exceptions where a corporation pursued basic research was Bell Laboratories. Here researchers from a number of disciplines were able to carry out basic research either within a disciplinary orientation or by integrating into interdisciplinary teams. However, this opportunity may have largely been an artifact of the peculiar legal status of American Telephone and Telegraph Company; its recent court-ordered reorganization has removed much of this support and what remains of the research efforts seem to be directed in a more applied vein. Thus, basic research seems still to be most common in the university setting, and here the institutional obstacles to cross-disciplinary endeavors do not seem to have any totally satisfactory answer.

At the level of professional societies and journals, on the other hand, there are far fewer obstacles to developing feasible cross-disciplinary arrangements. There are numerous examples of professional societies which deliberately try to draw members from several disciplines whose interests connect (sometimes as a result of sharing a common domain). These societies often put on their own conferences and publish journals that cross disciplinary lines. A clear example is the Cognitive Science Society, which puts on an annual conference and publishes the journal *Cognitive Science*. The society and journal are explicitly committed to fostering interdisciplinary communication between cognitive psychology, artificial intelligence, linguistics, neurobiology, and philosophy. The existence of such organizations is not to suggest that they are a panacea for those interested in cross-disciplinary endeavors in the area of focus. Even when practitioners from a variety of disciplines all espouse an interest in a common phenomenon like cognition, they need not understand it in the same way. Moreover, they may carry over their standards for doing research to the interdisciplinary arena, and so be hypercritical of those from other disciplines who do not adhere to the same standards of research. This makes the task of organizing conferences and editing interdisciplinary journals a particularly risky endeavor, for one can easily offend and alienate those disciplines with whom one is trying to interact.

Despite the difficulties confronting these various organizational and institutional structures for cross-disciplinary research, it is clear that there is an alternative to the institutional and organizational model of traditional disciplines. We can speak of these arrangements of cross-disciplinary committees and institutes in universities, interdisciplinary professional organizations and interdisciplinary journals as characterizing "interdisciplinary research clusters". While these clusters may not provide all the support that are provided by more traditional disciplinary units, they nonetheless do allow for research collaboration and cross-disciplinary communication. As well, they can also provide a focal point for funding. To return to the example of cognitive science, there has been a rapid development of cross-disciplinary academic programs in this area, especially in the United States, which have been fostered in part by the targeting of funds for this purpose by the Department of Defense and by private institutions like the Sloan Foundation.

To illustrate further the idea of an interdisciplinary research cluster, we can consider one of the historical examples of a cross-disciplinary project discussed in this volume, the evolutionary synthesis. The evolutionary synthesis contrasts with that of the other historical case considered, that of biochemistry, in that in this case no new unified discipline was established to which practitioners transferred their disciplinary identity. The disciplines incorporated into the synthesis, primarily transmission genetics, population genetics, systematics, and paleontology, and secondarily, cytology, embryology, and morphology,[16] remain distinct, having their own departments in some larger institutions and, in any case, having their own identities as specialties as a result of having professional organizations and journals. What happened is that, alongside the established departments and professional organizations, new ones emerged that served the interdisciplinary function. Thus one finds interdepartmental committees (The Committee on Evolutionary Biology at the University of Chicago), interdisciplinary societies like the Evolution Society, and multi-disciplinary journals like *Evolution.*

One final question that needs to be addressed is whether one can identify factors that determine whether the social and institutional arrangements in any particular interdisciplinary endeavor will take the form of a new discipline or of an interdisciplinary research cluster. Several factors affect this decision. One is whether there is a sufficiently unified domain that could provide the focus for a discipline. This may have been one of the factors that led to the formation of a new discipline in biochemistry. While evolutionary phenomena do constitute a domain, it is both an extremely broad domain and one that includes almost all of the domains of the disciplines constitutive of that interdisciplinary research cluster. This made it impractical for evolutionary sience to constitute itself as a single discipline. A second factor is the degree to which those interested in the cross-disciplinary area can maintain their status within their host disciplines. In the case of biochemistry, the case has been made that those researchers interested in questions at the border of chemistry and physiology could not acquire significant status and appointments in either discipline, motivating the development of autonomous departments and disciplinary organizations (see Kohler, 1982). Cambrosio and Keating present a similar view of Halberg's drive to create a separate discipline of chronobiology as a result of the inability to gain status within established disciplines. When, on the other hand, researchers can achieve status within their own discipline, the alternative organizational and institutional arrangements provided by

---

[16]   See Mayr and Provine (1980) for a discussion of how these various fields fit into the synthesis.

interdisciplinary research clusters may suffice and the demand to form separate disciplines may be less urgent.

There are also significant reasons for practitioners in interdisciplinary areas to maintain allegiances to their initial discipline. Often the particular problems and modes of approach to problems, even in the interdisciplinary area, are brought from one's original discipline and so one continues to have strong ties to other members of one's discipline not involved in the interdisciplinary activity. Then researchers may have no incentive to break free from their own disciplines, but only to enrich their endeavors by contact with those working on related problems who happen to come from other disciplinary backgrounds.

While it may be too early to make reliable prognoses about the contemporary cases of cross-disciplinary endeavors discussed in this volume, we can make tentative projections about whether they are likely to follow the tradition of forming interdisciplinary research clusters or of forming separate disciplines. In none of the cases do the ingredients appear to be present that would dictate the formation of new disciplines. For one, the domains involved are not tightly unified. The domain of cognitive and animal ethology, for example, would have to include much of the domain of cognitive science (minus, for the most part, those components primarily directed at language activity of the kind found in humans) and animal ethology. The other cases involve possible extensions to two already functioning interdisciplinary research clusters, one in evolutionary biology and one in cognitive science. Adding to an existing cluster is unlikely to produce a more consolidated disciplinary structure. What may be more likely is that the proposed expansions would make the clusters so large that they will have to divide. However, it seems most likely that the products of such divisions would still remain interdisciplinary clusters, not structured disciplines. Thus, one should not expect the formation of a new discipline for any of the contemporary cases examined in this volume.

5. Conceptual Products of Cross-Disciplinary Endeavors

By far the topic that is of most concern in the papers that follow concerns the kind of conceptual connections that are involved in or result from attempts to link research in different disciplines. The historical sessions have focused on the kinds of cognitive structures that developed in the case of biochemistry and the evolutionary synthesis while the authors of the papers focusing on current cross-disciplinary endeavors try to lay out conceptual frameworks through which cross-disciplinary research might proceed. The reason for this focus is obvious. One of the chief problems those engaged in interdisciplinary research readily recognize is that researchers from different fields approach problems with different conceptual tools and conceptual orientations. Kuhn (1970) spoke of these as "paradigms" and "disciplinary matrices" and, while there is room for plenty of disagreement as to the precise character of these conceptual frameworks, there is no doubt that investigators from different disciplines do construe their problems differently, use different theories to characterize the problem domain and possible solutions, and use different research techniques to deal with the problems. Given these differences, researchers coming from different disciplines may find serious problems in communicating and working together.[17] As a result, most meta-scientists who have examined interdisciplinary projects stress the need to develop integrating conceptual frameworks in which cross-disciplinary research can proceed.

Often appeals for scientific integration have been accompanied by elaborate schemes in terms of which all the disciplines of knowledge could be brought together. Two of the best known of these in the 20th century are the program of unified science proposed by the Logical Positivists and general systems theory of Von Bertalanffy. Common to both programs is the idea of a general integrating framework in which all human knowledge can be linked. A principal endeavor of the Unity of Science Project was to produce an "International Encyclopedia of Unified Science" which, according to Neurath (1938), was

> "to integrate the scientific disciplines, so to unify them, so to dovetail them toegther, that advances in one will bring about advances in the others" (p. 24).

The program was never carried out, but the strategy that was to be used to accomplish this integration is clear, especially from the contribution of Carnap (1938) to the original project. The tool for unification was theory reduction. As celebrated as that model of unification is, it is interesting that not a single paper in this volume employs the theory reduction model, or any of the subsequent

---

[17] In their long-term study of a forest ecology project, Barmark and Wallen (1978 and 1980) noted that even those who shared an ecological perspective had trouble integrating their work because of the differences in their research orientations. Wallen (1981) provides examples of the problems: "In addition to the self-evident differences in background knowledge, some concepts had different meanings in different disciplines. This is not a linguistic problem; rather they had different explanations of phenomena in reality. For instance "mineralization" could be regarded as a mainly chemical phenomenon or as a biological one. It is a basic characteristic of science to simplify and isolate certain objects for study. But every level in the ecosystem is studied on its own premises and it is difficult to overcome these borders. The plant physiologists for instance work with small parts and their models are fast (one day), their methods have been laboratory oriented and exact. Studies of plant growth and plant ecology used models encompassing 1–100 years and they are more field-oriented and have other standards for their data." See Gold and Gold, 1983, for further discussion of these problems.

modifications that have been proposed. One might well be puzzled by this inasmuch as a number of the writers are philosophers of science. To try to resolve that puzzle and to show that kind of connections actually are salient in these cases, I will begin by considering the deficiencies of the formal theory reduction model as a basis for cross-disciplinary research and then turn to alternative models of cognitive integration.

The model of theory reduction portrays the theories of different disciplines as being related deductively. (See Nagel, 1961, for the classical presentation of theory reduction, Schaffner, 1967, for some important modifications, and Ruse, 1973, for an attempt to use theory reduction as a unifying scheme for biology). Causey (1977) provides a recent account of what unity of science via theory reduction would involve, which I will employ as the basis for this discussion.[18] Causey views nature as consisting of levels, where entities at higher levels are strutured wholes comprised of entities at lower levels. At the higher level one has a theory that characterizes the behavior of the structured wholes. The first taks in performing a reduction is to provide a specification of these structured wholes in terms of their composition from lower level entities. Then one can identify the terms referring to objects at the higher level with lower level terms specifying their composition. One must similarly identify predicates in the upper level theory with those whose extension includes the structured wholes specified in lower-level terms. As long as one is willing to allow that entities apparently similar at the upper level may have diverse composition, and that similar lower level units may be components of different higher level units, the ability to carry out this much of the program is unproblematic. But it is a project that is, in itself, uninteresting. All it shows is that we can characterize the entities of the higher level in terms of their constitution of lower level entities.

The interest in the reductionist program comes with the requirement that the law statements about structured wholes (as stated in the lower-level theory) be derivable from law statements about their components and statements of prevailing environmental conditions. It is with the last condition that the program of unification of science through microreduction becomes problematic, for it is far from obvious that the necessary derivation will always be possible. Without offering a detailed response to Causey (see McCauley, 1981, for such a response), I will

---

[18]   There is an alternative sense of reduction, that between predecessor and successor theories, that does not require a deductive relationship between theories at different levels. The whole idea of deductive relationships is inappropriate in this context since the replacement theory is presumably changing and improving on the old theory. Rather, the endeavor is to show the similarities and differences between the theories which is of particular use in evaluating the improvements brought by the new theory. See Nickles, 1973, Wimsatt, 1976, and McCauley, in press, for further discussion of the differences between these modes of reduction, which have often been confused in the literature.

simply note a couple of the factors that make the ability to develop the necessary derivation of properties of structured wholes from properties of components problematic.

Although Causey's formal treatment does not require this, in his discussion he proposes that those engaged in a microreduction research program would identify the properties of the components of structured wholes by studying them when they are not incorporated into the structured wholes (Causey's non-bound condition), and then derive their behavior in the structured whole from this plus statements describing the organized structure in which they are bound and prevailing environmental conditions. It seems by no means obvious that one will learn all the properties of the parts when one studies them in the non-bound condition; in particular, one may question whether one will learn how they function when bound into structured wholes. If the non-bound and the bound properties of components are different, then we will have to incorporate knowledge of their behavior in bound conditions into the lower level account. This raises the question as to how much one is permitted, in the course of carrying out a microreduction, to modify the account of the components so as to support a derivation of the behavior of structured wholes from it. Causey is concerned about this problem, since unlimited revisibility can lead to trivializing the problem of reduction by simply allowing us to incorporate all the laws of higher level science as additional laws at the lower level. However, he has no specific proposals as to limit the acceptable modifications.

Hooker (1981) is much more open to fundamental revisions in the lower level theory in the course of developing a reduction. Once one has modified the lower level theory sufficiently to allow for the derivation, the issue becomes whether one has a truly unified theory at the lower level, or just a variety of theories stated in one vocabulary. In particular, the issue is whether the theoretical statements accounting for the behavior of parts in the non-bound condition are integrated with those introduced to explain their behavior in bound conditions. Causey sets forth elaborate conditions for theories being unified, but these conditions do not seem to exclude the possibility of there being two sets of statements in the theory that do not interact with each other. If that occurred, we would not really have a unified theory at the lower level, but merely a unified vocabulary. In this event, reduction has not really advanced the cause of unifying science.

There is yet a further problem with treating the reductionist program as a unification program. Throughout his book, Causey defers questions about the origin of particular structures to an analysis of evolutionary theories, which he promises to sketch at the end. When he gets to that sketch, Causey proposes to explain how structures come into existence by applying basic or derived dynamic laws (to which the reductionistic account had been directed) to specific environmental conditions (laws specifying selection might be of such a kind). This makes the question of whether the causal interaction between objects and their environment is correctly charcterized at the lower level critical. It seems plausible that these causal relations will be between the structured wholes. Causey will find this requirement unproblematic as long as the laws governing structured wholes can be derived from laws governing their components. Numerous people have argued, however, that structured wholes which behave identically at the upper level may in fact have widely different internal composition (see Fodor, 1974). Causey's response (and Hooker's) is to bifurcate the upper level kinds in such cases, treating structured wholes as different if they have different compositional characteristics, even if their interactive behavior is identical. The result of this, however, may not be to unify science, but splinter it, for we will end up with different accounts of cases that initially appeared to be the same. (Pylyshyn, 1984, has used the argument that one can state generalizations in the language of folk psychology that cannot be captured in the language of neuroscience or behaviorism to argue for the legitimacy of that level of discourse despite that fact it cannot be reduced in a simple fashion to lower level laws. His argument that these generalizations capture real features of nature despite the differences that may exist between different instances that fulfill the generalization seems a good reason to treat the higher level account as the more unified for that domain.)

The point of the previous two paragraphs is to question whether the program of reduction really would produce a unified science. Even if it did provide for a unified science, however, there are additional reasons why such a program may not serve the interests of scientists actually engaged in integrative research. Causey presents himself as describing a research program of scientists, but it is unclear to what extent scientists are engaged in such a program or what they could hope to gain from it. Causey presents such a program as providing explanation as well as ontological simplification and unification. Considering first the ontological simplification and unification, there are times when recognizing ontological connections can help the endeavor of a scientist. For example, recognizing that genes were units on chromosomes advanced research insofar as it revealed additional facts about the entities in question. But the kind of simplification and unification envisaged in the reductionists' program does not seem to offer any explanatory advantages to scientists. They do not expand their scope of explanation by showing that the theories they have developed can be derived from other theories at more basic levels. Moreover, the objective of most scientists working across disciplinary boundaries has not been to achieve ontological simplification and unification. The reasons are fairly obvious. Nickles comments:

" 'Reduction' *means* 'elimination,' 'trimming down,' 'consolidation' " (1973, p. 183).

Few scientists want to consolidate with others in a way that "eliminates" or "trims down" their own theories, since, as I noted above, having specialized theories is one of the defining characteristics of a discipline. Reducing one's theories to those of another discipline reduces oneself to an applied practitioner of that other discipline.

The other goal of reductionistic research programmes is potentially more significant, that of gaining explanatory power. This, however, depends on accepting a particular conception of scientific explanation, one according to which explanation involves deriving a statement describing what is to be explained from other statements. This conception has been seriously questioned in the literature (see Salmon, 1972 and 1984, Scriven, 1962, and Bromberger, 1965). This is not the place to continue that debate, but it should be noted that it is not obvious that a scientist in one discipline extends his or her explantory power by deriving the set of laws used in that discipline from those of another. This is not to say that there are not sometimes very good reasons to go to another level in nature for explanation. Richardson (1980), argues that one of Donnett's major insights is to show that scientists frequently turn to other levels when a system they are studying does not behave in the way the principles thought to explain that system predict. It is deviations from expectations that need explaining, and going to another level may sometimes provide the needed explanation. However, such explanations do not require a deductive argument from lower level facts to the higher level generalization; they only require a demonstration of how the parts of the system equip the whole to behave in a certain manner.

Perhaps the most general difficulty with the formal model of theory reduction as a basis for interdisciplinary work is that it works with completed, formally presented theories, not with theories still under underdevelopment. But researchers engaged in interdisciplinary work are generally engaged in the ongoing process for discovery, not with the attempt to systematize what is known. In rejecting such various formal models, including the model of theory reduction, as bases for interdisciplinary research, Gusdorf (1977) makes an important comment:

"Interdisciplinary learning should be a logic of discovery, a reciprocal opening up of barriers, a communication between the different realms of knowledge, a mutual fertilization–not a formalism that cancels out all meaning and bars all outlets. ... This grand design [of a formally unified science] presupposes the possibility of reducing all kinds of knowledge to unity and projecting them on to the same epistemological dimension, without denying the specificity of each. But the realm of knowledge has many dimensions. ... The fact that there are many different disciplines of knowledge entails a diversity of approaches, none of which can claim to incorporate all the others. The idea of interdisciplinarity does not mean a search for a lower common multiple or a highest common factor; it is concerned with the entire epistemological space within which the separate kinds of knowledge are deployed like so many paths through the unknown'' (595–597).

So, while granting the need for integrating frameworks in order to integrate the resources of different disciplines, we must also disown the kind of formalism as that represented by theory reduction and seek a model of integration more compatible with an ongoing process of discovery.

In giving up the search for a formal integrating scheme for unifying all sciences, we need to find some other means of bringing together the conceptual frameworks of different disciplines so as to provide a basis for integrating their research. A number of years ago, Darden and Maull (1977; see also Maull, 1977) argued, using historical studies, for a quite different conception of how to unify science without reduction. They argued for what they termed "interfield theories." (Although, as I noted in section 1, Darden and Maull's conception of a field was more limited than the broad notion of a discipline being employed here, the same kinds of cross-disciplinary relations may be found between disciplinary units that do not fit their specific definition of a field.) One of the important features of interfield theories is that such theories have evolved to serve actual explanatory ends of scientists; in particular, to solve problems that could not be solved in one field of inquiry alone. Another feature, and one that distinguishes an account of integrating science in terms of interfield theories from a reductionist account, is that the end product is typically just one theory that spans fields, not two theories related by a derivation relation.[19]

Interfield theories characterize the relations between the entities or phenomena studies in different fields. Darden and Maull introduce several kinds of relations that may be considered in interfield theories: identifying in one field the physical location of an entity or process discussed in another, frequently showing a part-whole relation between the two; finding in one field a description of the physical nature of an entity or process charcterized in the other theory; discovering in one

---

[19]   As Steve Fuller has pointed out (personal communication), there remains a danger, once interfield theories have been developed, cf one field (e.g., genetics) seeking to dominate another (e.g., cytology). The notion of an interfield theory does not bar the kind eliminative reduction proposed by Churchland (1979) and Rosenberg (19XX), wherein one science (neuroscience or evolutionary biology) comes to supplant another (cognitive psychology or sociology). However, such a result would not be a necessary product of an interfield theory and would be due more to the social character of the interacting disciplines than to the creation of an interfield theory.

field the structure underlying a function described in the other theory; and finding in one field the cause of an effect noticed in another field. Darden and Maull analyze a number of examples of interfield theories: the chromosomal theory of Mendelian heredity, which linked cytology and genetics; the operon theory, which related genetics and biochemistry, and the theory of allosteric regulation, which connected biochemistry and physical chemistry.

To show some of the character of an interfield theory as Darden and Maull conceived it (none of the cases considered in this volume show exactly this pattern), I will briefly summarize their discussion of the chromosomal theory. By 1903, geneticists had recognized Mendelian factors as the unit of heredity, but had not identified their physical location. Independently, cytologists had discovered chromosomes and had determined that they were involved in hereditary functions, but could not explain their role in producing individual hereditary characteristics. In this context, Boveri and Sutton developed an interfield theory by postulating that Mendelian factors are located on or in chromosomes. Although Darden and Maull point to ways in which the chromosomal theory modified ideas in both disciplines, it was genetics that primarily benefitted. The chromosomal theory provided the foundation for the classical genetics program of the Morgan school, which worked out a detailed account of the location of genes on chromosomes. The other cases considered by Darden and Maull differ from this in the kinds of relations they posit between the entities or processes in different fields. Through the operon theory, biochemistry provided a mechanism for regulation of gene expression, a phenomenon already identified by geneticists. In the theory of allosteric regulation, a physical-chemical cause is provided for the biochemically observed alteration in the level of protein activity. Common to all their examples, though, is one field filling in missing information about a phenomenon that was already partially understood in the other field.

In concluding their paper, Darden and Maull call on others to investigate additional cases of interfield theorizing in science so as to provide a better basis for understanding the ways in which scientists cross between fields of research. I have recently discussed a case (Bechtel, 1984) that differs from those studied by Darden and Maull in a couple of respects. The cases considered by Darden and Maull all involved one field posing a problem to another field. In the case I analyzed, in contrast, the interfield connection linking vitamin research with metabolism research (B vitamins being constituent parts of respiratory coenzymes) was discovered fortuitously. Once discovered, it helped to illuminate further research in the fields where each entity had first been discovered. The case is also distinguished from Darden and Maull's cases in that the linkage between the fields was discovered only after a critical reconceptualization occurred in each field separately.

Nonetheless, this case is like the cases discussed by Darden and Maull in that the researchers did not endeavor to reduce one theory (the theory of coenzymes) to the other (the theory of vitamins). The integration of research that was important for the scientists was accomplished without reduction. It involved identifying relationships between entities that had been studied independently that allowed researchers in each field to learn new information about the entities that were of primary interest to them.

Many cases of interfield theories involve relations between levels and illustrate an alternative to reductionism as a way of relating levels. These interfield theories make appeals to lower level entities to explain features of higher level entities without necessary providing a full account of the upper level entity. However, in some cases there will be reason to go the other direction and appeal to higher level entities to account for lower level behavior (Campbell, 1974). In particular, there will be reason to go up when selection forces are operating on higher level entities that determine the continued existence or replication of lower level entities. In this case, one can take a teleological perspective and view lower level entities as serving functions defined by higher level selection forces (Wimsatt, 1972). Machamer (1977) proposes that this kind of relation between different levels provides another important alternative to reduction. What is involved is an interfield connection which involves the interaction between the processes described in two theories, one of which explains how a system operates and the other tries to account for its existence. (I have discussed this model for interlevel theorizing further in Bechtel, 1985.)

Several of the case studies presented in this volume further take up the call of Darden and Maull and explore other kinds of theoretical connections between disciplines. Because any analysis of these cases must draw heavily on the specifics of the cases, I will postpone detailed comments until the Editor's Commentary following each set of papers. At this point I will merely note some of the patterns of cross-disciplinary connections that were revealed. The first such pattern involves developing sufficient conceptual links between disciplines so as to use perspectives developed in one discipline to modify the perspective adopted in another related disciplines, without developing major theoretical structures that subsume the disciplines in question. This mode of cross-disciplinary theorizing applies to Richardson's discussion of the use of linguistic frameworks to re-analyze neurological deficits in aphasics and to Kauffman's and Wimsatt's proposals of how to use developmental considerations to reconceptualize some aspects of evolutionary theorizing. A second pattern involves the recognition of a new level of organization with its own set of processes to solve problems unsolved within existing fields. This pattern is illustrated in two cases, the development of biochemistry and the development of the synthetic theory of evolution. The new theories developed at the new levels were designed to interact with the perspective previously adopted at other levels, but it was the framework developed at the new level that was decisive in solving previously unsolved problems. Beatty's analysis of the evolutionary synthesis shows a third pattern of interfield theorizing, wherein one uses research techniques developed in one discipline to help elaborate a theoretical model in another. A fourth pattern involves taking a theoretical framework from one domain and modifying and extending it in order to apply it in another domain where independently researchers think such extension is plausible (Mason's and Anderson's proposals for taking cognitive perspectives adopted in human psychology to understand animal behavior). A final pattern involves developing a new theoretical framework that will reconceptualize research in now separate domains as it tries to integrate them. Rumbaugh and Sterritt exemplify this when they introduce their control theory as a vehicle for integrating the study of intelligence amongst humans and other animals. Savage–Rumbaugh and Hopkins' focus on the idea of communicative intention in order to overcome inadequacies of both the ethological and learning perspectives for dealing with animal communication further illustrate this pattern. This last case offers the prospect of further modifying another discipline, human cognitive psychology, insofar as the framework of communicative intention is found to be useful in dealing with human linguistic behavior.

All of the cases have in common that disciplines are being brought together and integrated so as to help solve identifiable problems. To this extent, Darden and Maull's concept of interfield theory is applicable. However, there is variation in the degree to which one well-developed theory is being proposed to bridge the disciplines. In the historical cases, such theories are more apparent, but one should not expect such well-worked out theories at the initiation of a cross-disciplinary research endeavor. Rather, one should expect suggestive frameworks that will need to be elaborated and tested. Some of the proposals in the three contemporary cases are more developed than others, but all are still at the stage of working proposals for developing interfield theoretical connections.

6. Conclusion

Within this introduction I have sketched a framework in which to examine interdisciplinary research. I have considered first various ways we might understand the units of science that can be labelled disciplines and then considered how these units have become isolated from each other so as to make interdisciplinary

48

endeavors problematic. I have also tried to indicate some of the factors that actually motivate scientists to engage in cross-disciplinary research so that our focus is on the real activities of science, not those of the *a priori* theorist of knowledge. Finally, I have explored some of the social and institutional as well as cognitive arrangements that are employed in cross-disciplinary endeavors. The analysis offered here is not intended as a definitive statement about what interdisciplinarity does or should involve, but only to provide a framework for further development. Some of that development is offered by the cases that follow, but they are only a small selection of possible cases for consideration in the life sciences. The topic of cross-disciplinary research is ripe for further inquiry both by scientists engaged in such efforts and by historians, philosophers, sociologists and others interested in meta-science. With such detailed studies we can hope to get past the idealizations that have frequented discussions of interdisciplinarity in science and come to a sound understand of what cross-disciplinary research entails, what it can accomplish, and how we can improve the potential for its success.

Acknowledgement

I am most grateful to Adele Abrahamsen, who has given me extensive help both in writing this introduction and in compiling this volume. Rita Anderson, Lindley Darden, and Steve Fuller have also made many very helpful comments on earlier drafts of this paper, of which I am most appreciative. Some of the ideas presented here were developed during a Fellowship for Independent Study and Research, awarded by the National Endowment for the Humanities, which I held during 1983.

References

Barmark, Jan and Wallen, Goran (1978). Knowledge production in interdisciplinary groups. Report number 37, second series, Department for Theory of Science, University of Gothenburg.

Barmark, Jan and Wallen, Goran (1980). The development of an interdisciplinary project. In Karin Knorr, Roger Krohn, and Richard Whitley (Eds.), *The social process of scientific investigation. Sociology of the sciences, Volume IV*. Dordrecht: Reidel.

Barnes, Barry (1977). *Interests and the growth of knowledge*. London: Routledge and Kegan Paul.

Barth, Richard T. and Steck, Rudy (1979). *Interdisciplinary research groups: Their management and organization*. Vancouver: International Research Group on Interdisciplinary Programs.

Bechtel, William (1984). Reconceptualizations and interfield connections: The discovery of the link between vitamins and coenzymes. *Philosophy of Science, 51*, 265–292.

Bechtel, William (1985). Teleological functional analyses and the hierarchical organization of nature. In N. Rescher, ed., *Teleology and natural science*. Landham, MD: University Press of America.

Ben-David, Joseph and Collins, Randall (1966). Social factors in the origins of a new science: The case of psychology. *American Sociological Review, 31,* 451–465.

Birnbaum, Philip H. (1978). Academic context of interdisciplinary research. *Educational Administration Quarterly, 14,* 80–97.

Birnbaum, P. H. (1983). "Predictors of long-term research performance. In S. R. Epton, R. L. Payne, and A. W. Pearson, eds., *Managing interdisiplinary research.* New York: John Wiley and Sons.

Bloor, David (1976). *Knowledge and social imagery.* London: Routledge and Kegan Paul.

Blumenthal, Arthur L. (forthcoming). The emergence of psycholinguitstics.

Bourdieu, P. (1975). The specificity of the scientific field and the social conditions of the progress of reason. *Social Science Information 14:* 19–47.

Bradley, D.; Garrett, M.; and Zurif, E. (1980). Syntactic deficits in Broca's aphasia. In D. Caplan (ed.), *Biological studies of mental processes.* Cambridge: MIT Press.

Brandon, Robert (1982). The levels of selection. In P. Asquith and T. Nickles (Eds.), *PSA 1982.* Volume 1. East Lansing: Philosophy of Science Association.

Bromberger, S. (1965). An approach to explanation. In R. J. Butler, ed., *Studies in Analytical Philosophy.* Second Series. Oxford: Blackwell.

Cambrosio, Alberto and Keating, Paul (1983). *The disciplinary stake: The case of chronobiology. Social Studies of Science, 13,* 323–353.

Campbell, Donald T. (1969). Ethnocentrism of disciplines and the fish-scale model of omniscience. In M. Sherif and C. W. Sherif, eds., *Interdisciplinary relationships in the social sciences.* Chicago: Aldine Publishing Company.

Campbell, Donald T. (1974). 'Downward causation' in hierarchically organized biological systems. In F. Ayala and T. Dobzhansky, eds., *Studies in the philosophy of biology.* Berkeley: University of California Press.

Carnap, Rudolf (1938). Logical foundations of the unity of science. In O. Neurath, R. Carnap, and C. Morris (Eds.), *International ecncyclopedia of unified science.* Volume I. Chicago, The University of Chicago Press.

Causey, Robert L. (1977). *Unity of science.* Dordrecht: Reidel.

Chubin, Daryl E. (1976). The conception of scientific specialties. *The Sociological Quarterly, 17,* 448–476.

Chubin, D. E. (1982). *Sociology of sciences: An annotated bibliography on invisible colleges, 1972–1981.* New York: Garland.

Churchland, Paul M. (1979). Scientific realism and the plasticity of mind. Cambridge: Cambridge University Press.

Coleman, William (1985). The cognitive basis of the discipline: Claude Bernard on physiology. *Isis, 76,* 49–70.

Crane, Diana (1969). Social structure in a group of scientists: A test of the invisible college hypothesis. *American Sociological Review, 34,* 335–352.

Crane, Diana (1972). *Invisible colleges.* Chicago: University of Chicago Press.

Cutler, Robert S. (1979). A policy perspective on interdisciplinary research in U.S. universities. In Richard T. Barth and Rudy Steck (Eds.), *Interdisciplinary research groups: Their management and organization.* Vancouver: International Research Group on Interdisciplinary Programs.

Darden, Lindley, and Maull, Nancy (1977). Interfield theories. *Philosophy of Science, 44,* 43–64.

Edge, D. (1979). Quantitative measures of communication in science: A critical review. *History of Science, 17,* 102–134.

Epton, S. R.; Payne, R. L.; and Pearson, A. W. (1983). *Managing interdisciplinary research.* New York: John Wiley and Sons.

Fodor, Jerry (1974). Special sciences (Or: Disunity of science as a working hypothesis. *Synthese, 28,* 97-115.

Garfield, E. (1979). *Citation indexing.* New York: Wiley.

Gold, S. E. and Gold, H. J. (1983). Some elements of a model to improve productivity of interdisciplinary groups. In S. R. Epton, R. L. Payne, and A. W. Pearson (Eds.), *Managing interdisciplinary research.* New York: John Wiley and Sons.

Grodzinsky, Y.; Swinney, D.; and Zurif, E. (1983). Agrammatism: Structural deficits and antecedent processing disruptions. In M. L. Keane, ed., *Agrammatism.* New York: Academic Press.

Gusdorf, Georges (1977). Past, present, and future in interdisciplinary research. *International Social Science Journal, 29,* 580-600.

Hagstrom, W. O. (1965). *The scientific community.* New York: Basic Books.

Hooker, C. A. (1981). Towards a general theory of reduction. *Dialogue, 20,* 38-59; 201-236; 496-529.

Hull, David (1978). A matter of individuality. *Philosophy of Science, 45,* 335-360.

Hull, David (1982). Exemplars and essences. Unpublished manuscript.

Knorr, Karin D. (1981). *The manufacture of knowledge. Toward a constructivist and contextual theory of science.* Oxford: Pergamon.

Knorr-Cetina, Karin D. and Mulkay, Michael (1983). Introduction: Emerging principles in social studies of science. In K. D. Knorr-Centina and M. Mulkay, eds., *Science observed.* London: Sage.

Kohler, Robert E. (1977). Rudolf Schoenheimer, isotopic tracers, and biochemistry in the 1930's. *Historical Studies in the Physical Sciences, 8,* 257-297.

Kohler, Robert E. (1982). *From medical chemistry to biochemistry.* Cambridge: Cambridge University Press.

Krantz, David and Wiggins, L. (1973). Personal and impersonal channels of recruitment in the growth of knowledge. *Human Development, 16,* 133-156.

Krohn, Roger (1980). Introduction: Toward an empirical study of scientific practice. In K. D. Knorr, R. Krohn, and R. Whitley, eds., *The social process of scientific investigation. Sociology of the sciences, Volume IV.* Dordrecht: Reidel.

Kuhn, Thomas (1970). *The structure of scientific revolutions.* Second Edition. Chicago: University of Chicago Press.

Lakatos, Imre (1970). Falsification and the methodology of scientific research programmes. In I. Lakatos and A. Musgrave, eds., *Criticisms and the growth of knowledge.* Cambridge: Cambridge University Press.

Latour, Bruno. and Woolgar, Steve. (1979). *Laboratory life. The social construction of scientific facts.* Beverly Hills: Sage Publications.

Laudan, Larry (1977). *Progress and its problems.* Berkeley: University of California Press.

Law, John (1973). The development of specialties in science: The case of X-ray protein crystallography. *Science Studies, 3,* 275-303.

Lemaine, Gerald; MacLeod, Roy; Mulkay, Michael; and Weingart, Peter (1976). Introduction: Problems in the emergence of new disciplines. In *Perspectives on the emergence of scientific disciplines.* The Hague: Mouton.

Machamer, Peter (1977). Teleology and selection processes. In Robert G. Colodny, ed., *Logic, laws and life: Some philosophical complications.* Pittsburgh: University of Pittsburgh Press.

Machlup, Fritz (1982). *Knowledge: Its creation, distribution, and economic significance. Volume II. The branches of learning.* Princeton: Princeton University Press.

Maull, Nancy (1977). Unifying science without reduction. *Studies in the History and Philosophy of Science, 8,* 143-162.

Mayr, Ernst and Provine, William B. (1980). *The evolutionary synthesis. Perspectives on the unification of biology.* Cambridge: Harvard University Press.

McCauley, Robert (1981). Hypothetical identities and ontological economizing: Comments on Causey's program for the unity of science. *Philosophy of Science,* **48,** 218–227.

McCauley, Robert (in press). Intertheoretic relations and the future of psychology. *Philosophy of Science.*

Merton, Robert K. (1973). *The sociology of science.* Chicago, The University of Chicago Press.

Mullins, Nicholas C. (1972). The development of a scientific specialty: The phage group and the origins of molecular biology. *Minerva,* **10,** 51–82.

Nagel, Ernst (1961). *The structure of science.* New York: Harcourt, Brace and World.

Neurath, Otto (1938). Unified science as encyclopedic integration. In O. Neurath, R. Carnap, and C. Morris (Eds.), *International encyclopedia of unified science.* Volume I. Chicago, The University of Chicago Press.

Nickles, Thomas (1973). Two concepts of intertheoretic reduction. *The Journal of Philosophy,* **70,** 181–201.

Polanyi, Michael (1966). *The tacit dimension.* New York: Doubleday, 1966.

Popper, Karl (1972). *Objective knowledge.* Oxford: Oxford University Press.

Price, Derek J. de Solla (1961). *Science since Babylon.* New Haven: Yale University Press.

Pylyshyn, Zenon (1984). *Computation and cognition.* Cambridge: MIT Press/Bradford Books.

Reber, Arthur S. (forthcoming). The rise and (surprisingly rapid) fall of psycholinguistics.

Richardson, Robert C. (1980). Intentional realism or intentional instrumentalism. *Cognition and Brain Theory,* **3,** 125–135.

Rosenberg, Alexander (1980). *Sociobiology and the preemption of social science.* Baltimore: Johns Hopkins University Press.

Rosenberg, Charles (1979). Toward an ecology of knowledge: On discipline, context, and history. In A. Oleson and J. Voss, *The organization of knowledge in modern America: 1860–1920.* Baltimore: Johns Hopkins University Press.

Rossini, F.; Porter, A. L.; Chubin, D. E.; and Connolloy, T. (1983). Cross-disciplinarity in the biomedical sciences: A preliminary analysis of anatomy. In S. R. Epton, R. L. Payne, and A. W. Pearson (Eds.), *Managing interdisciplinary research.* New York: John Wiley and Sons.

Ruse, Michael (1973), *The philosophy of biology.* London: Hutchinson.

Russell, M. G. (1983). Peer review in interdisciplinary research: Flexibility and responsiveness. In S. R. Epton, R. L. Payne, and A. W. Pearson (Eds.), *Managing interdisciplinary research.* New York: John Wiley and Sons.

Salmon, Wesley (1971). *Statistical explanations and statistical relevance.* Pittsburgh: University of Pittsburgh Press.

Salmon, Wesley (1984). *Scientific explanation and the causal structure of the world.* Princeton: Princeton University Press.

Saxberg, Borje O., Newell, William T., and Mar, Brian W. (1979). The integration of interdisciplinary research with the Organization of the University. In Richard T. Barth and Rudy Steck (Eds.), *Interdisciplinary research groups: Their management and organization.* Vancouver: International Research Group on Interdisciplinary Programs.

Saxberg, B. O. and Newell, W. T. (1983). Interdisciplinary research in the university: Need for managerial leadership. In S. R. Epton, R. L. Payne, and A. W. Pearson (Eds.), *Managing interdisciplinary research.* New York: John Wiley and Sons.

Schaffner, Kenneth (1967). Approaches to reduction. *Philosophy of Science,* **34,** 137–147.

Scriven, Michael (1962). Explanations, predictions, and laws. In H. Feigl and G. Maxwell (Eds.),

Minnesota studies in the philosophy of science. Minneapolis: University of Minnesota Press.

Shapere, Dudley (1974). Scientific theories and their domains. In F. Suppe (Ed.), *The structure of scientific theories.* Urbana: University of Illinois Press, pp. 518–565.

Shapere, Dudley (1984a). Remarks on the concepts of domain and field. In D. Shapere (Ed.), *Reason and the search for knowledge.* Dordrecht: Reidel.

Shapere, Dudley (1984b). Alteration of goals and language in the development of science. In D. Shapere (Ed.), *Reason and the search for knowledge.* Dordrecht: Reidel.

Shapin, Steven, (1982). History of science and its sociological reconstructions. *History of Science,* **20,** 157–211.

Teich, Albert H. (1979). Research centers and non-faculty researchers: Implications of a growing role in American universities. In Richard T. Barth and Rudy Steck (Eds.), *Interdisciplinary research groups: Their management and organization.* Vancouver: International Research Group on Interdisciplinary Programs.

Toulmin, Steven (1972). *Human understanding.* Princeton: Princeton University Press.

Wallen, Goran (1981). The interaction between the development of knowledge and organization in the Swecon project. Unpublished internal report.

Whitley, Richard (1976). Umbrella and polytheistic scientific disciplines and their elites. *Social Studies of Science,* **6,** 471–497.

Whitley, Richard (1980). The context of scientific investigation. In K. D. Knorr, R. Krohn, and R. Whitley (Eds.), *The social process of scientific investigation.* Dordrecht: Reidel.

Whitley, Richard (1982). The establishment and structure of the sciences as reputational organizations. In N. Elias, H. Martins, and R. Whitley (Eds.), *Scientific establishments and hierarchies. Sociology of the sciences, Volume VI.* Dordrecht: Reidel.

Whitley, Richard (1984). The rise and decline of university disciplines in the sciences. In R. Jurkowich and J. H. P. Paelinck (Eds.), *Problems in interdisciplinary studies.* Hampshire: Gower Publishing Company.

Wimsatt, William C. (1972). Teleology and the logical structure of function statements. *Studies in the History and Philosophy of Science,* **3,** 1–80.

Wimsatt, William C. (1976). Reductionism, levels of organization, and the mind-body problem. In G. Globus, G. Maxwell, and I. Savodnik (Eds.), *Consciousness and the brain: A scientific and philosophical inquiry.* New York: Plenum Press.

PART I
THE COMING TOGETHER OF BIOCHEMISTRY

Introduction

AHMED ABDELAL
*Department of Biology, Georgia State University, Atlanta, Georgia
30303–3083, U.S.A.*

In our current period the status of biochemistry as an energetic cross-disciplinary endeavor has been largely obscured. This is partly due to the fact by the 1930s biochemistry had coallesced into a well defined discipline, with graduate departments, professional organizations, and journals. It is also no doubt due to the fact that in more recent decades molecular biology has, in part, consciously, defined itself in contrast to biochemistry. In the process, biochemistry has sometimes been viewed as a dry discipline that is no longer a focus of interest. This perception of classical biochemistry stems from the fact that by the 1940s many of the major theories of biochemistry were well worked out. As a result, the vital nature of the early days of the discipline, when the pathways of intermediary metabolism were still very much in doubt and where controversy reigned, has been obscured from view. Many people no longer recognize that in those early days it was not yet clear that there would be a separate discipline of biochemistry and that those who contributed the foundational work to what we now call biochemistry were actually involved in research in numerous different disciplines.

During the past two decades, it has been increasingly recognized that significant further progress in dealing with fundamental biological questions requires the integration of a number of disciplines, particularly biochemistry (now treated as a discipline), genetics, physiology, and biophysical chemistry. The term "molecular biology" has been frequently employed to describe such interdisciplinary studies, and in the process biochemistry has been more narrowly defined. As a result, exciting developments in the biological sciences have become associated with molecular biology rather than biochemistry.

The papers that follow return to the early gestational days of biochemistry and explore the process by which a distinctive discipline evolved. Both papers focus on the critical period from 1900 to 1940. However, it is useful to remember that the idea of chemical explanations of biological phenomena has a much longer ancestory. While speculative proposals as to the chemical events occurring in the

Bechtel, W (ed), Integrating Scientific Disciplines. ISBN 90-247-3242-5.
© *1986, Martinus Nijhoff Publishers, Dordrecht. Printed in The Netherlands.*

body can be traced back to the earliest days of physiology and medicine, the modern period is conveniently treated as beginning with the achievements of Lavoisier just before the beginning of the 19th century (see Holmes, 1985). Early in the 19th century the basic chemical constitution of the body was analyzed and by the 1840s concerted efforts were underway to trace the chemical conversions within plants and animals. The work of Justus Liebig (see his 1842) in many respects marks a major episode in this development.

During this period there was considerable controversy over the possibility of explaining all the physiological processes in organisms in chemical terms. While some researchers advocated the attempt to analyze all events in chemical terms, others claimed a special status for living organisms and denied the capacity of chemistry to ever account for the most fundamental processes of life. While the controversy is often cast as vitalism versus mechanism, this terminology can be rather misleading. As Lenoir (1982) has shown, many of those who opposed purely chemical explanations were not vitalists, they only insisted on the importance of physiological organization and that it had to be taken as a primitive in explaining the events occurring in living organisms. It was the fact that this organization was neglected in ordinary chemistry that made it impossible to develop purely chemical accounts of vital phenomena. This controversy came to a climax in the work of Bernard (1865), who tried to develop a program in which chemical explanations could be advanced for physiological processes that would still recognize the role of organization as critical to the realization of vital phenoman (see Holmes, 1974 and Coleman, 1985).

The process of fermentation provided one of the focal points for this controversy, following the proposal by Schwann (1837) that alcoholic fermentation is a physiological function of the yeast cell. Two of the leading chemists of the time, Berzelius (1836) and Liebig (1842), opposed the view that fermentation is the result of microbial activity on the basis of their belief that such a process can be explained in terms of the catalytic powers of certain chemical agents. The controversy was joined by Pasteur (1858) as he began his studies of fermentations between 1857 and 1876. It was these classical studies that convinced the scientific community that all fermentations were indeed the result of microbial activities.

The development of knowledge about the chemical basis of fermentation began only following the discovery by Buchner (1897) that fermentation could be accomplished in extracts in which the whole living cells had been destroyed. As Kohler (1973) has argued, this was a major factor in the development of the enzyme theory that played a guiding role in the development of biochemistry in the early 20th century. The detailed analysis of alcoholic fermentation led to the understanding that this physiological function of microorganisms can be explained in terms of a series of chemical reactions, each catalyzed by a specific enzyme. The premise that all biological phenomena can be explained in physiochemical terms provides a basis for modern molecular biological studies.

The following two papers deal with the development of biochemistry in the late 19th and early 20th centuries. The two papers have somewhat different foci and provide different perspectives on the major events in the development of biochemistry. Frederick L. Holmes is trained as an historian and has previously worked extensively on the development of physiological chemistry in the 19th century and is now involved in a detailed study of the events leading up to the development of the citric acid cycle in the 1930s. His focus is on how researches from different disciplines coallesced onto a common problem of detailing the chemical events involved in metabolism. He considers how, through a variety of different pursuits, researchers came to recognize some compounds as playing a central role in different metabolic processes and began to put together from a variety of perspectives the pieces of the metabolic mechanisms that are central to all living processes. William Bechtel, trained as a philosopher, also began work in the 19th century, focusing on transformations in the conception of the cell. In his paper he explores a reason why the result of interdisciplinary work on intermediary metabolism was the development of an independent discipline of biochemistry.

References

Bernard, Claude (1865). Translated as *An introduction to the study of experimental medicine*. New York: Dover, 1957.

Berzelius, Jons Jacob (1836). Einige Ideen 'über eine bei der Bildung organischer Verbindungen in der lebenden Natur wirksame, aber bisher nicht bemerkte Kraft. *Jahresbericht 'uber die Fortschritte der Chemie, 15*, 237–245.

Buchner, Eduard (1897). Alkoholische Gahrung ohne Hefezellen. *Berichte der deutschen chemischen Gesellschaft, 30*, 117–124.

Coleman, William (1985). The cognitive basis of the discipline: Claude Bernard on physiology. *Isis, 76*, 49–70.

Holmes, Frederic Lawrence (1985). *Lavoisier and the chemistry of life*. Madison: University of Wisconsin Press.

Holmes, Frederic Lawrence (1974). *Claude Bernard and animal chemistry*. Cambridge: Harvard University Press.

Kohler, Robert E. (1973). The enzyme theory and the origin of biochemistry. *Isis, 64*, 181–196.

Lenoir, Timothy (1982). *The strategy of life*. Dordrecht: Reidel.

Liebig, Justus (1842). *Animal chemistry or organic chemistry in its application to physiology and pathology*. Translated by W. Gregory. Cambridge: John Owen. Reprinted: New York: Johnson Publishing Company, 1964.

Pasteur, Louis (1858). Mémoire sur la fermentation appelée lactique. *Annales de chimie et de physique*

(3rd series), **52**, 404–418. Portions translated in James Bryant Conant, *Harvard case histories in experimental science*. Cambridge: Harvard University Press, 1957.

Schwann, Theodor (1837). Vorläufige Mitteilung betreffend Versuche 'über die Weingarung und Faulnis. *Poggendorf's Annalen der Physik und Chemie,* **11**, 184–193.

Intermediary Metabolism in the Early Twentieth Century

FREDERIC L. HOLMES
*Section of the History of Medicine, Yale University, School of Medicine,
New Haven, Connecticut 06510–8015, U.S.A.*

During the first third of this century, the subject of "intermediary metabolism" emerged as an active, clearly defined area of scientific research. By the end of that period intermediary metabolism appeared to constitute one of the principal divisions of biochemistry. The sub-field was well delineated, however, before biochemistry itself has solidified as a scientific discipline. Those who made contributions to this growing investigative stream between 1900 and 1930 published from departments of physiology, chemistry, organic chemistry, physiological chemistry, biochemistry, agricultural chemistry, botany, internal medicine, pathology, pathological chemistry, and others. Papers entered the literature from research institutes, the laboratories of hospital clinics, even from breweries. The parent field of the individual investigator often influenced the niche within the general problem area of intermediary metabolism from which he or she entered the sub-field. A paper emerging from a botanical or an agricultural institution, for example, was more likely to be about alcoholic fermentation than about muscle glycolysis. An investigator from a department of medicine might well approach intermediary metabolism from the vantage point of metabolic disorders, such as diabetes. Such contributions coalesced with others from institutions with other objectives, however, to form a common problem area that did not coincide with the boundaries of any of these fields. From whatever direction investigators came to a set of problems which lay at the intersection of their respective professional disciplines, in order to become recognized participants in the development of intermediary metabolism they had to adopt certain common concepts and problems, and to conform to certain methodological standards, as well as criteria for acceptable solutions. Each of the parent disciplines itself contributed some share to the intellectual framework and the methodological criteria, but none controlled them. A concensus gradually grew out of the nature of the problem itself. To some extent each of the disciplines involved may have helped to shape some aspects of the problem; but in the long run it was the intrinsic shape of the problem which

Bechtel, W (ed), Integrating Scientific Disciplines. ISBN 90-247-3242-5.
© *1986, Martinus Nijhoff Publishers, Dordrecht. Printed in The Netherlands.*

60

gradually reshaped the relationships between the disciplines.

The conceptual boundaries of intermediary metabolism were already clearly drawn in 1900 by the prior development of the field of general or quantitative metabolism. During the nineteenth century, chemists and physiologists developed methods for measuring accurately the food intake, the excretions, and the respiratory exchanges of humans and animals. With these methods they were able to establish the absolute rates and the relative proportions in which the three main classes of foodstuffs – proteins, fats, and carbohydrates – are consumed and decomposed to end products, the most important of which are carbon dioxide, water and urea. Many of those who carried out such studies assumed that there must exist extended series of step-by-step oxidations within the organism, connecting these starting and end points. Until late in the century, however, these intermediate stages, known by then to occur mainly within the minute cells of the tissues, seemed inaccessible to direct investigation. In the last decade of the century, those interested in metabolism began to probe more persistently the question of what takes place between the initial and final stages of these processes. Although the specific steps involved were mostly blank at the turn of the century, the highly developed knowledge of the end points provided sharply defined boundaries for the problem itself – boundaries that structured both the approach to the problem and the nature of the solutions that would be acceptable. Even as this new research area opened up previously unexplored investigative spaces, therefore, it presented from the beginning a set of what Herbert Simon (1977) defines as "well-structured problems."

It was not only the conceptual boundaries of intermediary metabolism that were set in advance by the earlier era of general metabolism, for the broad contours of the experimental means used to explore the new domain also followed those that had been developed in the older field. That field had grown out of the merger of two types of measurement: first, large respirometers that could determine the quantities of oxygen a human or animal consumed, and of carbon dioxide and water it exhaled; and second, a set of chemical methods for identifying quantitatively the input of each class of alimentary substance and the output of the products of metabolism in the excreta. These two components of the measurements made on intact animals in the nineteenth century were adapted in the twentieth century to corresponding measurements made on the component parts of organisms – first by perfusing isolated whole organs, then by developing more delicate means to measure the metabolic exchanges of isolated tissues. The bulky apparatus necessary to measure the respiratory exchanges of whole animals in the old era was represented in miniature in the new, by manometers that could measure with precision the relatively minute respiratory exchanges of tiny bits of tissue. The analytical methods for determining the comparatively bulk quantities of substances ingested and excreted by intact animals were represented by a more subtle array of micromethods for detecting the very small quantities of substrates that isolated tissues consumed from and discharged into the fluid media in which they were studied. Also derived from the quantitative metabolism of the previous century was the fundamental axiom of those who investigated intermediary metabolism in the new century: that it was through the *rates* at which substances are consumed in the tissues that one can best draw inferences concerning their respective roles in the metabolic processes.

This transition in the physiological level at which metabolic processes were investigated, from animals to tissues, was paralleled by a transition in the chemical level at which the central metabolic processes were defined and studied. Near the end of the nineteenth century the work of Emil Fischer and others began to reveal the molecular architecture of proteins and carbohydrates, showing them to be made up of smaller units linked through characteristic types of readily hydrolysable bonds. As strong evidence emerged that the foodstuffs are actually dissociated into these smaller molecules during digestion, the main thrust of investigations of metabolic processes could be transferred to these molecules – as Frederick Gowland Hopkins expressed it in 1913, from "complex molecules which elude ordinary chemical methods," to "simple substances undergoing comprehensible reactions" (Hopkins, 1947, p. 137; see also Fruton, 1974, pp. 108–120 and Holmes, 1979). Thus the question of protein metabolism could be sharpened down to the degradation and synthesis of the amino acids; that of carbohydrate metabolism to the transformations of the simple sugars and their derivatives; and that of fat metabolism to the fatty acids and glycerol. The structure of the field remained, however, heavily oriented around the original three categories. In the first decades of the twentieth century, the primary subdivisions of intermediary metabolism were still protein, carbohydrate, and fat metabolism, with only subordinate attention given to the metabolism of substances, such as nucleic acids, that did not fit into this old classification.

The major obstacle to attempts to identify intermediate metabolic compounds during the era of feeding experiments on whole animals was that these substances rarely accumulated in the excretions, or even the blood of the animals. From their very absence, however, investigators inferred that intermediates must gener a generally be decomposed as rapidly as they are formed. That was what one would expect if the intermediary processes formed reaction chains leading step-by-step from the foodstuffs to the final decomposition products. This same obstacle was also one of the reasons that some of the early contributions to the study of normal intermediary metabolism originated in medical research; for in some diseases, most notably diabetes, substances appeared in the urine that are under normal circumstances not excreted. The discovery in the 1880's that in the extreme condition known as acidosis, diabetics excrete -hydroxybutyric acid (CH₃ . CHOH . CH₃ . COOH), and acetoacetic acid (CH₃ . ĊO-CH₃ . COOH), and acetone (CH₃ Ċ-CH₃) became particularly important in this regard. These compounds, called collectively the "ketone bodies," soon came to be regarded as partial decomposition products that were excreted because diabetics could not further oxidize them (Minkowski, 1884a and b and Rosenfield, 1906). Much subsequent research in intermediary metabolism was directed at trying to link up the oxidation of carbohydrates and fatty acids with these compounds.

One way to obtain partial decomposition products from intact normal animals was to feed them not the natural compounds composing their foodstuffs, but analogous molecules that the animals might not be able to oxidize fully. In 1904 Franz Knoop achieved a major success through such a trick. After feeding dogs aromatic derivatives of fatty acids, he was able to identify in their urine aromatic acids that were always shorter by an even number of carbon atoms than those that the animal had ingested. Assuming that ordinary straight-chain fatty acids would undergo analogous reactions, Knoop inferred that the fatty acids of the body and of the normal nourishment are decomposed by the successive removal of two carbon atoms at a time. He named the process "β-oxidation." Subsequently other investigators confirmed Knoop's theory using organ perfusion methods. The β-oxidation theory provided the first sequence of intermediary reactions in a major metabolic pathway to be established with strong experimental evidence (Knoop, 1904 and 1905; see also Fruton, 1972, pp. 450–456). During the following three decades Knoop became the leading spokesperson for the point of view that the central goal of physiological chemistry ought to be to acquire

> "a knowledge of the course of the decompositions and oxidations of the building materials and nutrient substances in the animal organism which would leave no gaps" (Knoop, 1904, p. 3).

Between 1906 and 1908 Gustav Embden was able to link up Knoop's theory of β-oxidation with the appearance of the ketone bodies. Perfusing the isolated livers of dogs with fatty acids, he showed that when he added even-numbered fatty acids to the fluid entering the organ, there was an increase in the quantity of ketone bodies in the fluid emerging from the organ, whereas odd-numbered acids had no such effect. The β-oxidation theory could readily explain this result, since the successive removal of two atoms at a time from an even numbered fatty acid would eventually give rise to a four carbon fatty acid which would be converted, during the course of the oxidation of its next β-carbon atom, to β-hydroxybutyric acid. Embden showed by similar perfusion methods that β-hydroxybutyric acid was converted in turn to acetoacetic acid. Embden's fundamental investigation thus appeared to move a long way toward the goal of an unbroken series of metabolic reactions (Embden and Kalberlah, 1906; Embden, Salomon and Schmidt, 1906; Embden and Marx, 1908; and Embden and Engel, 1908).

To complete this picture, one would need to know how the ketone bodies themselves are normally further decomposed. Embden also took up that question. The natural intermediate to suspect was acetic acid (CH_3 COOH), because it was the expected product of one further β-oxidation of β-hydroxybutyric acid, and because when fed to animals it was readily metabolized. Embden was unable to obtain conclusive evidence for this view, but his research helped to bring acetic acid into special prominence as a probable nodal intermediary substance.

Too little was known about the specific intermediary steps in metabolic processes, during the first decade of this century, to support extended discussion of the patterns into which such steps might be connected. From passing comments, however, we can ascertain that it was commonly assumed that each class of component molecules of the foodstuffs – sugars, amino acids, or fatty acids – would give rise to its own linear sequence of successive oxidations, leading gradually but directly to their respective final end products. It was evident that these sequences could not be strictly independent of one another, since animals were known to be able to convert carbohydrate, or protein, or both, to fat. The nature of the interconnections, however, was ill-defined. The mergence of acetic acid as a likely product of fatty acid metabolism presented the first clearly delineated departure from this vague pattern. As a breakdown product of amino acids as well, and as a product of the fermentation of sugar, acetic acid was beginning to appear as a common link in the sequences of reactions that all three classes of foodstuffs undergo. Acetic acid offered, therefore, the first concrete picture of three pathways that did not run in parallel oxidation sequences, but converged upon a single, relatively simple intermediary compound.

The pioneering studies of respiratory oxidations in isolated tissues were begun, independently and almost simultaneously, in 1906, by Federico Battelli and Lina Stern in Switzerland, and Thorsten Thunberg in Sweden. Battelli and Stern used minced muscle preparations, in which they assumed the cell structures to be intact, suspended in blood or saline media. They measured the oxygen consumption by connecting the flasks containing the respiring tissues to simple U-shaped manometers. They attained respiratory rates approaching those of normal warm-blooded animals by maintaining the flasks in a constant temperature bath and agitating them continuously. The simplicity of their apparatus allowed them to carry out many experiments, and to test extensively the effects on the respiratory rate of adding to the media various substances, including substances the cells might be expected to oxidize. Thunberg utilized a more complicated, less flexible apparatus, but one that permitted more precise measurements. Although his initial assumptions and objectives differed markedly from those of Battelli and Stern, their respective research programs converged during the next four years. After systematically testing the effects of a large number of organic acids on the respiratory rate, Thunberg discovered in 1909 that malic acid, ($COOH$. $CHOH$. CH_2 . $COOH$), citric acid ($COOH$. CH_2 . $C(OH)COOH$. CH_2 . $COOH$), and succinic acid ($COOH$. CH_2 . CH_2 . $COOH$) increased the quantity of carbon dioxide formed. Learning of Thunberg's results, Battelli and Stern found that in their system succinic acid dramatically increased the overall respiration rate. The succinic acid was not completely decomposed, however, but only oxidized, as they thought, to malic acid. They had, apparently, been able to identify one of the discrete, single steps within the pathways of respiratory oxidation. Subsequently they showed that citric, malic, and fumaric acid ($COOH$. $CH = CH$. $COOH$) are totally oxidized (Battelli and Stern, 1907, 1910, 1911a and b, and Thunberg, 1909a and b, 1911).

Battelli and Stern's discovery was not accepted immediately as of capital importance, because the compounds in question – three small dicarboxylic acids and a tricarboxylic acid – were not regarded as metabolically important substances. They were not chemically related in any evident direct manner to the three classes of foodstuffs whose oxidative decomposition was deemed the central object of intermediary metabolism (Dakin, 1912, pp. 45–6). Nevertheless, the fact that, out of the many compounds tested, only these four conspicuously increased the respiration of isolated tissue, could not be overlooked. During the next twenty years the question of their role often attracted and puzzled investigators of intermediary metabolism.

The positions of these four major contributors – Knoop, Embden, Battelli and Stern, and Thunberg – to the early formative period in the study of intermediary metabolism illustrate the interdisciplinary nature of the emerging research area. Knoop carried out his investigation of fatty acid decomposition as a student in the laboratory of Franz Hofmeister, holder of one of the few chairs in physiological chemistry, at the University of Strassburg (Th., 1948). Embden too worked for a time in Hofmeister's laboratory, but when he began the investigation summarized above, he was director of the chemical laboratory of a clinic in Frankfurt. In 1907 he was able to establish an independent institute for physiological chemistry in the municipal hospital there, but in the same year he habilitated in experimental pathology at the University of Bonn (Deuticke, 1933). Thunberg was Professor of Physiology at the University of Lund when he began his experiments on tissue respiration (Monnier, 1944) while Battelli and Stern were *Privatdozenten* in the physiology laboratory at the University of Geneva (Kahlson, 1976). The nature of the problem made it a part of physiology and of chemistry, and it was possible to approach the subject, intellectually as well as institutionally, from either direction. In addition the problems it raised were of obvious interest to pathology.

In the pre-World War period the dominant subject for the study of the intermediary steps in metabolism was, in spite of the auspicious investigations summarized above, not animals, but yeast. The success of Eduard Buchner in carrying out alcoholic fermentation in cell-free yeast extracts stimulated intensive investigations of this process, with special attention to finding a sequence of reactions that might account for the chemically complex transformation of a 6-carbon sugar to the 2-carbon molecule ethyl alcohol. One could also study fermentation in intact yeast cells, which, as microorganisms, were already equivalent to the isolated cells of animal tissue, far more easily than one could at the time track down intermediary processes within animals.

A number of reaction schemes were proposed between 1900 and 1913, by Buchner himself, by Alfred Wohl, and others. Most of these were modifications of a scheme postulated by the organic chemist Adolph Baeyer in 1870, mainly on the basis of his knowledge of the reaction mechanisms of organic compounds. The later schemes were also based in part on inferences from reactions familiar to organic chemistry, but included as well compounds that were suspected to be intermediates because they either appeared in the fermentation solution or, if added to it, diminished in the course of the reaction. During the first decade of the century, a more stringent criterion, that a substance to be considered an intermediate must be capable of fermenting as rapidly as glucose does, was accepted in principle by most investigators. Exceptions to this rule were repeatedly made, however, when there appeared to be compelling theoretical reasons to incorporate a particular compound into a reaction sequence (Baeyer, 1870; Wohl, 1907; Buchner and Meisenheimer, 1904; Slator, 1907; see also Kohler, 1972).

In 1913 Carl Neuberg offered a theory of the reaction steps in alcoholic fermentation that resembled some of the earlier schemes in general outline, but that included some specific features that were attractive enough so that it soon dominated the field:

1. $C_6H_{12}O_6 - 2H_2O = C_6H_8O_4$
   sugar   methylglyoxal-aldol

2. $C_6H_8O_4 = 2CH_2{:}COH{\cdot}CHO$ (or $2CH_3{\cdot}CO{\cdot}CHO$)
              methylglyoxal

66

3. $CH_2:C(OH) CHO + H_2$ $CH_2OH CHOH CH_2OH$ (glycerol)

$$=$$

    $CH_2:C(OH) CHO + O$ $CH_2:COH COOH$ (enol pyruvic acid)

4. $CH_3CO COOH = CO_2 + CH_3 CHO$
   pyruvic acid   acetaldehyde

5. $CH_3 CO CHO$ O $CH_3 CO COOH$
   methylglyoxal   pyruvic acid

          $+$ $=$

   $CH_3 CHO$ H_2 $CH_3 CH_2 OH$
   acetaldehyde   ethyl alcohol

The pyruvic acid left at the end of reaction 5 re-enters reaction 4, so that the end products are the alcohol of the last reaction and the CO2 formed in reaction 4. Part of this scheme was based on two capital discoveries in which Neuberg had played a major part. In 1911, almost simultaneously with Otto Neubauer, he had found that pyruvic acid can be fermented. Soon afterward he could make it ferment as rapidly as glucose, so that pyruvic acid became the first compound to meet the stringent test for an intermediate. Neuberg showed in addition that the fermentation of pyruvic acid produces acetaldehyde and CO2, a reaction he could explain by the simple equation: pyruvic acid = CO2 + acetaldehyde. The CO2 had been released by splitting the carboxyl group off from this -keto acid. The enzyme that Neuberg inferred had caused this reaction he designated a ''carboxylase,'' emphasizing that it was the prototype for a general class of enzymatic decarboxylations. In addition to identifying a type of reaction which potentially could account for the shortening of a carbon chain in numerous metabolic situations, Neuberg brought pyruvic acid into the center of attention. Like acetic acid, pyruvic acid soon came to appear as a nodal intermediate expected to occupy the crossroads connecting various metabolic pathways. In contrast to the solid experimental evidence Neuberg had for including pyruvic acid and acetaldehyde, he placed methylglyoxal in the center of his scheme in spite of his failure to show that it could be fermented. He did so not only because it could be fitted into the scheme very nicely in accordance with known organic reaction mechanisms, but also because he could represent it as undergoing a special ''dismutation'' reaction which appeared to have great biological significance. In such a reaction half of the molecules are oxidized to carboxylic acids, while the other half are reduced to the corresponding alcohols. This reaction seemed particularly significant because it provided a mechanism for explaining biological oxidations that can take place without molecular oxygen. Incorporating the most exciting new developments in the field, Neuberg's fermentation theory was widely accepted for nearly twenty years (Neuberg and Hildescheimer, 1911; Neuberg and Tir, 1911; Neuberg and Karczag, 1911; Neuberg, 1913; Parnas, 1910).

Those who participated in the study of alcoholic fermentation in the early twentieth century also belonged to diverse scientific disciplines. Because the organism in which the phenomenon occurs is a plant, and because yeast was commercially important, those who worked in the area tended to come from different fields than those who worked on human and animal metabolism. They were less often fields associated with medical institutions, and more often with agricultural or industrial interests. Among those who made significant contributions, Alfred Wohl worked in the organic chemistry laboratory of the technical *Hochschule* in Danzig, and had connections with brewers; Buchner and Jakob Meisenheimer carried on research in the chemical laboratory of an agricultural *Hochschule* in Berlin; Leonid Iwanoff worked in the botanical section of a forestry institute in St. Petersburg; S. Kosteytschew worked in another botanical laboratory in St. Petersburg; and Peter Boysen-Jensen did research in a plant physiology laboratory of the University of Copenhagen. An exception to the non-medical orientation of the institutional affiliations of those who investigated fermentation were Arthur Harden and William Young, who worked in the biochemical laboratory of the Lister Institute of Preventive Medicine.[1] Neuberg's career epitomizes the interdisciplinary cast of the subject area in which he achieved prominence. Trained as an organic chemist, he worked successively in the Chemical Division of a Department of Pathology and the Chemical Unit of a Department of Animal Physiology in Berlin, before he became assistant director of the Kaiser Wilhelm Institute for Experimental Therapy in 1913. Retrospectively he is viewed as an outstanding biochemist, but he did not become one within the institutional framework of an established scientific discipline (Gottschalk, 1956).

Up until the time that Neuberg's theory became predominant, the investigators of alcoholic fermentation disagreed on many of the specific features of the process, and promoted rival variations on a basic outline of the reactions involved. They reached a common concensus, however, regarding the general methods appropriate to the study, the central problems to be solved, and the types of solution that were acceptable. These standards were imported, in part, from the fields from which the investigators came to the problem; but they were selected by, and grew up around, the demands of the problem itself. The nature of the problem created an informal community of investigators cutting across institutionalized disciplinary lines.

---

[1]   The institutional affiliations of these researchers were taken from the heading of articles they published during this period.

Moreover, because the approaches that appeared successful in the study of fermentation became a model for application to other metabolic problems, these standards spread in turn to investigations of human and animal metabolism within such fields as physiology, biochemistry, and experimental pathology. When Neuberg proposed his theory of alcoholic fermentation, it was already assumed that the anaerobic phase of carbohydrate metabolism in animal tissue was very similar to fermentation. With minor changes necessary to make the sequence of intermediary reactions end with lactic acid in place of alcohol, Neuberg's scheme prevailed in this research area also, until the end of the 1920's.

In the area of oxidative metabolism, there were also efforts, in the period between 1910 and 1930, to form a coherent picture of the way in which the pathways may be organized. The person who led the way was again Thunberg. In 1913 Thunberg had expressed the view that

> "The oxidative processes in the living cell must be thought of as forming chain reactions, a series of reactions connected to one another in such a way that, by and large, none of the links in the reaction chain can proceed more rapidly than the others."

In 1916 he found a new means to begin identifying these links. Hans Einbeck (1914), seeking to confirm Battelli and Stern's discovery that succinic acid is oxidized to malic acid in tissues, found instead that fumaric acid is produced. To Thunberg the conversion of succinic acid (COOH . CH$_2$. CH$_2$. COOH) to fumaric acid (COOH . CH = CH . COOH) suggested that the theory of Heinrich Wieland, according to whom in biological oxidations two hydrogen atoms are simultaneously removed from substrate molecules, was applicable to this first link in the chain of cellular oxidations to be identified experimentally. In order to test his interpretation, Thunberg devised a method to carry out the reaction anaerobically, in an evacuated test tube containing methylene blue to act as an acceptor for the hydrogen atoms that should, according to the theory, be released. When he mixed these substances with an extract from muscle tissue, the solution turned colorless, an indication that the dye had been reduced to the leuco-form by absorbing hydrogen. He concluded:

> "The biological oxidation of succinic acid must take place in the following way. First the succinic acid is dehydrogenated through the action of the muscle enzyme. If oxygen is then present, it exerts an oxidizing action on the hydrogen. But the hydrogen can also be transported to other substances which readily accept hydrogen" (Thunberg, 1913 and 1916).

By 1920 Thunberg had tested dozens of substances in his system, and found that the same four acids that had accelerated the respiration of isolated tissues in the earlier experiments he and Battelli and Stern had carried out – that is, succinic, fumaric, malic, and citric acid – all reacted strongly. In addition, the ordinary fatty acids, formic, butyric, and caproic, as well as lactic and several other acids were "unequivocal activators." In a long paper which became the standard point of reference for the study of oxidative intermediary metabolism, for the next decade, Thunberg set forth the principles that he believed should guide future endeavours to construct out of such results the sequences of the oxidative reaction chains. The substances that reduced methylene blue were not automatically to be regarded as participants in the metabolic chains, despite the fact that these reactions were enzymatic ones, taking place only in the presence of tissue extracts. To narrow down the possibilities, he argued that an activator of such a reaction is most likely to be an intermediate if the product of the dehydrogenation is itself a compound that undergoes a further reaction in the oxidative chain. At this stage such a test had to be made on theoretical grounds, because Thunberg's experiments showed only that the compounds in question reduce methylene blue. The product of the reaction had to be deduced from Wieland's dehydrogenation theory (Thunberg, 1920).

Applying these principles to the particular substances that he had found to reduce methylene blue, Thunberg inferred the reactions that could meet his criteria. Malic acid, for example, must give rise either to oxaloacetic or oxyfumaric acid:

$$COOH.CHOH.CH_2.COOH - 2H \rightarrow COOH.CO.Ch_2.COOH \text{ or } COOH.COH = CH.COOH$$

malic acid $\qquad\qquad$ oxaloacetic acid $\qquad\qquad$ oxyfumatic acid

For the case of acetic acid, Thunberg's guidelines brought him to an unusual and highly significant conclusion. Its activity in his system reinforced the existing view of its importance as an intermediate. Yet the elimination of two of its hydrogen atoms could not result in a known compound. He proposed instead

"a reaction in which two acetate molecules are simultaneously each deprived of one hydrogen atom, with the joining of their carbon atom chains into one. The substance which must thereby form is succinic acid.

$$2CH_3.COOH \rightarrow COOH.CH_2.CH_2.COOH + H_2$$

The transformation of acetic acid would in this way slip into the pathway opened by the conversion of succinic acid."

Noting that biochemists had long found it difficult to understand how acetic acid could be further decomposed, Thunberg now thought he could both explain this difficulty and establish a connecting link between the isolated segments of reaction chains that had been previously identified (Thunberg, 1920, pp. 31–34 and 54–55).

Three years later a student in Thunberg's laboratory, Gunnar Ahlgren, published a hypothesis which showed how Thunberg's methodological principles could connect such partial sequences into a closed system:

> If one sets out from the hypothesis formulated by Thunberg, according to which succinic acid is formed through the dehydrogenation of acetic acid (2 molecules of acetic acid − H_2 = 1 molecule of succinic acid) one can imagine the following circulation within the decomposition process:

1 mol. succinic acid-fumaric acid-malic acid-oxaloacetic acid-pyruvic acid-acetic acid

1 mol. succinic acid-fumaric acid-malic acid-oxaloacetic acid-pyruvic acid-acetic acid

$\left.\right\}$ 1 mol. succinic acid + $2CO_2$ + $2H_2$

Behind Ahlgren's somewhat awkward representation we can see that the hypothesis accounted for the complete oxidation of succinic acid, shown in overall form on the right, by a continuous chain of reactions. This chain joined Thunberg's theoretical synthesis of succinic acid from acetic acid with Battelli and Stern's respiratory oxidation of succinic acid as modified by Einbeck, and extended the chain through further dehydrogenations and hydrolyses to include all four of the closely related dicarboxylic acids. Then, by means of two successive decarboxylations based on the mechanism that Carl Neuberg had introduced from his studies of fermentation (in his discussion Ahlgren inserted an additional step between pyruvic acid and acetic acid, which was Neuberg's original fermentation reaction, pyruvic acid acetaldehyde + CO_2), he brought the hitherto isolated phenomenon of the respiratory activity of these acids into the broader picture of the decomposition of foodstuffs, by linking the former with the two acids – pyruvic and acetic – that had come to be viewed as pivotal to the intersecting paths of carbohydrate and fatty acid metabolism. Ahlgren specified that the connection to the -oxidation mechanism for fatty acids was probably through acetoacetic acid, an activator in the methylene blue system, which he thought could be divided by hydrolysis into two molecules of acetic acid (Ahlgren, 1923).

At nearly the same time that Ahlgren published this scheme, Franz Knoop included a nearly identical series of reactions in a more general discussion of the mechanisms of intermediary metabolism, (Knoop, 1923). Henrich Wieland too apparently drew independently from Thunberg's paper of 1920 the same inferences that Ahlgren and Knoop had reached. In 1925 Wieland presented in Oppenheimer's standard handbook of biochemistry a sequence of reactions for the oxidation of succinic acid that closely resembled those postulated by Ahlgren and by Knoop (Wieland, 1925; see also Wieland, 1922). The three versions of this hypothesis differed somewhat in emphasis. Ahlgren focused on Thunberg's synthetic reaction of acetic acid and the "circulation" that it closed. Knoop embedded the same reactions less conspicuously within a broader network of pathways of decomposition and synthesis. Wieland depicted the five reactions beginning with succinic acid as a biological manifestation of his dehydrogenation mechanism, relegating the synthesis of succinic acid to a subordinate and uncertain position. In spite of these nuances, however, contemporaries merged these three versions of a common hypothesis into what became known as the Thunberg–Knoop–Wieland scheme.

We may recall that Ahlgren was a student in a physiological laboratory, Knoop was a physiological chemist, and Wieland was an organic chemist. The differences in the priorities which the three gave respectively to certain aspects of this scheme may reflect personal orientations that can be attached to the fields from which they came. The striking feature of the situation, however, is that they converged on schemes that were so alike. The fact that they did so illustrates again that intermediary metabolism had become a coherent investigative stream, shared by several disciplines, and lapping over the conventional boundaries of each.

The Thunberg–Knoop–Wieland scheme provided a picture of the pathways of intermediary metabolism far different from the vague image of parallel linear decompositions that had prevailed at the beginning of the century. One could now envision well-defined reaction sequences forming closed cycles, chains of oxidative decomposition reactions interrupted by synthetic steps, and intersections that were beginning to link up the paths of carbohydrate, amino acid, and fatty acid metabolism into an integrated network. During the rest of the decade, those who directed themselves to the goal of filling in the steps of intermediary pathways found this scheme highly attractive. Some of them wove other, even more comprehensive loops around the basic cycle (Kuhnau, 1928). Others provided stronger evidence that individual reactions postulated in the scheme actually occur in biological tissues (Hahn and Haarmann, 1927, 1928, 1929a and b, 1930). The one reaction they could not demonstrate, however, was the most crucial one of all – the postulated synthesis of succinic acid from two molecules of acetic acid. As the 1930's began, therefore, the Thunberg–Knoop–Wieland scheme remained an enticing but unconfirmed hypothesis.

Reviewing the topic of "oxidations in the animal body" in 1931, Franz Knoop maintained as staunchly as he had in 1904, that "the final goal of physiological chemistry" [is to] "present a scheme that puts together an unbroken series of equations of all of the reactions from the foodstuffs which continuously supply to the organism its energy needs, all the way to the slag that again leaves the organism as energyless final oxidation products."

The problem had proven so difficult, however, he acknowledged thirty years after he had taken it up, that one could still provide only an "*a priori* conception of what such a scheme would have to look like" (Knoop, 1931, pp. 7 and 9).

From the foregoing account of those thirty years we can see that the *a priori* conception of which Knoop spoke had by then taken on a very concrete form. Although no complete reaction chains were firmly established in detail, the outlines to which they would have to conform were specified in a number of ways that closely restricted the acceptable solutions to the remaining unsolved problems. The starting and ending points of the chains had long been fixed. It was now generally accepted that the intermediate stages must pass by way of certain compounds such as lactic, pyruvic, and acetic acid, the ketone bodies, and the dicarboxylic acids. Certain sequences, such as the progressive shortening of fatty acids by -oxidation; the dicarboxylic acid series; the cleavage of hexose sugars into one of several possible 3-carbon compounds; and the early deamination of amino acids, were taken for granted. It was almost universally assumed that a small set of characteristic reactions, including in particular dehydrogenations, decarboxylations, and hydrolyses, would occur repeatedly along the various reaction chains. In his synthetic overview of 1931 Knoop pointed out certain broader architectural features of the reaction pathways that must apply no matter what the details might turn out to be. The ability of organisms to compensate for variations in their food supply by converting one form of foodstuff to another was a basic premise that must be explained in terms of specific bridges connecting the pathways of carbohydrate, fat, and protein decomposition. On general chemical grounds it was evident that six-carbon sugars, long-chain fatty acids, and amino acids could, in general, not be directly interconverted, so that they must be linked through the simple products common to their deeper stages of decomposition (Knoop, 1931, pp. 8, 22–23, 27–29). The combination of all these considerations left little room for theoretical maneuver. Unless in their long search investigators had somehow overlooked an entire route, or class of organic reactions, the main task facing them would appear to be to determine which of the few available alternatives for building the necessary bridges between the established pieces of the metabolic picture were the correct ones.

In 1933 the reign of the Neuberg methylglyoxal scheme suddenly ended, when Gustav Embden demonstrated convincingly that in glycolysis, fructose-1,6-diphosphate is cleaved into the 3-carbon compounds glyceraldehyde-3 phosphate and dihydroxyacetone phosphate. With some modifications added by Meyerhof in 1935, the reaction sequence presented by Embden became the definitive pathway for the anaerobic phase of carbohydrate metabolism (Fruton, 1972, pp. 347-352).[2] In 1937 Hans Krebs proposed the "citric acid cycle" of oxidative carbohydrate metabolism (Krebs and Johnson, 1937) that proved to be what the Thunberg–Knoop–Wieland scheme had once tried to be. These were brilliant discoveries, and they have come to be seen as foundations upon which the modern field of intermediary metabolism is built. Only the Knoop β-oxidation theory of fatty acid decomposition survived intact from the previous era as one of the major metabolic pathways. Although the investigations of the thirties supplanted most of the coherent pathways previously proposed, they were not revolutionary. The Embden–Meyerhof pathway and the Krebs cycle did not result from sharp departures from the approaches that developed during the preceding decades. They were successful largely because they fulfilled superlatively the criteria for acceptable solutions to these problems that had been established by an extended era of active investigation.

In current parlance we would probably identify the Embden–Meyerhof pathway and the Krebs cycle as the paradigms around which intermediary metabolism coalesced as a sub-field of biochemistry. These achievements probably did function similarly to the way Thomas Kuhn postulates that a paradigm organizes further investigations in what it constitutes as a field of normal science. If that is so, however, then the order of events is the reverse of those depicted by Kuhn as typical. It was not the success of the paradigms that supplied the methods, the criteria, and hidden assumptions that enabled "scientists to investigate some part of nature in a detail and a depth that would otherwise be unimaginable" (Kuhn, 1970, p. 24). Criteria were set out and guided scientists through years of detailed investigation of a special part of nature before these investigators attained any success dramatic enough to qualify as a paradigm. The paradigm achievements grew out of the general acceptance of these methods, criteria, and assumptions, rather than the other way around.

It should also be clear that the methods and standards shared by those who participated in the study of intermediary metabolism during the first third of the century were not imposed by any single organized discipline. The stream of investigation flowed, during this period, through channels that were not fully

---

[2] A sociological interpretation might invoke the "gatekeepers" of scientific disciplines, the editors of the journals, as setting these standards for what they would accept for publication in this area. The burden would be on such an interpretation to show that such editors imposed standards that differed from concensus views of the investigators in this area.

contained in any of the scientific fields formally established through institutional structures and labels. Although the conceptual and methodological criteria that defined intermediary metabolism by 1930 were drawn from the contributing fields, especially organic chemistry, physiology, and physiological chemistry, those that prevailed were the ones that had proven effective over the preceding decades by interactions among the participants in this area itself.

This case suggests that when we talk about disciplines as though they divide scientific inquiry into discrete territories, we may be taking a metaphor too literally. Perhaps we also view "intellectual territory" too much from the perspective of our capitalistic culture, assuming that domains of inquiry, like the terrain filling our physical landscapes, must be someone's property. The story told here suggests that a research area may thrive even though it is not in the possession of any of the disciplines that contribute to its progress. The relation between research areas and disciplines is fluid and shifting, and discipline boundaries can be very porous. A progressive research area at the intersections of two or more fields can attract investigators from each of them, and these individuals can comprise a community of shared interests and practices that exists athwart the domains of formally organized disciplines. Such a condition need not be a mere transition stage in the formation of a new discipline. As an interdisciplinary area of investigation, intermediary metabolism lasted for at least thirty formative years. By hindsight we may treat those years as the ones in which biochemistry was acquiring the institutional foundations that would ultimately enable it to absorb the "territory" of intermediary metabolism.[3] Even so, the period before it did so lasted for as long as the subsequent period in which intermediary metabolism was predominantly a subfield of biochemistry. More recently still, further shifts in the fluid relation between this research area and institutionally defined disciplines have occurred, so that the problems of intermediary metabolism extend almost as deeply into the domains of cellular biology, molecular biology, physiology, and other fields, as they do within biochemistry. Scientific problem areas are more natural than, and often more stable than, the socially constructed disciplines which lay claim to them.

References

Ahlgren, G. (1923). Sur le champ d'action des hydrogénases musculaires. *Acta Medica Scandinavica*, **57**, 508–510.

Baeyer, A. (1870). Ueber die Wasserentziehung und ihre Bedeutung für das Pflanzenleben und die Gährung. *Berichte der deutschen chemischen Gesellschaft*, **3**, 63–75.

[3]   For a detailed account of the acquisition of these institutional foundations, see Kohler (1982).

Battelli, F. and Stern, L. (1907). Recherches sur la respiration élémentaire des tissus. *Journal de Physiologie et de Pathologie générale,* 1–16.

Battelli, F. and Stern, L. (1910) Oxydation de l'acide succinique par les tissus animaux. *Comptes Rendus Société de Biologie,* 301–303.

Battelli, F. and Stern, L. (1911a). Oxydation der Bernsteinsäure durch Tiergewebe. *Biochemische Zeitschrift,* **30,** 172–178.

Battelli, F. and Stern, L. (1911b). Die Oxydation der Citronen-, Apfel- und Fumarsäure durch Tiergewebe. *Biochemische Zeitschrift,* **30,** 478–502.

Buchner, E. and Meisenheimer, J. (1904). Die chemischen Vorgänge bei der alkoholische Gährung. *Berichte der deutschen chemische Gesellschaft,* **37,** 417–418.

Dakin, H. (1912). *Oxidations and reductions in the animal body.* London: Longmans, Green.

Deuticke, H. J. (1933). Gustav Embden. *Ergebnisse der Physiologie und experimentelle Pharmakologie,* **35,** 32–49.

Einbeck, H. (1914). Über das Vorkommen der Fumarsäure im freschen Fleische. *Hoppe Seylers Zeitschrift für physiologische Chemie,* **90,** 303–307.

Embden, G. and Kalberlah, F. (1906). Über Acetonbildung in der Leber: Erste Mitteilung. *Beiträge zur Chemie, Physiologie und Pathologie,* **8,** 121–128.

Embden, G., Salomon, H. and Schmidt, Fr. (1906). Über Acetonbildung in der Leber: Zweite Mitteilung: Quellen des Acetons. *Beiträge zur Chemie, Physiologie und Pathologie,* **8,** 129–155.

Embden, G. and Marx, A. (1908). Über Acetonbildung in der Leber. *Beiträge zur Chemie, Physiologie und Pathologie,* **11,** 318–319.

Embden, G. and Engel, H. (1908). Über Acetessigsäurebildung in der Leber. *Beiträge zur Chemie, Physiologie und Pathologie,* **11,** 323–326.

Fruton, J. (1972). *Molecules and Life: Historical Essays on the interplay of chemistry and biology.* New York: Wiley-Interscience.

Gottschalk, A. (1956). Prof. Carl Neuberg. *Nature,* **178,** 722–723.

Hahn, A. and Haarmann, W. (1927). Ueber die Dehydrierung der Bernsteinsäure. *Zeitschrift für Biologie,* **87,** p. 107.

Hahn, A. and Haarmann, W. (1927). Ueber die Dehydrierung der Apfelsäure. *Zeitschrift für Biologie,* **87,** 465–471.

Hahn, A. and Haarmann, W. (1930). Ueber Dehydrierungsvorgänge im Muskel. *Zeitschrift für Biologie,* **89,** 563–572.

Holmes, F. L. (1979). Early theories of protein metabolism. In P. R. Srinivasan, J. S. Fruton, and J. T. Edsall (eds.), *The origins of modern biochemistry: A retrospect on proteins.* New York: New York Academy of Science.

Hopkins, F. G. (1947). The dynamic side of biochemistry. In J. Needham and E. Baldwin (eds.), *Hopkins and biochemistry.* Cambridge: Heffer.

Kahlson, G. (1976). Thorsten Ludvig Thunberg. In C. C. Gillispie (ed.), *Dictionary of scientific biography,* Volume 13. New York: Scribner's.

Knoop, F. (1904). *Der Abbau aromatischer Fettsäuren im Tierkörper.* Freiburg: Kuttruff.

Knoop, F. (1905). Der Abbau aromatischer Fettsäuren im Tierkörper. *Beiträge zur chemischen Physiologie,* **6,** 150–162.

Knoop, F. (1923). Wie werden unsere Hauptnährstoffe im Organismus verbrannt und Wechselseitig ineinander 'übergeführt? *Berliner Klinische Wochenschrift,* **2,** 60–63.

Knoop, F. (1931). *Oxydationen im Tierkörper.* Stuttgart: Enke.

Kohler, R. E. (1972). The reception of Eduard Buchner's discovery of cell-free fermentation. *Journal of the History of Biology,* **5,** 327–353.

Kohler, R. E. (1982). *From medical chemistry to biochemistry: the making of a biomedical discipline.* Cambridge: Cambridge University Press.

Krebs, H. A. and Johnson, W. A. (1937). The role of citric acid in intermediate metabolism in animal tissues. *Enzymologia, 4,* 148–156.

Kuhn, T. S. (1970). *The structure of scientific revolutions.* Chicago: University of Chicago Press.

Kühnau, J. (1928). Über den Abbau der Beta-Oxybuttersäure durch Fermente der Leber. *Biochemische Zeitschrift, 200,* 29–60.

Minkowski, O. (1884). Ueber das Vorkommen von Oxybuttersäure im Harn bei Diabetes mellitus. *Archiv für experimentelle Pathologie und Pharmakologie, 18,* 35–48.

Minkowski, O. (1884). Nachfrag über Oxybuttersäure im diabetischen Harne. *Archiv für experimentelle Pathologie und Pharmakologie, 18,* p. 150.

Monnier, M. (1944). Frédérick Battelli. *Ergebnisse der Physiologie und experimentelle Pharmakologie, 45,* 12–15.

Neuberg, C. and Hildesheimer, A. (1911). Über Zucherfreie Hefegärung, I. *Biochemische Zeitschrift, 31,* 170–172.

Neuberg, C. and Tir, L. (1911). Über zucherfreie Hefegärungen, IV. *Biochemische Zeitschrift, 36,* 60–67, 68–75.

Neuberg, C. (1913). Der Zucherumsatz der Zelle. In C. Oppenheimer (ed.), *Handbuch der Biochemie des Menschen und der Tiere,* Suppl. Vol., 581–582. Jena: Fischer.

Parnas, J. (1910). Ueber fermentative Beschleunigung der Cannizaroschen Aldehydumlagerung durch Gewebesäfte. *Biochemische Zeitschrift, 28,* 274–287.

Rosenfield, G. (1906). Fett und Kohlenhydrate. *Berliner Klinische Wochenschrift, 43,* 978–981.

Simon, H. A. (1977). *Models of Discovery.* Dordrecht: Reidel.

Slator, A. (1907). Über Zwischenprodukte der alcoholischen Gärung. *Berichte der deutschen chemischen Gesellschaft, 40,*p. 123.

Thunberg, T. (1909a). Ein Mikrorespirometer: ein neuer Respirationsapparat, um den respiratorischen Gasaustausch kleinerer Organe und Organismen zu bestimmen. *Skandinavisches Archiv für Physiologie, 15,* 74–85.

Thunberg, T. (1909b). Studien über die Beeinflussung des Gasaustausches des überlebenden Froschmuskels durch verschiedene Stoffe. *Skandinavisches Archiv für Physiologie, 22,* 406–427.

Thunberg, T. (1911). Studien über die Beeinflussung des Gasaustausches des überlebenden Froschmuskels durch verschiedene Stoffe. *Skandinavisches Archiv für Physiologie, 24* (1911): 22–61.

Thunberg, T. (1913). Zur Kenntnis einiger autoxydabler Thioverbindungen. *Skandinavisches Archiv für Physiologie, 30,* 289–290.

Thunberg, T. (1916). Über die vitale Dehydrierung der Bernsteinsäure bei Abwesenheit von Sauerstoff. *Zentralblatt für Physiologie, 31,* 91–93.

Thunberg, T. (1920). Zur Kenntniss des intermediären Stoffwechsels und der dabei wirksamen Enzyme. *Skandinavisches Archiv für Physiologie, 40,* 9–91.

Th., K. (1948). Franz Knoop zum Gedächtnis. *Hoppe–Seyler's Zeitschrift für physiologische Chemie, 282,* 1–8.

Wieland, H. (1922). Über den Mechanismus der Oxydationsvorgänge. *Ergebnisse der Physiologie, 20,* 498–500.

Wieland, H. (1925). Mechanismus der Oxydation und Reduktion in der lebendedn Substanz. In C. Oppenheimer (ed.), *Handbuch der Biochemie des Menschen und der Tiere, II.* Jena: Fischer.

Wohl, A. (1907). Die neueren Ansichten über den chemischen Verlauf der Gärung. *Biochemische Zeitschrift, 5,* 46–49.

Biochemistry: A Cross-Disciplinary Endeavor That Discovered a Distinctive Domain

WILLIAM BECHTEL

*Department of Philosophy, Georgia State University, Atlanta, Georgia
30303–3083, U.S.A.*

Introduction

While the quest to explain physiological phenomena in chemical terms has a long history, the development of a distinct cross-disciplinary research area in physiological chemistry was a product of the later part of the 19th century that came to full fruition only at the beginning of the 20th century. There are several factors that led to biochemistry becoming a flourishing area of science at this time. One of these was the intense effort directed toward discipline building by some of the original pioneers (see Kohler, 1982). Another was the development of a clear conception of the kind of process that was thought to be involved in intermediary metabolism. The idea that intermediary metabolism consisted of a sequence of basic chemical reactions opened up a domain of inquiry that was of significance to researchers from a variety of disciplinary orientations. In the previous paper Holmes has explored this issue in some depth, showing how the idea of a metabolic pathway emerged and researchers committed their efforts to identifying the intermediate reactions constitutive of metabolic processes.

While biochemistry became an active research area during this period, its ultimate state and the way in which it would relate to chemistry and physiology were not yet determined. Two options existed. One was that it could continue at the crossroads between the two general disciplines of physiology and chemistry (each of which possessed multiple specialties), being an applied area of organic chemistry and a foundational part of physiology. The other was that it could become a discipline in its own right. By now the outcome is clear – biochemistry shows all the signs of an independent discipline. This is a result that is of interest to the student of cross-disciplinary research and in this paper I will explore one of the reasons for it. The reason I will focus on is cognitive – I will show how developments in the understanding of the subject matter of biochemistry provided a basis for its autonomous status. By focusing on this aspect of biochemistry's development,

Bechtel, W (ed), Integrating Scientific Disciplines. ISBN 90-247-3242-5.
© *1986, Martinus Nijhoff Publishers, Dordrecht. Printed in The Netherlands.*

though, I do not mean to suggest that it provides the whole explanation and that other factors do not matter. But I will argue that this was one significant factor in the development of biochemistry.

What I shall be arguing is that the process that became the subject matter of classical biochemistry, intermediary metabolic processes like glycolysis, biological oxidation, and the like, occur at a level of organization that is neither chemical nor physiological. These reactions involve more than a series of chemical reactions, yet they occur in the cell below the level of physiologically distinct units. (Only after the time period I will be examining were sub-cellular structures such as mitochondria generally recognized as fundamental units involved in these processes.) Thus, the subject matter is not part of either organic chemistry or of physiology, but lies between. (See Kohler, 1975, for similar claims.)

This level of organization, however, was only discovered in the 1930s. Prior to then researchers had limited knowledge of the organized nature of biochemical processes and generally viewed the cell as what Herbert Simon (1980) calls "a nearly decomposable system". In a nearly decomposable system, each component of the system carries out its operations largely in isolation from other components so that the interaction between the components will be less significant than the autonomous operations of the components. While the output of one component may provide the input for another component, the activities within each component will be otherwise unaffected by those occurring in other components of the system. With the rise of the enzyme theory in the early years of the 20th century, it became common to think of the major active elements in intermediary metabolism as the enzymes that catalyzed the various reactions. To study the process, one could decompose the system into its separate enzyme systems. Coupled with this conception of the cell as a system nearly decomposable into its component enzymes was the perspective that it was a linear system in which substances were metabolized in a sequence of enzyme catalyzed steps until only waste products remained.

Had this view of metabolism as involving simply a linear set of reactions that were each accomplished by a discrete enzyme been substantiated, biochemistry could well have continued to be simply a hybrid discipline or simply an applied part of organic chemistry. The one feature that would have distinguished biochemistry would have been that it had to identify the particular set of enzymes included in the cell that directed metabolism along one sequence of reactions rather than other possible ones. Such pathways, however, would have been nothing more than the sequencing of several discrete enzymes processes.

A different conception of the domain of biochemistry emerged in the 1930s. Basic biochemical processes like glycolysis were seen to involve not just a sequence of reactions but highly integrated chemical systems. The significance of the

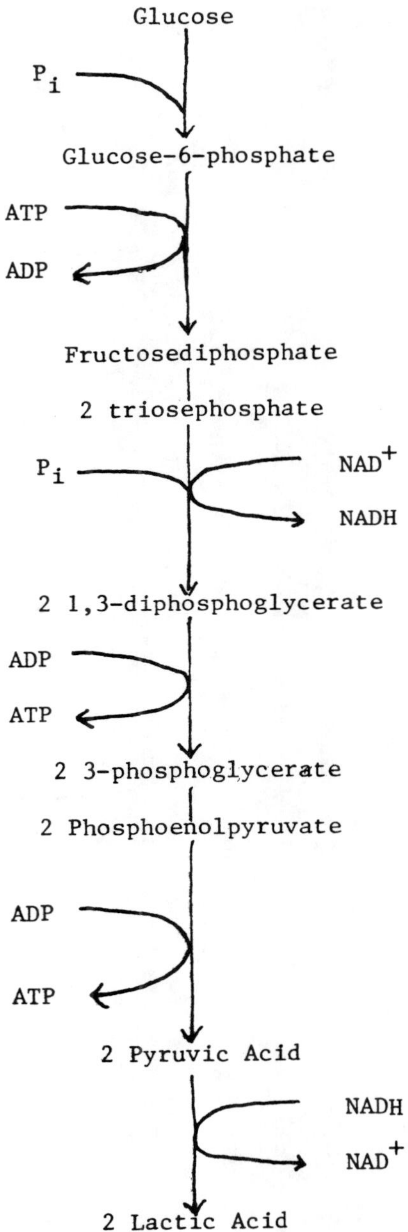

Figure 1. Major reactions in the glycolytic pathway (muscle glycolysis).

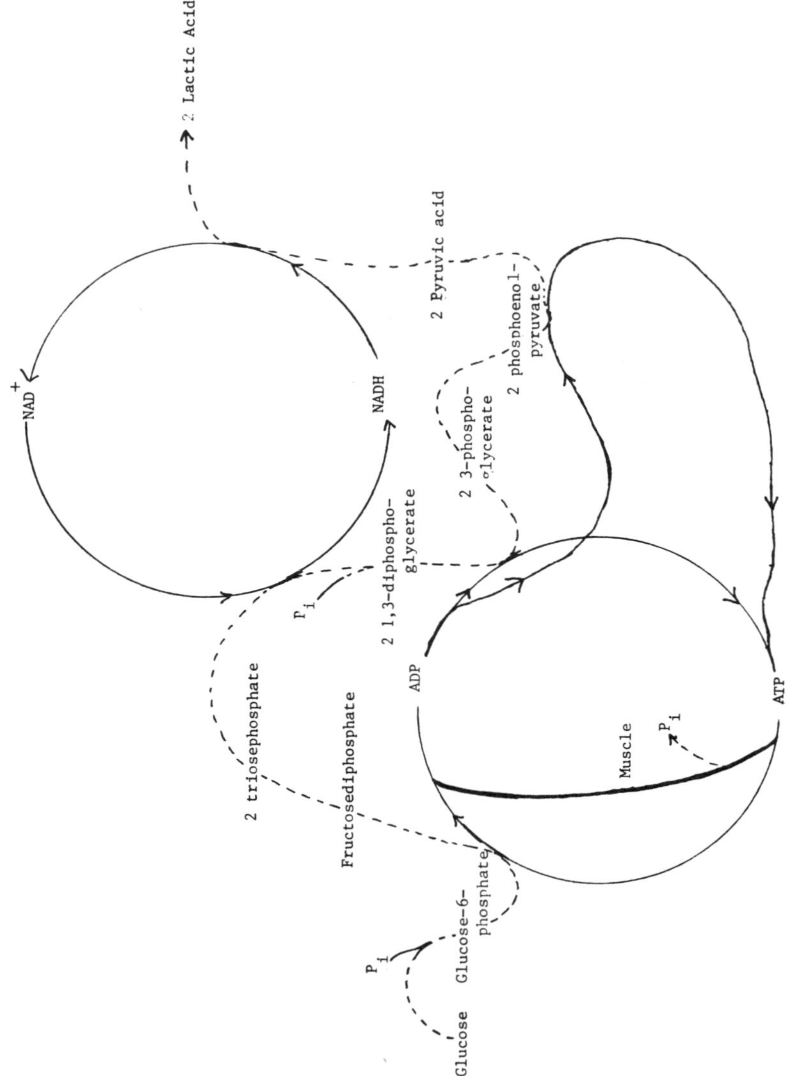

Figure 2. Same reactions as shown in Figure 1. The difference is that the coenzyme cycles are pictured as complete cycles, with the pathway of glycolysis passing through this established architecture.

development of this perspective within biochemistry is not fully grasped today. While the fundamental exemplars that students learn in basic biochemistry courses (e.g., the citric acid cycle, the process of energy transfer through "high energy phosphate bonds") are means by which this integration is achieved, the importance of this integration is partly obscured by the fact that many of the processes in intermediary metabolism are still commonly portrayed in terms of a sequence of reactions along a main reaction pathway. For example, glycolysis can be characterized as in Figure 1. However, the coenzymes, whose function is commonly portrayed (as here) only by loops indicating where they interface with the main pathway, actually serve to integrate the various reactions along the pathway. To recognize the integration that exists one has to notice that these coenzymes function cyclicly so that a product formed in one reaction (NADH or ATP) is then broken down in other reactions. This cyclic activity of the coenzymes links together different reactions with the result that one reaction may affect other reactions in the cell. As a result of this form of integration, the overall metabolic process cannot so readily be decomposed into discrete chemical reactions. (See Figure 2 for a presentation of the glycolytic pathway that shows the integrative function of the coenzymes.) Since the significance I attach to the discovery of coenzyme functions in the 1930s is not generally recognized, this picture of the history of biochemistry may strike some as novel and revisionist. Yet, it is a picture that, as I will try to show, both is supported by the historical record and helps to explain the current status of biochemistry as a discipline. (One researcher who did recognize at the time that biochemical processes constitute a more complex, integrated system was Needham. See Needham, 1937, where she describes an integrated phosphate cycle and predicts where new pieces will be discovered that will fit into that cycle.)

To make the difference between the two conceptions of the cell more concrete, I will present in the first section examples of the theories of intermediary metabolism that were prevalent during the first quarter of this century. In the second section I will contrast these with accounts developed in the 1930s, in which the function of various coenzymes was discovered and their role in integrating metabolic processes recognized. Some qualifications need to be made about the kind of contrast I will be drawing. In characterizing the biochemists in the early period as adopting a nearly decomposable view of the cell, I am not saying that they made this decision consciously or were strongly committed to it. Rather, it was the simplest hypothesis, and so the natural starting point for their research. Far from criticizing these early researchers, I would maintain that they played a critical role. Moreover, the advances that were made in the 1930s did not stem from other researchers adopting a different philosophical perspective. Rather, the modes of integration that were discovered resulted from investigators recognizing difficulties in the initial, simplified conception that treated intermediary metabolism as a more decomposable process.

In previous writings I have charcterized the view of metabolism that emerged in the 1930s as providing a partial vindication of the claims of those who opposed the reductionistic endeavor of the earlier generation of biochemists (Bechtel, 1984a), but it must be remembered that it was not those who protested the enzymologists who produced the major breakthroughs. Rather, it was the work of those adopting the reductionistic conception of biochemistry that would have made it into "applied chemistry" that led to the discovery of the integrating function of coenzymes and consequently to the conception of a higher level of organization involved in intermediary metabolism. My remarks in this paper are thus not intended to criticize those who assumed a nearly decomposable conception of intermediary metabolism but rather at recognizing the importance of an ontological reconceptualization that occurred subsequently.

The Early View of Metabolism as a Linear, Nearly Decomposable Process

The acceptance of the conception of metabolism as involving linear, nearly decomposable processes was itself revolutionary. Prior to the 1890s many physiological chemists thought that the mode of organization found in living systems radically distinguished these systems from non-living ones. Thus, Pflüger (1875) postulated the notion of a protoplasm molecule which, as a result of its structure, was able to accomplish various reactions that were not otherwise possible (e.g., oxidation at temperatures where oxidation was not typically possible). Several developments during the 1890s, foremost among which was Buchner's demonstration that cell extracts in which the cell structure had been destroyed could still carry out fermentation, helped to rekindle a view that was more commonly held earlier in the century. That view held that physiological processes were primarily chemical reactions. Thus, Buchner (1897) interpreted his results as showing "that the fermentive power of yeast juice is due to the presence of a dissolved substance" which he labelled "zymase". Developments like this encouraged biochemists to adopt the assumption stated by Hopkins that

> "in ... the intermediate process of metabolism we have to deal, not with complex substances which elude ordinary chemical methods, but simple substances undergoing comprehensible reactions" (Hopkins, 1913).

Although Buchner initially attributed the whole process of metabolism to a single enzyme catalyzing a single reaction, most investigators quickly recognized that several reactions were involved in carrying out physiological processes. Thus, Buchner and Meisenheimer (1904) revised Buchner's original proposal to argue that fermentation involved two enzymatically catalyzed reactions. The first, performed by zymase, produced lactic acid from sugar, while the second, accomplished by another enzyme named "lactacidase," converted lactic acid into alcohol. As Holmes has discussed in the preceding paper, Knoop's scheme of β-oxidation of fatty acids provided an exemplar of how a stepwise sequence of basic chemical reactions could accomplish an overall metabolic process. This model of a sequence of basic reactions received considerable support from other investigators like Dakin and Neubauer. Out of these numerous investigations the requirement developed that the component reactions must each involve a known chemical process like oxidation, decarboxylation, or deamination. (Note that this was not true of Buchner.) Thus, Knoop set as the goal for biochemistry the acquisition of

> "knowledge of the course of the decompositions and oxidations of the building materials and nutrient substances in the animal organism that would leave no gaps" (Knoop, 1904, p. 3).

The linear, nearly decomposable conception of intermediary metabolism articulated by Knoop was to guide research during the next three decades. As I noted above, researchers did not have a particular commitment to simple, linear chains. They merely offered the simplest possible models and so provided a useful starting point for investigations. Thus, researchers working on carbohydrate metabolism took as their task the endeavor to trace the sequence of decomposition and oxidation from glucose to either lactic acid (in the case of fermentation) or carbon dioxide and water (in the case of oxidation). In addition to trying to build models that employed already well-known basic chemical reactions, researchers had at their disposal a number of physiological techniques they could use to ensure that the models were biologically realistic. If one postulated that a substance was an intermediary, one could try to show that it actually occurred in living systems and was metabolized by them. It was often rather difficult to show that a potential intermediate actually appeared in biological systems since, if the intermediate met the second requirement of being metabolized by the system, it would not remain long in the system once it appeared. To overcome this problem, elaborate techniques were devised to trap potential intermediaries or to poison the system in such a way that metabolism would not continue beyond formation of the intermediary in question.

To see the kinds of theoretical frameworks developed during this period, consider Neuberg and Kerb's (1913) theory of fermentation, summarized in formulae 1–3 in Figure 3. This was the most widely accepted of various schemes of fermentation that were advanced prior to the 1930s. (See Harden, 1932, for a description of several other theories.) The basic mechanism proposed is that a sugar molecule is scissioned to form two molecules of methylglyoxal. Neuberg proposed that the methylglyoxal was then oxidized by oxygen from a molecule of water to form pyruvic acid. The hydrogen from the water molecule he supposed served to reduce an aldehyde, which itself resulted from the decarboxylation of the pyruvic acid formed in an early round of the reaction. (Neuberg proposed this scheme when oxidation was thought to involve the uptake of an atom of oxygen. The process whereby a single molecule of water was thought to oxidize one substrate and reduce another was referred to as a Cannizaro reaction or a dismutation.)

The proposal of this particular set of reactions grew out of numerous empirical investigations. Even prior to Buchner's success with cell-free fermentation, several organic chemists had used alkalis to catalyze the decomposition of sugar. These efforts yielded small quantities of alcohol and larger quantities of several three carbon products – dihydroxyacetone, glyceraldehyde, and methylglyoxal. This suggested the possibility that these substances were themselves intermediaries, a possibility which researchers tried to investigate by developing techniques by which they could fix and remove these substances from living cells. Neuberg himself pursued a long sequence of such investigations. For example, he employed sodium sulfate to prevent the reduction of acetaldehyde and show its occurrence in fermenting material (Neuberg and Reinfurth, 1920).

Besides trapping possible intermediates, one had to show how they fit together in a coherent scheme. It was this that seemed to make the strongest case for the scheme that Neuberg proposed, for the scheme utilized only fairly common reactions like the Cannizaro oxidation-reduction and decarboxylation. Thus the proposed sequence of reactions seemed well supported on both chemical and physiological grounds. It faced just one outstanding difficulty: methylglyoxal failed the test of itself being fermentable. Neuberg, however, offered an explanation for this apparent difficulty with this theory. He argued that there were many forms of methylglyoxal and proposed that only certain forms were fermentable. While many found this to be a stumbling block with his theory, no other theory seemed equally compatible with the available data.

For the most part, Neuberg's theory fit the conception of metabolism as involving a linear sequence of separable catabolic reactions. Two features of Neuberg's theory, however, violate this pattern and so are particularly noteworthy. First is the fact that the final product of the reaction, ethyl alcohol, is in a lower oxidation state than the intermediate product, pyruvic acid. The inclusion of pyruvic acid in the sequence is anomalous, since it requires an endothermic reaction to produce the more reduced substance ethyl alcohol and thus violates the assumption that the process involved simply a sequence of catabolic reactions. This exception to the pattern, however, was supported on empirical grounds. Neubauer had shown that it was readily metabolized by yeast cells (Neubauer and Fromhertz, 1911) and fixation studies had revealed its occurrence in fermenting cells.

A second notable feature of this pathway is that a later step in the reaction, the reduction leading to ethanol, is linked with the early oxidation step, making the process not strictly linear. This introduction of a non-linear feature in the pathway was born of chemical necessity. At the beginning of the reaction, before any aldehyde was available to be reduced by the hydrogen from the water in the Cannizaro reaction, Neuberg proposed that it was another molecule of methylglyoxal that was reduced in the Cannizaro reaction, forming glycerol (according to formula 4 in Figure 3). However, this would lead to a build up of glycerol, so Neuberg proposed that subsequently it was the now formed aldehyde that participated in the reaction. The two features of the Neuberg model that do not fit the general pattern of a linear, nearly decomposable system are simply the result of difficulties in fitting the data into the simpler pattern. They represent the beginning of the move to more complex models, but are not of the order of those to be developed in the 1930s.

$$2\ H_2O$$

(1.) $C_6H_{12}O_6\ \longrightarrow\ 2\ C_3H_4O_2$
    Hexose   Methylglyoxal

(2.) $C_3H_4O_2\ +\ C_2H_4O\ +\ H_2O\ \longrightarrow\ C_3H_4O_3\ +\ C_2H_5OH$

    Methylglyoxal Aldehyde   Pyruvic Acid Alcohol

(3.) $C_3H_4O_3\ \longrightarrow\ C_2H_4O\ +\ CO_2$
    Pyruvic Acid   Aldehyde

(4.) $2\ C_3H_4O_2\ +\ 2H_2O\ \longrightarrow\ C_3H_8O_3\ +\ C_3H_4O_3$
    Methylglyoxal   glycerol   pyruvic acid

Figure 3. Steps in glycolysis as proposed by Neuberg and Kerb (1913). Note that the aldehyde produced in step 3 is utilized in step 2. Step 4 represents a preliminary reaction that is used instead of step 2 until enough aldehyde is formed to permit the reaction of step 2.

The conception of metabolism as a linear, non-decomposable process extended beyond work on fermentation. The research on oxidation via dehydrogenation discussed by Holmes (previous paper) offers another clear example where researchers adopted this perspective. In this case, Thunberg (1916) proposed to account for the metabolism of the various dicarboxylic acids in terms of a sequence of decarboxylations. As Holmes describes, Thunberg departed from this scheme in the case of acetic acid. Since it could not be further dehydrogenated, he proposed a synthetic step whereby two acetic acid molecules were synthesized and then dehydrogenated to form succinic acid. Otherwise, Thunberg's scheme involved a linear model of discrete reactions. Thunberg himself appealed to an analogy with an assembly line to describe the type of process involved:

> "I consider the catabolism of the food stuffs to take place in a series of continuous dehydrogenations, carried out by a series of dehydrogenases. This procedure, to which the complicated food-stuff molecules are thus subjected, might be compared to what happens in modern factories where a piece of metal glides along on rails from workman to workman each of whom has his special task to carry out in the course of the work until the metal piece leaves their hands as a finished product" (Thunberg, 1930, p. 327).[1]

Thus, like Neuberg, the basic conception Thunberg adopted was of the metabolic system as a linear, nearly decomposable system. The complications he introduced were a result of chemical necessity.

Discovering the Integration of Reactions within the Cell

In this section I will focus on how researchers came to revise this conception of metabolism as a linear, nearly decomposable process as a result of discovering the functions of various coenzymes during the 1930s, but particularly adenosine triphosphate (ATP). Before describing how the discovery of these coenzymes led to a reconceptualization of metabolism, though, it is worth noting that the occurrence of coenzymes had been discovered much earlier than the 1930s. As a result of the linear, nearly decomposable conception of metabolism they adopted, however, researchers were unsuccessful in recognizing the role they played.

The need for coenzymes in reactions like fermentation was established shortly

---

[1]   Other remarks of Thunberg show that he thought each of the reactions in this sequence could be studied in isolation. For example, in response to the objection that breaking up the cell structure disrupts the normal biological reactions he argued: "what the cell structure seems to be able to effect in preference to a simple physical-chemical system, is a harmonic course of the continuous links of a long reaction chain and, further, the transformation of chemical energy into other forms of energy, e.g. mechanical energy (via surface tension changes or imbibation changes), electrical energy, osmotic energy, etc. But there is no reason to suppose that in order to take place a simple chemical process would require the presence of a cellular structure" (Thunberg, 1930, quoting an unspecified earlier publication).

after Buchner's original work. Harden and Young (1906) showed that in order to sustain fermentation in cell extracts they had to add boiled yeast juice. Boiling the yeast juice made it incapable of sustaining fermentation itself (this was correctly attributed to the heat incapacitating the thermolabile enzymes), but it seemed that even the boiled yeast juice carried something needed to sustain the reaction of the non-boiled yeast juice. They called this substance a "co-ferment" while other variously spoke of it as a "coenzyme," "cozymase," or "cofactor."

During the next quarter of the century numerous studies were done to learn the composition of the coenzyme and to determine its function in fermentation. In part, this endeavor was complicated by the fact that what Harden and Young took to be one cofactor was comprised of at least two substances, ATP and nicotinamide adenine dinucleotide (NAD). Despite this obstacle, significant progress was made in determining the basic molecular composition of the cofactor and points where "it" figured in the fermentation pathway. Various researchers claimed "it" facilitated either what was generally recognized as a necessary initial phosphorylation of the glucose or the oxidation- reduction step of the reaction. Subsequent inquiry showed that they were right on both of these counts, since ATP is involved in phosphorylation while NAD functions in the oxidation-reduction step. However, they had no clue as to how the coenzymes figured in such functions so that as late as 1932 Harden was led to comment: "The function of this coadjuctor is still unknown" (p. 17). Generally, until the 1930s the cofactor was simply thought to play some ancillary or facilitating role in the action of one of the enzymes. This, in fact, is the only kind of function that could be envisaged within the conception of metabolism as a linear, nearly decomposable process. The cofactor had to figure at *one* of the points of action.

What was discovered during the 1930s was that the different substances that make up the cofactor do not function at just one step in the pathway but serve to connect steps by transporting products formed from the substrate in one reaction to another reaction in which they are required. In this manner, the coenzymes serve to integrate the various reactions in the metabolic pathways. One example of this integrative function involves NAD, which, as the research of Warburg and his colleagues (Warburg and Christian, 1936) showed, functions by receiving the hydrogen removed in the course of oxidizing triose phosphate in one reaction and releasing it in a reduction of pyruvate. What NAD does, therefore, is to link these two reactions together so that each is dependent on the other. If the reduction of pyruvate is blocked, the oxidation of the triose phosphate is also inhibited. (I discuss the integrative function performed by NAD and other respiratory coenzymes further in Bechtel, 1984b.) In a sense, this integrative function assigned to NAD was already accounted for in the Neuberg and Kerb model of fermentation by the fact that they linked the oxidation of (in their scheme) methylglyoxal with the reduction of pyruvate in one reaction that they spoke of as a dismutation. These are, however, different reactions catabolized by distinct enzymes. In fact, the nature of the reduction is different in fermentation and muscle glycolysis due to the different enzymes involved. By treating them as one reaction Neuberg not only missed this relevant distinction but also did not notice the general pattern of integration contributed by the coenzymes.

In terms of its role in integrating functions in fermentation, ATP is even more important. Moreover, understanding its role required an even more radical reconceptualization of the metabolic process. The significance of ATP is that it serves as the conduit of energy within the cell. Prior to the 1930s, however, no role for such a conduit was envisioned for it was not thought that energy was stored and transported chemically. Hence, it was assumed that the issue of how energy was stored and transported could be safely ignored in devising a model of the chemical reactions. (Generally, it was assumed that the energy was transfered in the form of heat, and so numerous researchers did use the evidence about temperature changes as a guide in attempting to trace the disruption (comprehension and production) were orthogonal to the set of distinctions which linguists take as central (phonology, syntax, semantics, pragmatics, and lexical structure). Initially, the endeavor was to recast the aphasia research in terms of the linguistics distinctions (so as to treat Broca's aphasia as involving a syntactic deficit – see Bradley, Garrett, and Zurif, 1980). Now the aphasia work is suggesting that a further modification may be required in the linguistics categories so as to focus on a distinction between automated and non-automated functions (Grodzinsky, Swinney, and Zurif, 1983). Thus, a dialectic is occurring between different approaches to distinguishing cognitive function, with ideas from each discipline being modified to fit the other.

The discovery of the linkages between these processes began with the discovery in the late 1920s of two substances in the cell – creatine phosphate (identified by Fisk and Subbarow, 1927) and adenosine triphosphate (ATP) (identified by Lohmann, 1929). Both of these substances possessed a notable property – the phosphate bonds within them had a very high heat of hydrolysis. This suggested that these substances might play a role in muscle contraction. Lundsgaard established the reality of this possibility in 1930 when he showed that even when glycolysis is blocked, muscle contraction can continue as long as a supply of creatine phosphate remained. While ATP was equally marked by a high heat of hydrolysis, some of the early research on it focused on its role in the cofactor Harden and Young had established as necessary for alcoholic fermentation. Lohmann (1931) established that its presence was necessary for fermentation and, together with Meyerhof, went on to suggest that the breakdown of ATP to AMP (in actuality, only to ADP) at one stage in fermentation was coupled with the resynthesis of ATP later:

"the adenylpyrophosphate cycle maintains the lactic acid formation. The synthesis of phosphagen is therefore made possible ... by the cleavage energy of the adenylpyrophosphate, while the energy of lactic acid formation (from phosphate esters) serves to resynthesize the cleaved pyrophosphate" (Meyerhof and Lohmann, 1931, p. 576).

With this suggestion of a cycle, the framework for understanding the integrating role of ATP was beginning to take shape. However, piecing together a comprehensive account of the role of ATP took the remainder of the decade. A significant advance was made in 1934 when Lohmann established that ATP also functions in the hydrolysis of creatine phosphate. He treated it as a coenzyme that functioned by taking on a phosphate bond from the creatine phosphate and then surrendering it in turn. After providing evidence for this function of ATP, Lohmann comments on the use of ATP as a coenzyme both in fermentation itself and in the breakdown of creatine phosphate:

"Viewed teleologically, this dual function seems to be a very ingenious arrangement for insuring the orderly sequence of the chemical processes involved in the muscle twitch.... The contraction brings about a fission of adenylpyrophosphoric acid which in turn imposes a cleavage of creatine phosphate, thereby simultaneously reconstituting adenylpyrophosphoric acid; the latter can now interact as co-enzyme by mobilizing glycogen for lactic acid formation" (Lohmann, 1934a).

At this stage Lohmann was beginning to appreciate the integrating function of ATP, but another comment he made during the same year reveals what yet had to be learned:

"As regards the question of the chemical and energetic relationship of the breakdown and resynthesis of adenylpyrophosphate to the fundamental process of muscular contraction, it may be assumed that there is no direct relation" (Lohmann, 1934b).

Yet, this was precisely the kind of relation that was discovered in the later part of the decade. Engelhardt and Lyubimova (1939) established that muscle myosin was an enzyme for the hydrolysis of ATP and that this energy releasing reaction was coupled with the process of contraction. At the other end, researchers working on glycolysis were able to show first that it was phosphorylated substances like triose phosphate, not the unphosphorylated compound methylglyoxal, that figured in the fermentation process (Embden, Deuticke, and Kraft, 1932), and second, that the oxidation process was followed by a transfer of a phosphate group from phosphoglycerate to ADP (Parnas, Ostern, and Mann, 1934). (Parnas thought the transfer was actually to creatine phosphate, but this error in detail does not undermine the importance of the linkage he had found.) Subsequently it was established that the oxidation of triose phosphate is accompanied by the taking up

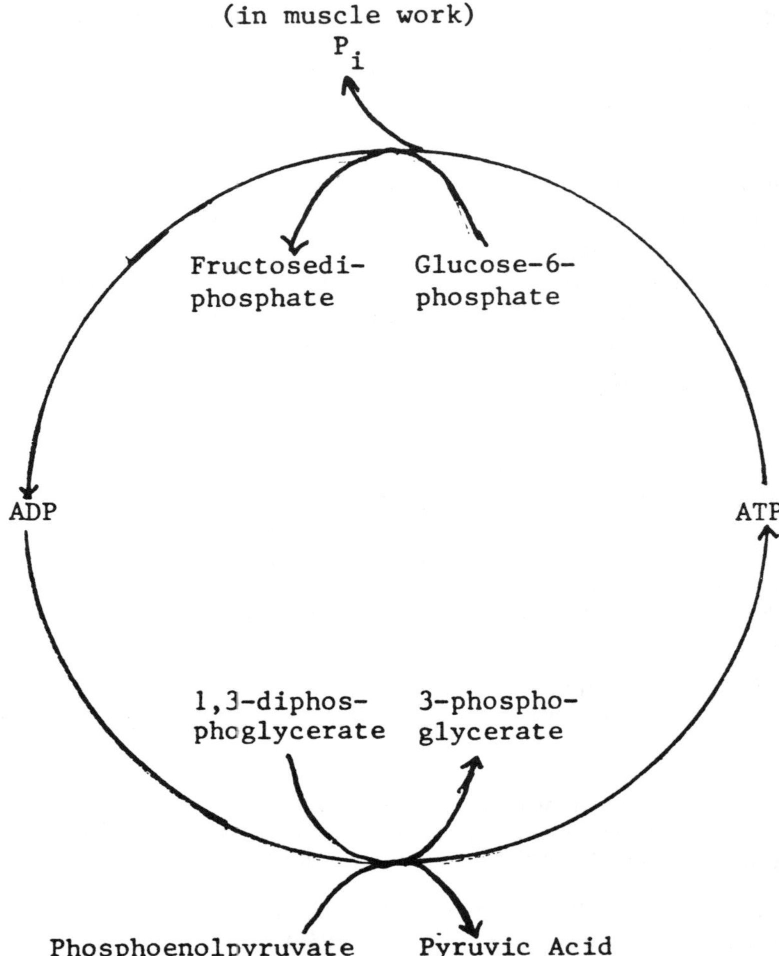

Figure 4. Role of ATP in transporting energy. Adding a phosphate bond to ATP requires significant energy, which is then liberated in the scissoning of the ATP molecule.

of an additional phosphate group to form a diphosphoglycerate, which then is transfered to ADP (Needham and Pillai, 1937, and Negelein and Bromel, 1939). Through these reactions, ATP was able to receive the energy produced by the oxidation reaction in fermentation. The stage was now set for Lipmann's introduction of the term "high-energy phosphate bond" for the bonds in ATP, diphosphoglycerate, and phosphoenolpyruvate and his comprehensive account of how these bonds provided for the storage and transfer of energy between cell reactions (Lipmann, 1941). (See Figure 4 for an overview of the role ATP serves in transfering energy between reactions. I have discussed the advances mentioned in this paragraph in more detail in Bechtel, in press.)

What resulted from the discovery of ATP was the recognition that the different energetic processes in the cell are coupled to each other. This coupling has the result that different reactions are not separable from one another. Whether glucose will be esterfied, which it must be if it is to undergo fermentation, depends on the availability of ATP, which is itself produced in later stages of fermentation (or in other reactions). Similarly, whether the triose phosphate, formed by scissoning the phosphorylated glucose, is fermented depends on the availability of ADP, to which the high-energy bond formed during the process can be transfered. This, in fact, explains the puzzling result, first noticed by Harden and Young, that in *in vitro* fermentation studies hexose diphosphate accumulated. They erroneously thought this showed that hexose diphosphate was not fermentable. However, the result was an artifact stemming from the fact that under *in vitro* conditions there was no ADP available to receive the high-energy phosphate bond. In this case, the assumption that the fermentation system was decomposable led to an artifact which confused researchers for nearly thirty years.

The discoveries of the coenzymes during the 1930s thus demonstrated that the cell was highly integrated. A particular reaction could be dependent on others not just for its substrate, but for providing coenzymes that were needed for that reaction. These discoveries further showed that if one did not recognize the modes of integration involved and assumed the system was nearly decomposable, one could end up producing artifacts and totally fail to understand how the reactions actually occur in the living organism.

Integrated Pathways as Defining a Level of Organization

In the previous sections I have contrasted the conception of intermediary metabolism researchers had before the 1930s with the one that emerged during that decade. What I will do in this section is show how the latter conception of intermediary metabolism provided biochemistry with a domain of its own at a distinct level of organization in nature, one that was at a higher level than the level of inorganic and organic chemistry but below that usually considered in physiology. In arguing that intermediary metabolism involved a higher level of organization than ordinary chemistry, I am rejecting the view that the integration of chemistry and physiology should be viewed as a reduction of physiological functions to chemical reactions. The key to rejecting the claim that a reduction occurred is the contention that metabolic reactions are the product of an *organized* system, not merely of independently operating components. Clarifying what is meant by the terms "organized system" and "level of organization" is critical to establishing this claim, since reductionists readily admit that there are complex systems that are built out of components and yet do not see this as creating any obstacles to a reduction. They will note that any time parts are brought together, there is some pattern of organization amongst the parts and they will proceed to treat organization as simply one of the boundary conditions that must be noted if one is to have a reduction of the kind described by Nagel (1961) or Causey (1977). What I will be arguing in this section is that it is the *type* of organization found in the systems of intermediary metabolism, not the mere occurrence of some mode of organization, that makes it inappropriate to view the relationship between physiology and biochemistry as one of reduction.

The question of mode of organization has become central in another recent controversy over reductionism, that concerning whether group selection can be reduced to individual selection. In analyzing the group selection controversy, Richardson (1982) has proposed a taxonomy of types of organized systems that will be useful in this context. His concern is to show what kind of organization rules out a reduction of group selection to individual selection. Thus, he is concerned to distinguish cases where a group trait must be treated as resulting from group selection from cases where it might be taken to be merely a consequence of selection operating on lower level traits.

In the course of his analysis, Richardson sets out a taxonomy of systems composed of interrelated components and evaluates each for whether it involves the kind of organization that requires an appeal to a higher level of organization. The initial parts of this taxonomy is based on three parameters (where a high value on a parameter indicates a lower grade of organization): aggregativity, intersubstitutivity, and decomposability. A system scores high in aggregativity if the output of the system depends not on the organization of the components but simply the quantity of components. It scores high on intersubstitutivity if the parts lack specialized functions and so can replace each other. Finally, a system scores high on decomposability if the functioning of parts does not depend on what occurs elsewhere in the system. Richardson presents his taxonomy such that each subsequent parameter is considered only after a low value is reached on the previous one. Thus, he establishes the hierarchical ordering of organized systems shown in Figure 5.

Richardson's hierarchy effectively captures several factors relevant to determining when an entity is at a higher level of organization and is not simply the product of the lower level components. We can best see this by proceeding up the hierarchy step by step. High aggregativity counts against appeal to a higher level

1. Equipotential systems (high intersubstitutivity, high decomposability)
    1.1. Aggregative systems (high aggregativity)
    1.2. Simple organized systems (low aggregativity; organization relevent to system function)
2. Composite Systems (low aggregativity, low intersubstitutivity)
    2.1. Component systems (high decomposability)
    2.2. Interlocking systems (low decomposability: component function partly organizationally determined)

Figure 5. A hierarchy of organized systems based on Richardson (1982).

entity, since we can explain the behavior of the whole aggregate simply by summing over the characteristics of the components. With simple organized systems organization begins to take on significance. The parts are intersubstitutable, but the contribution they make to the behavior of the whole system depends on how they are related to other parts. In this context, though, organization seems to be just a condition on the operation of the components and so the tendency to treat it as a background condition seems reasonable. As the components become less intersubstitutable because of their differing capacities, as they do in component systems, the potential arises for an interaction between the character of the components and the nature of the organization so that the reductionist attempt to differentiate the contributions of the parts and of organization becomes more suspect. However, as long as the system is decomposable so that we can examine each of the parts in turn and then add those functions together to determine the function of the whole, the reductionist can still propose to deal with the whole part by part. When the system ceases to be decomposable, though, the behavior of the parts can no longer be determined in isolation and this strategy fails. We must recognize the character of the whole system in developing explanations of its behavior.

Richardson's hierarchy helps us to explain how the discoveries of the 1930s showed that the metabolic systems in the cell constitute a higher level of organization. This hierarchy, however, cannot be interpreted strictly since even as researchers reduced their estimates as to how intersubstitutable its parts were and how decomposable it was, they continued to view these systems as highly aggregative. The conception of the cell as high in aggregativity may have been largely a reaction against earlier views that speculatively attributed the metabolic capacities of cells to the organization found in "living protoplasm". This legacy made many investigators extremely apprehensive of accounts that appealed to physical organization within the cell to account for the capacities of the cell.

Although there were deviations from this perspective (e.g., Warburg emphasizes on organized surfaces in his theory of oxygen transfer via Atmungsferment–see Kohler, 1973), it was only after the period I have been examining that the importance of organelles and mitochondria in providing a physical compartmentalization was discovered.

While still treating the cell as largely an aggregative system, researchers did reduce their estimates as to how intersubstitutable the parts were and how decomposable the system was. Intersubstitutability became a central issue after the identification of enzymes as responsible for catalyzing metabolic reactions. For example, after the development of the scheme of oxidation via dehydrogenation, some researchers proposed that one enzyme might be responsible for many different dehydrogenation reactions. Thunberg offered evidence (based on how heating and cooling affected different reactions) that the dehydrogenases were highly specific, but this was disputed until Quastel (1926) developed convincing arguments that at best the same dehydrogenase could work on different substrates with the same basic structure. During the time when the composition and function of the coenzymes was being developed, similar unclarity existed as to whether different coenzymes could intersubstitute for one another. The discovery of the different functions of ATP and NAD brought some clarity to this, but with the more closely related coenzymes, for example, flavin mononucleotide and flavin dinucleotide, opposing claims continued to be made during the 1930s about the specificity of function of each coenzyme.

The discussion of the previous section was directed primarily at showing how researchers came to change their understanding of how decomposable the cell was. My contention has been that the discovery of the functions of the different coenzymes revealed ways in which different reactions in the sequence were linked to one another so that one reaction could not continue unless the others with which it was linked functioned normally. Thus, the reactions in which ATP is synthesized could not be studied in isolation from those where it was broken down unless a fresh supply of ADP were continually provided. Because of the way these cyclic processes connect the different reactions in the metabolic pathways, the metabolic processes themselves must be attributed to the organized systems and not simply to their distinguishable components. Thus, the discovery of the coenzyme functions pushed the cellular metabolic system up to the highest level of Richardson's hierarchy as presented so far.

Richardson, though, adds an additional stage to his hierarchy, represented by what he calls an integrated system. He characterizes an integrated system as one involving functional subordination. Functional subordination is possessed by systems in which the functions of the components are defined only in terms of their contribution to the operation of the whole (or of a supersystem of which the system itself is a functional component). The question of how to define functions has been the subject of much recent philosophical discussion, but Richardson basically endorses the approach that identifies the function of something in terms of how it meets selective forces operating on it (see Wimsatt, 1972). What the notion of functional subordination does, therefore, is to distinguish components whose selection is dependent on the way they contribute to a larger whole from those who might be viewed as being selected independently. Whether the selection of a component must be viewed as so dependent on its contribution to a larger whole will depend on the way in which the parts of the system are interlocked. For example, if the system is appropriately interlocked such that a component performs its function only when activated by activity elsewhere in the system, then it would seem appropriate to construe the part not only as incidently serving a function for the system, but as subserving the system in performing that function.

Richardson's interest in this additional stage in the hierarchy of organized systems stems from his concern about when group selection can be demonstrated. While our concern is not with the group selection controversy, these same considerations can be applied to understanding the kind of organization found in the case of the metabolic pathways. The kind of linkage between reactions provided by coenzymes not only serves to make the system an interlocking one, but also allows the behavior of components to be subordinated to the needs of the whole system. For example, the concentration of ATP serves to govern the occurrence of the catabolic reactions involved in the release of energy, allowing them to occur only as needed to maintain cell functions. When all the available ADP has been turned into ATP, energy releasing reactions like oxidation and fermentation cease.

For the most part evolutionary evaluations of the functionality of the metabolic systems have been beyond the range of interest of biochemists. However, when biologists have discovered highly intricate systems like those involved in metabolism, some of them have been led to inquire as to why such systems exist. This curiousity is driven by the fact that organic chemists were aware of simpler reaction pathways that accomplished the same overall reactions. Krebs, who contributed to the discovery of what is perhaps the best known metabolic pathway, the citric acid cycle, later took up the inquiry as to why the cell adopted more complex modes of organization such as that found in the tricarboxylic acid cycle, the phosphate cycle, and in the cycles involving FAD and NAD, rather than the simpler modes identified by organic chemists. He suggested a variety of respects in which these more complex pathways serve to coordinate the overall reactions so that these reactions serve the needs of the whole cell. He noted, for example, that the cell must use non-reversible reactions if it is to release energy from the substrates. However, many of these reactions occur very rapidly so that if they were not

96

regulated in this way, enormous food supplies would be needed to restore the system. He argues that the regulation achieved through ATP, for example, makes it possible for the cell to use an overall sequence that is non-reversible and yet not destroy the "dynamic equilibrium" (Krebs borrows the phrase from Hopkins) of the cell by carrying all the reactions to completion. Krebs also suggests a second function of this more complex design involving ATP: by releasing the energy in small reactions it becomes possible for the energy to be released and stored in quantities useable elsewhere in the cell (Krebs, 1946).

What Krebs is proposing is that these modes of organization provide an evolutionary advantage. Krebs, in fact, recommends looking at biochemical phenomena in terms of how they facilitate cell functions as a general biochemical research strategy:

> "One of the general working hypotheses in biochemical research, which has often proved correct, is the assumption that chemical substances and chemical reactions occurring in living matter are as a rule not accidental but serve some purpose. If, then, living matter often employs cyclic mechanisms we may assume that these are best suited for the requirements of living matter" (Krebs, 1946, p. 98).

There are reasons today to reject this kind of inference as universally valid: we recognize that evolution is not an optimizing process and that many characteristics of organisms are not due to their being selected for their adaptive character (Gould and Lewontin, 1978). Yet, in this case it is extremely plausible to construe the mode of organization provided by the coenzymes as directly providing advantage to the cell by coordinating the individual reactions catalyzed by cellular enzymes so as to function as needed. Thus, it seems that the metabolic pathways are examples of the kind of integrated system manifesting functional subordination as characterized by Richardson.

Even if the claim concerning the adaptiveness of these mechanisms cannot be fully supported, though, we could at least recognize that these pathways manifest a mode of organization that brings particular reactions under the control of the whole system. The occurrence of such a mode of organization is sufficient to undermine the reductionist programme and to show that these metabolics systems constitute organized systems that are at their own level of organization and must be studied in their own right. This point can be established by noting a critical evolutionary consideration that does not depend on embracing the adaptationist view that every feature of a system is present because it is well adapted. It depends only on recognizing how the results of the historical evolutionary process determine boundary conditions under which chemical laws must be applied to biological conditions.

The reductionist approach emphasizes the applicability of basic laws in explaining higher phenomena. It appears as though a reduction is accomplished once it is shown that a description of the phenomena in question can be derived from the laws of the basic science given appropriate boundary conditions. Within the reductionistic approach, it should be recognized that boundary conditions are codeterminants of the behavior of the components within the system. If there were no order to the boundary conditions, though, the boundary conditions would not assume any special significance. They would probably be treated in some statistical manner. But, insofar as the boundary conditions cohere into stable structures that are heritable they acquire a significant status and must be accommodated in any general endeavor to describe the course of events. After they arise, some of these stable structures may be perpetuated because of the adapative advantage they confer onto the overall system, while others may be perpetuated without confering any such advantage. In any case, once it is recognized that these organizational structures are the result of an historical process, the significance of any attempt to give a reductionistic explanation is radically reduced. To complete the reduction, one must fill in the details of the boundary conditions as they have historically arisen, a task that cannot be completed with just the laws of the basic theory (Mercer, 1981).

In the case at hand, the structure provided by the transport functions of the coenzymes is a critical boundary condition which cannot be ignored in applying the basic rules of chemistry to the biological situation. I have tried to dramatize this point in this paper by looking historically at the development of biochemistry so as to see how the discovery of the functions of the coenzymes significantly altered the accounts biochemists offered of the processes of fermentation and oxidation. The difference made by this discovery of the mode of organization imposed by the coenzymes thus shows the limited utility of attempting to reduce these biological functions to organic chemistry. While these reactions accord with basic organic chemical laws, to understand them one also needs knowledge of the structured pathways in the cell. These structured pathways provided the subject matter of biochemistry. As a result, biochemistry cannot be viewed as simply an applied area of chemistry but a discipline with it sown subject matter. After the period that has been the focus of this paper, structured pathways were found for many other cell functions, especially the various synthetic processes by which the cell manufactures its basic constituents on the basis of information stored in the genome. Although there are ongoing questions about whether some of these processes should be treated by a separate discipline of molecular biology, it has not been proposed that they constitute simply an applied domain of organic chemistry. The discovery that these processes involve highly organized systems is sufficient to differentiate this domain from that of organic chemistry. In this respect, then, the discoveries of the 1930s provided the conceptual foundation for construing biochemistry as a discipline with a domain of its own.

Acknowledgement

I thank Frederic L. Holmes for his helpful comments on earlier versions of this paper. Work on this project was partly supported by a research grant from the National Endowment for the Humanities, which is gratefully acknowledged.

References

Bechtel, W. (1984a). The evolution of our understanding of the cell: A study in the dynamics of scientific progress. *Studies in the History and Philosophy of Science,* **15**, 309–356.

Bechtel, W. (1984b). Reconceptualization and interfield connections: The discovery of the link between vitamins and coenzymes. *Philosophy of Science,* **51**, 265–292.

Bechtel, W. (in press). Building interlevel pathways: The discovery of the Embden–Meyerhof pathway and the phosphate cycle. In Dorn, J. and Weingartner, P. (eds.), *Foundations of Physics and Biology.*

Buchner, E. (1897). Alkoholische Gährung ohne Hefezellen. *Berichte der deutschen chemischen Gesellschaft,* **30**, 117–124. Translated in Friedman, H. C. (ed.), *Enzymes.* Stroudsburg, PA: Hutchinson Ross Publishing Company, 1981.

Buchner, E. and Meisenheimer, J. (1904). Die chemischen Vorgänge bei der alkoholische Gährung. *Berichte der deutschen chemischen Gesellschaft,* **37**, 417–418.

Causey, R. (1977). *The unity of science.* Dordrecht: Reidel.

Embden, G., Deuticke, H. J., and Kraft, G. (1933). Uber die intermediaren Vorgänge bei der Glykolyze in der Muskulatur. *Klinische Wochenschrift,* **12**, 213–215. Translated in Kalckar (1969), pages 67–72.

Engelhardt, V. A. and Lyubimova, M. N. (1939). Myosine and adenosinetriphosphatase. *Nature,* **144**, 668–669.

Fiske, C. and Subbarow, Y. (1927). The nature of the 'inorganic phosphate' in involuntary muscle. *Science,* **65**, 401–403.

Gould, S. J. and Lewontin, R. (1978). The spandrels of San Marco and the Panglossian paradigm: A critique of the adaptationist programme. *Proceedings of the Royal Society,* **B77**, 405–420.

Harden, A. and Young, W. J. (1906). The alcoholic ferment of yeast-juice. *Proceedings of the Royal Society of London,* **B77**, 405–420.

Harden, A. (1932). *Alcoholic Fermentation.* Fourth edition. London: Longmans and Green.

Hopkins, F. G. (1913). The dynamic side of biochemistry. *Nature,* **92**, 213–223.

Kalckar, H. M. (1969). *Biological Phosphorylation.* Englewood Cliffs, NJ: Prentice Hall.

Knoop, F. (1904). *Der Abbau aromatischer Fettsäuren im Tierkörper.* Freiburg: Kuttruff.

Kohler, R. E. (1973). The background to Otto Warburg's conception of the *Atmungsferment. Journal of the History of Biology,* **6,** 171–192.

Kohler, R. E. (1975). The history of biochemistry: A survey. *Journal of the History of Biology,* **8,** 275–318.

Kohler, R. E. (1982). *From medical chemistry to biochemistry: the making of a biomedical discipline.* Cambridge: Cambridge University Press.

Krebs, H. A. (1946). Cyclic processes in living matter. *Enzymologia,* **12,** 88–100.

Lipmann, F. (1941). Metabolic generation and utilization of phosphate bond energy. *Advances in Enzymology,* **1,** 99–162.

Lohmann, K. (1929). Über die pyrophosphatfraktion im Muskel. *Naturwissenschaften,* **17,** 624–625.Translated in Leicester, H. (ed.) *Source Book in Chemistry 1900–1950.* Cambridge: Harvard University Press, pages 367–369.

Lohmann, K. (1931). Darstellung der Adenylpyrophosphorsäure aus Muskulatur. *Biochemische Zeitschrift,* **233,** 460.

Lohmann, K. (1934a). Über den Chemismus der Muskel Kontraktion. *Naturwissenschaften,* **22,** 409. Reprinted in Kalckar, 1969.

Lohmann, K. (1934b). Über die enzymatische Aufspaltung der Kreatinphosphorsäure; augleich ein Beitrag zum Chemismus der Muskelkontraktion. *Biochemische Zeitschrift,* **271,** 264–277.

Lundsgaard, E. (1930). Untersuchungen über Muskelkontraktion ohne Milchsäurebildung. *Biochemische Zeitschrift,* **217,** 162–177.

Mercer, E. H. (1981). *The foundations of biological theory.* New York: Wiley-Interscience.

Meyerhof, O. and Lohmann, K. (1931). Über die Energetik der anaeroben Phosphagensynthese ("Kreatinphosphorsäure") im Muskelextrakt. *Naturwissenschaften,* **19,** 575–576.

Nagel, E. (1961). *The structure of science.* New York: Harcourt, Brace and World.

Needham, D. (1937). Chemical cycles in muscle contraction. In Needham, J. and Green, D. E. (eds.), *Perspectives in biochemistry.* Cambridge: Cambridge University Press, pages 201–214.

Needham, D. and Pillai, R. K. (1937). Coupling of oxidations and dismuttions with esterification of phosphate in muscle. *Biochemical Journal,* **31,** 1837–1851.

Negelein, E. and Brömel, H. (1939). R-Diphosphoglycerin säure, ihre Isolierung und Eigenschaften. *Biochemische Zeitschrift,* **303,** 132–144. Translated and reprinted in Kalckar, 1969.

Neubauer, O. and Fromherz, K. (1911). Über den Abbau der Aminosäuren bei der Hefegärung. *Zeitschrift für physiologische Chemie,* **70,** 326–350.

Neuberg, C. and Kerb, J. (1913). Über zucherfreie Hefegärungen. XII. Über die Vorgänge der Hefegärung. *Biochemische Zeitschrift,* **53,** 406–419.

Neuberg, C. and Reinfurth, E. (1919). Weitere Untersuchungen über die korrelative Bildung von Acetaldehyd und Glycerin bei der Zucherspaltung und neue Beiträge zur Theorie der alkoholischen Gärung. *Biochemische Zeitschrift,* **52,** 1677–1703.

Parnas, J. K., Ostern, P., and Mann, T. (1934). Über die Verkettung der chemischen Reaktion der Muskel. *Biochemische Zeitschrift,* **272,** 64–70. Translated in Kalckar (1969), pages 74–79.

Pflüger, E. (1875). Ueber die physiologische Verbrennung in den lebendigen Organismen. *Pflügers Archiv für die gesamte Physiologie des Menschen und der Tiere,* **10,** 251–367.

Quastel, J. H. (1926). Dehydrogenations produced by resting bacteria. IV. A theory of the mechanism of oxidations and reductions in vivo. *Biochemical Journal,* **20,** 166–193.

Richardson, R. C. (1982). Grades of organization and the units of selection controversy. In Asquith, P. and Nicles, T. (eds.) *PSA 1982.* Volume 1. East Lansing: Philosophy of Science Association. Pages 324–340.

Simon, H. (1980). *The sciences of the artificial.* Cambridge: MIT Press.

Thunberg, T. (1916). Über die vitale Dehydrierung der Bernsteinsäure bei Abwesenheit von Sauerstoff. *Zentralblatt für Physiologie,* **31,** 91–93.

Thunberg, T. (1930). The hydrogen activation enzymes of the cell. *Quarterly Review of Biology,* **5,** 318–347.

Warburg, O. and Christian, W. (1936). Pyridin, der wasserstoffübertragende Bestandteil von Gärungsfermenten. *Biochemische Zeitschrift,* **287,** 291–328. Portions translated in Kalckar (1969), pages 86–97.

Wimsatt, W. C. (1972). Teleology and the logical structure of function statements. *Studies in the History and Philosophy of Science,* **3,** 1–80.

Editor's Commentary

This case of interdisciplinary research is different from those to be considered in the following parts of this volume in that it resulted in the formation of a separate discipline. We can see this by measuring the biochemistry against the various characteristics of disciplines considered in the introduction to this volume. Biochemistry has its own domain set off at its own level of inquiry–the various chemical reactions involving macromolecules that perform physiological functions. It has developed its own theoretical schemes, perhaps the best known of which is the citric acid cycle; it also has its own set of research problems and a variety of accepted research techniques. Finally, the term "biochemistry" frequently figures in the name of academic departments, albeit often in conjunction with another disciplinary name, and it has its own professional organizations (e.g., the Biochemical Society in Britain and the American Society of Biological Chemists), complete with sub-societies for specialties within biochemistry, and journals (e.g., *The Biochemical Journal, The Journal of Biological Chemistry, Zeitschrift für Physiologische Chemie* and Biochemische Zeitschrift).

However, as Holmes' warns us in the concluding paragraphs of his paper, one must be careful not to over-reify disciplines, since the problem areas in science are often fluid and fall within the purview of reseachers from numerous disciplines. Thus, biochemistry is not a homologous unit. Recent decades have, for example, seen sometimes acrimonious conflict between those who have adopted the banner of "molecular biology" and those who maintain that molecular biology is just one sub-area of biochemistry. Robert Kohler (1975) provides a useful way to think about the disciplinary status of biochemistry when the draws an analogy to a biological species to guide his own treatment of the history of biochemistry:

> "The idea is to use central theories or themes to trace out the changing center or subcenters of the discipline community, a diverse population of partly overlapping, competing groups, yet having certain concerns recognizably in common – like a biological species" (p. 288).

Bechtel, W (ed), Integrating Scientific Disciplines. ISBN 90-247-3242-5.
© 1986, Martinus Nijhoff Publishers, Dordrecht. Printed in The Netherlands.

As long as one does not insist on a defining trait for a species, but recognizes them as historical, developing entities, this analogy provides a useful model.

There is one critical respect, though, that biochemistry departs from the pattern commonly found in biological species. It is most common for new species to be formed by splitting of current species. However, as the papers in this unit have tried to show, biochemistry emerged from a bringing together of efforts from several different disciplines. One point on which Holmes and I have differed is in the way we have identified the disciplines that were brought together in the founding of biochemistry. The disciplines Holmes considers are such disciplines as physiology, organic chemistry, agricultural chemistry, botany, internal medicine, pathology, and pathological chemistry. Each of these disciplines have their own principal problem areas and objectives in dealing with the problems in those areas (e.g., pathologists were interested in explaining and curing diabetes, while agricultural chemists were concerned to improve fermentation processes). What Holmes has shown is that during the early decades of this century each of these areas began to consider and work on parts of a common problem, that concerning the intermediate chemical processes in metabolism. A fitting metaphor for the picture he presents is that biochemistry is a stream with tributaries in a number of different disciplines. I, on the other hand, following a lead from Hopkins (1936), have construed biochemistry as involving the integration of biological disciplines with chemical disciplines. I then tried to explicate further the idea of Kohler that there is

> "a particular level of biological organization between the chemical molecule and the unit cell, which is the particular domain of the biochemist, a domain lying between that of the organic chemist on the one hand and the biologist on the other" (1975, p. 288).

I have tried to show that it was only somewhat later in its history, during the 1930s, that this level became clearly defined.

Given this difference in how our two papers have characterized the disciplines that were integrated into biochemistry, the accounts we offer of the integration process naturally differ. Holmes tries to show first how the common problem of intermediary metabolism arose from the various endeavors occurring in these different disciplines. Many of these disciplines were focusing on the catabolism of particular substances in an organism; what brought them together was often the discovery that similar intermediaries (e.g., acetic acid and pyruvic acid) were found to be involved in several seemingly different metabolic processes. He further tries to show how, as researchers came to recognize that solving each of their individual problems depended on solving the common problem of discovering the processes of intermediary metabolism, there emerged a common understanding of what type of explanatory model needed to be developed, a common set of research techniques that could be used, and a set of criteria to evaluate proposed models.

I, on the other hand, having focused on the different levels at which physiological and chemical explanations were developed, considered the process by which biochemistry arrived at a new level of explanation when it discovered the highly organized pathways involved in metabolism. My concern was to show that the product of the endeavor whose history Holmes' presented was not just a linking together of chemical and physiological theories so that one might try to reduce physiology to chemistry or even create an interfield theory of the kind portrayed by Darden and Maull (1977). Rather, it was the discovery of a new level of theorizing, with its own distinctive domain. While I did not emphasize it, accompanying the discovery of such a level was the development of research techniques that were appropriate to phenomena at that level.

As a result of this difference in focus, I reached a conclusion that appears, at least on the surface, to be inconsistent with that of Holmes. Towards the end of his paper, Holmes contends that the theoretical models developed in the 1930s, the Embden–Meyerhof pathway and the citric acid cycle, which now provide the basic exemplars of biochemical explanations, were not radically different from those advanced in the preceding three decades. Both in my paper here and in Bechtel (in press) I emphasize the differences between the models of metabolism developed in the 1930s and those offered earlier, arguing that it was the more complex modes of organization recognized in the later models that made them distinctively different from the models of organic chemistry and provided biochemistry with its own specific domain. This difference, however, may be more of a matter of the focus Holmes and I have taken than a substantive disagreement. I certainly agree with Holmes that if one concentrates on the processes by which theories were developed, there is a clear continuity. The research approaches of Embden, Meyerhof, Warburg, Krebs, Lipmann, and others who contributed to the new standard models of metabolism that were developed during the 1930s, differed little from those of their predecessors like Knoop, Dakin, Neuberg, and Thunberg. My focus is on the difference in the conceptual framework that resulted from recognition of the inadequacy of the simpler models and the inclusion of more complex modes of organization, especially those that integrated different steps in the metabolic processes. It was these discoveries that made the difference between organic chemistry and biochemistry ultimately irresolvable and secured a separate status for biochemistry.

Both Holmes and I have focused on the way in which the conceptual framework of biochemistry emerged during its formative period. However, conceptual activity does not occur in a social vacuum and recently Robert Kohler (1982) has produced a detailed study of the way in which the institutional features of biochemistry, in

104

particular academic departments, emerged during the last decades of the 19th century and especially the early decades of the 20th century. He argues for a quite different explanation of the formation of a separate discipline of biochemistry:

> ... intellectual achievement or the lack of it is not the reason why biochemists failed to build a discipline in nineteenth-century Germany or why they succeeded in America, a provincial backwater if judged by research output. Differences in achievement cannot explain why the timing, location, and character of discipline building differed so markedly in the United States, Britain, and Germany. These patterns have to do with the political and economic support system of science: movements for reform of universities and medical school, changing hospital practice, expanding markets for scientific professionals, and evolving division of labor among disciplines (p. 4).

Kohler provides a very probing analysis of the factors that militated against the formation of new chairs in biochemistry in Germany (of which one of the most significant factors is the fact that universities were no longer expanding) and that contributed to the development of new departments in England and especially in the United States (where the endeavors to reform medical education played a major role in the development of academic departments for biochemistry). This dispute over whether cognitive factors or political and economic factors were most decisive in the creation of a discipline of biochemistry, while it may be useful heuristically in bringing to the fore a variety of factors hitherto ignored, probably should not be maintained. There was a role for both the domain and cognitive factors Holmes and I have focused on, and the organizational and institutional factors Kohler has brought to light; it would seem that both needed to be in place for biochemistry to achieve and maintain some measure of identity as a separate discipline.

Since Kohler has dealt so ably with the social and institutional factors involved in the establishment of biochemistry, I will not consider further how they developed. Rather, it will be useful to conclude the discussion of this example of interfield theorizing by considering more generally the character of the theories or explanatory models developed in biochemistry. While Holmes and I have differed some in our assessments of what is most significant in these models, common basic features can be noted. What was involved was the development of a common conception of the chemical processes underlying different metabolic activities. This involved the development of a unified picture of metabolism according to which basic chemical reactions like decarboxylations and dehydrogenations were linked together so as to create a pathway through which a metabolite was processed. This picture became more robust as some elements were found to be common to different pathways and as various processes served to connect later parts of the pathway with either earlier parts of the same pathway or with other pathways (an example of the later is the Pasteur effect according to which some organisms are able to change between using respiration and fermentation depending on oxygen supply – see Lipmann, 1941). What emerged from the pursuits of different inquiries was the recognition of a hitherto unanticipated domain which provided the point of integration for their efforts. It was to characterize the processes in this newly discovered domain that the explanatory models of biochemistry were developed.

In this respect of developing theories for a new domain, the case of biochemistry provides an example of interfield theorizing that is distinctively different from those characterized by Darden and Maull (1977). Part of the emphasis of their work was to show that interfield connections need not have the form of a reduction, which they did in part by emphasizing the development of relationships between fields of analysis as opposed to domains of phenomena (this emphasis is particularly evident in Maull, 1977). In this case, however, one equally sees the lack of a reduction through the discovery of a new domain for analysis. While the development of theories for this new domain allowed one to see the relationship between individual chemical events and physiological phenomena, these theories were not themselves about those connections, but about the pathways that constitute the domain of biochemistry itself.

Once these theories were developed within biochemistry, they did serve to integrate the previously independent disciplines. For example, one could often determine through these models how particular chemical factors, which were previously known to be involved in particular medical problems, actually produced their pathological effects. This allowed for a more useful passing of information between disciplines and for the development of particular interfield theories that are more like those of Darden and Maull. For example, as I have discussed elsewhere (Bechtel, 1984), the discovery of the fact that many B vitamins are important constituents of respiratory coenzymes allowed one to explain why particular chemical deficiencies in the diet manifested themselves in the avitaminosis pathologies. Here is an example of an interfield connection that is quite similar to that between genes and chromosomes which Darden and Maull discuss. While the development of biochemistry did permit the creation of such interfield theories, the point of central importance here is that the central explanatory models of biochemistry, while they emerged from cross-disciplinary research efforts, were not themselves interfield theories, but theories for a new domain. Darden's discussion of the evolutionary synthesis in the following section will provide another example of how cross-disciplinary research may lead to such discoveries of new domains about which theorizing can then develop.

106

References

Bechtel, W. (1984). Reconceptualization and interfield connections: The discovery of the link between vitamins and coenzymes. *Philosophy of Science, 51,* 265–292.
Bechtel, W. (in press). Building interlevel pathways: The discovery of the Embden–Meyerhof pathway and the phosphate cycle. In Dorn, J. and Weingartner, P. (eds.), *Foundations of physics and biology.*
Darden, L. and Maull, N. (1977). Interfield theories. *Philosophy of Science, 44,* 43–64.
Hopkins, F. G. (1936). The influence of chemical thought on biology. *Science, 84,* 255–260.
Kohler, R. E. (1975). The history of biochemistry: A survey. *Journal of the History of Biology, 8,* 275–318.
Lipmann, F. (1941). Pasteur effect. In *A symposium on respiratory enzymes.* Madison: University of Wisconsin Press.
Maull, N. (1977). Unifying science without reduction. *Studies in the History and Philosophy of Science, 8,* 143–162.

PART II
Dobzhansky's Contribution to the Evolutionary Synthesis

Introduction

ROBERT BRANDON
*Department of Philosophy, Duke University, Durham, North Carolina
27706, U.S.A.*

During the 1930s and 1940s research efforts from a variety of disciplines, including population biology, genetics, systematics, paleontology, and morphology were brought together in what has become known as the "evolutionary synthesis." There have already been numerous analyses of the synthesis (see Mayr and Provine, 1980, for a variety of perspectives and references to many earlier accounts). There is considerable disagreement about when precisely the synthesis occurred; this dispute depends in part on whether the synthesis involved the integration of Mendelian models and accounts of natural selection in the population genetics of Fisher, Haldane, and Wright (this seems to be the view of Allen (1978)), or whether it primarily involved the development of models by Dobzhansky and Mayr to account for macroevolutionary phenomena like speciation (this is the view of Mayr (1982)). Since the synthesis itself was a very broad undertaking, the following papers are focusing on one component of it: the role played by Theodosius Dobzhansky, particularly through his seminal 1937 book *Genetics and the Origin of Species*. In focusing on the contribution of one scientist, we hope to gain a perspective on how individuals are able to develop cross-disciplinary programs through their own research. In this respect, this case offers a useful contrast to the case of biochemistry discussed in the previous section, where a group of researchers coming from different disciplines fed into the same research enterprise.

To provide a framework in which to approach the contribution of Dobzhansky, it will be useful to provide a brief sketch of the historical background. At the outset of the 20th century a bitter controversy existed between the biometricians who favored Darwin's mechanism of natural selection as the explanation of evolution and the Mendelians who sought in genetic mutations the source of new species and credited selection with only limited efficacy. The biometricians focused primarily on the statistical distribution of traits within species and treated evolution in gradualist terms. The Mendelians, particular de Vries, focused on the potential for new species arising through macro-mutations and so offered a saltationist view of evolution.

Bechtel, W (ed), Integrating Scientific Disciplines. ISBN 90-247-3242-5.
© 1986, Martinus Nijhoff Publishers, Dordrecht. Printed in The Netherlands.

The gap between the competing conceptions of evolution was gradually narrowed over subsequent decades. One factor responsible for this was the recognition (due in large part to the work of the Morgan school) that the genome actually consisted of a large array of genes that each accounted for only a small part of the character of the phenotype, rather than consisting of only a few genes responsible for macro traits. This provided the foundations for the population genetics of Fisher, Haldane, and Wright, who tried to model how natural selection and other factors like drift could affect the genic distribution in the species. While these population geneticists left many questions about species unanswered, they did point the way to a bridging of Mendelism and Darwinism, with Mendelism providing the mechanism for replication and variation on which the Darwinian mechanism of natural selection could work.

The two contributors to this section, Lindley Darden and John Beatty, are trained in both the history and philosophy of science. Nearly a decade ago Darden, together with Nancy Maull, developed the notion of an interfield theory. Darden's analysis of Dobzhansky's contribution to the synthesis involves a further development of the concept of an interfield theory. She distinguishes different levels at which theories had been worked out for the mechanism of inheritance and for changes within populations, and portrays Dobzhansky as first recognizing the need for considering a third level of analysis and then developing a theory that related the processes at all three levels. Beatty has, in recent years, been exploring the utility of the semantic conception of theories to understanding evolutionary theory and his construal of Dobzhansky's role can be seen as an extension of that work. According to the semantic view of theories, a theory is a formal model. A formal model may or may not apply to some part of the real world. In particular, population genetic models based on the Hardy–Weinberg law allow for various possible modes of evolutionary change. Which modes are actual depends on the actual values of certain variables (e.g., mutation rate, strength of selection, etc.). According to Beatty, Dobzhansky's major contribution to the synthesis was to employ experimental studies and field studies of natural populations so as to determine the values of variables needed to apply the theories of population genetics to actual biological cases. These two papers thus offer distinctly different views of Dobzhansky's contribution to integrating disciplines in the evolutionary synthesis.

References

Allen, Garland (1978). *Life science in the twentieth century*. Cambridge: Cambridge University Press.

Dobzhansky, Theodosius (1937). *Genetics and the origin of species*. New York: Columbia
  University Press.
Mayr, Ernst (1982). *The growth of biological thought*. Cambridge: Harvard University Press.
Mayr, Ernst and Provine, William (eds.) (1980). *The evolutionary synthesis*. Cambridge:
Harvard University Press.

Relations Among Fields in the Evolutionary Synthesis

LINDLEY DARDEN
*Committee on the History and Philosophy of Science, University of
Maryland, College Park, Maryland 20742, U.S.A.*

Introduction

The synthetic theory of evolution is a multi-level theory that serves to synthesize
knowledge from fields at different levels of organization. It provides a solution to
the problem of the origin of species. Biologists attempted to solve this problem for
years, during which time the key fields emerged and developed to the point that the
synthesis was possible. Prior to the synthesis, debate occurred about what
components were necessary to solve the problem and alternative theories were
proposed. Had any of the prior theories been correct, then the fields that exist
within evolutionary studies would have been different. What fields exist is a
contingent fact about the nature of the world and the way we study it. What fields
are synthesized to solve certain problems is contingent on the nature of the solution
and the stage of development of various fields at the time the solution is proposed.

Relations among levels of organization, fields of study, and the problem of the
origin of species are the subject of this paper. The focus is on the synthetic theory
of evolution as proposed by Dobzhansky in 1937, the multiple levels within it, the
knowledge from different fields that it synthesized, and various of its predecessors.
A final section will discuss the concept of a synthetic theory and contrast that with
previously-studied interfield theories.

The Synthetic Theory

Theodosius Dobzhansky published the seminal book outlining the evolutionary
synthesis in 1937: *Genetics and the Origin of Species.* Although others contributed
to the synthesis, this discussion will focus primarily on Dobzhansky's 1937 book.
In it he clearly distinguished between evolution (change of species) and the causal
mechanisms of evolutionary change. It is worth quoting him at length on this topic:

Bechtel, W (ed), Integrating Scientific Disciplines. ISBN 90-247-3242-5.
© *1986, Martinus Nijhoff Publishers, Dordrecht. Printed in The Netherlands.*

114

> The theory of evolution asserts that the beings now living have descended from different beings which have lived in the past; that the discontinuous variation observed at our time-level, the gaps now existing between clusters of forms, have arisen gradually, so that if we could assemble all the individuals which have ever inhabited the earth, a fairly continuous array of forms would emerge; that all these changes have taken place due to causes which now continue to be in operation and which therefore can be studied experimentally. (Dobzhansky, 1937, p. 7)

Dobzhansky thus stated the three components that are usually referred to as common descent, gradualism, and uniformitarianism. The explanation of common descent and the gaps now existing between forms was the key problem that he addressed in his book. Note that adaptations were not mentioned. Since Darwin combined the explanation of adaptations and the origin of new forms in his theory of natural selection, these two problems are often lumped together. Although Dobzhansky advocated selection as a means of producing adaptive change, it was not the only mechanism he postulated for the production of new forms. Furthermore, the other theories we will be considering were directed more to solving the problem of the origin of new forms than to the explanation of adaptations.

Dobzhansky clearly stated the three levels that constitute the "mechanisms of evolution." These constitute the "synthetic theory of evolution," though Dobzhansky did not use that term.

First:

> "Mutations and chromosomal changes are ... the first stage, or level, of the evolutionary process, governed entirely by the laws of the physiology of individuals."

Secondly, the populational level:

> "A mutation may be lost or increased in frequency ... without regard to the beneficial or deleterious effects of the mutation. The influences of selection, migration, and geographical isolation then mold the genetic structure of populations into new shapes in conformity with the secular environment and the ecology, especially the breeding habits, of the species. This is the second level of the evolutionary process, on which the impact of the environment produces historical changes in the living population."

Third, the species level:

> "Finally, the third level is a realm of fixation of the diversity already attained on the preceding two levels.... A number of mechanisms encountered in nature (ecological isolation, sexual isolation, hybrid sterility, and others) guard against a ... fusion [due to interbreeding]." (Dobzhansky, 1937, pp. 12–13).

Knowledge from three levels of organization was synthesized: the level of genes and chromosomes, the level of the population, and the level of the species. These levels constitute an inclusive hierarchy: genes are parts of chromosomes which make up individual organisms which make up populations which make up species. Dobzhansky as the geneticist focused on the organism's genes and chromosomes; the individual organism was not accorded a separate level in his hierarchy. Additionally, a direction of causal influence characterizes the relations among the levels: mutational changes provide raw material for changes in populations; populational changes, in turn, are necessary for the production of new species.

Different fields provided the key knowledge about mechanisms operating at each level: Mendelian genetics for mutations; cytology for chromosomal abnormalities; mathematical population genetics and experimental and field studies of populations for the populational level. Finally, the study of isolating mechanisms in speciation was a new area of study that emerged as a result of the synthesis. Dobzhansky was aware that the final level was the least studied as of 1937: "The origin and functioning of the isolating mechanisms constitute one of the most important problems of the genetics of populations." (Dobzhansky, 1937, p. 14). It is clearly too simplistic to say that the synthetic theory synthesized only Mendelism and Darwinism; numerous other components were involved. Additionally a new level was established as separate from the previously existing ones. Dobzhansky drew on extensive knowledge from these numerous fields as well as delineating problems for the new level of organization. These he combined into one comprehensive explanation of the origin of new species.

It is sometimes asked why the synthesis did not occur shortly after the rediscovery of Mendel's laws in 1900. The stage of development of the fields at the time of the synthesis was crucial. Mendelism by 1937 was a very different field than Mendelism in 1900. By 1937, Dobzhansky could draw on extensive knowledge of mutations, modifying genes, genes that produce sterility – none of which was available in 1900. Thus, fields must be seen as developing entities, providing shifting information at different times to problem-solving occurring in other fields. The stage of development determines whether appropriate knowledge is available. We will now examine alternative explanations of the origin of new species prior to the synthetic theory and see how the various fields were related within those theories.

Other Theories, Other Levels

We will begin with Darwin's theory of natural selection, examine Hugo de Vries's mutation theory, and the objections of some of the early Mendelians to Darwinian selection. This analysis will show how prior theories focused on different levels or organization as the key to the origin of species. Had any of the alternatives been correct, then the fields supplying knowledge would have been different. Levels themselves are not just "given," but must be discovered to exist as causally efficacious.

In the *Origin of Species* of 1859 Darwin laid out his theory, which explained the origin of new species and the origin of adaptations that organisms exhibited. Darwin argued that his theory of natural selection was sufficient to explain the origin of all new species. The theory may be stated as a deductive argument:

I. In all organic forms, heritable variations occur.
II. Organic forms multiply at a greater rate than their sources of sustenance, therefore, there is a struggle for existence.
Conclusion 1: Those organisms which possess variations advantageous in the struggle for existence will tend to survive.
Conclusion 2: A permanent and adaptive change in organic forms will be effected, as over time, the new forms replace the old. (modified from Vorzimmer, 1970.)

The levels of organization found in Darwin's theory are the level of variation in individual organisms and the population in which selection occurs. No separate level of isolating mechanisms to produce new species was postulated; in fact, Darwin emphasized the continuity between varieties and species. Thus, if we were still operating with Darwin's theory we would have fields for the study of variation and the field of population studies but no separate study of speciation mechanisms. That a separate level existed in order to move from the population to the species and that separate study of isolating mechanisms was needed was one of the additions of the synthesis.

Darwin amassed much empirical evidence for the existence of variations, both in domesticated forms and in nature. However, in the *Origin* he did not have a theory to explain the origin or inheritance of variations. Darwin attempted to supply such a theory in 1868 – his unsuccessful theory of pangenesis. He postulated that cells produced hereditary units called "gemmules." The gemmules varied in quantity or quality, and modified cells produced modified gemmules. Since Darwin believed that most hybrids were intermediate in form between the two parental forms, hybrids were postulated to produce hybridized gemmules. The gemmules circulated throughout the body and collected in the reproductive areas, were passed on to offspring, and grew into cells in the developing embryo. (Darwin, 1868, Ch. 27). Darwin's theory of pangenesis was thus both particulate and accounted for blending inheritance. (Unfortunately particulate and blending are often contrasted as two different types of hereditary theories. To use later terminology, blending at the phenotypic level may or may not be correlated with discrete units at the genotypic.)

Darwin's theory of pangenesis was not confirmed when tested (Galton, 1871). The problem of variation remained unsolved in the late nineteenth century. Hugo de Vries took up Darwin's theory of pangenesis and modified it. As a result of his experiments on variations, de Vries discovered that Darwin was wrong: hybrid forms do not produce hybrid units; they carry one or other of the pure parental units in an unmodified form (de Vries, 1900; Darden, 1976, but see Kottler, 1979). This is Mendel's law of segregation, namely that hereditary units separate or segregate in a pure form in the formation of pollen and egg cells so that each sex cell receives one or the other of a pair of units for a trait. After de Vries (and Correns) published this finding in 1900, the field of Mendelism began.

Even though Mendelism emerged as a field in 1900, as late as 1908, Vernon Kellogg in *Darwinism Today* said that one of the chief problems in biology was the "origin, the causes and the primary control" of variations (Kellogg, 1908, p. 30). The relation between Darwinian selection, with its stress on small frequently occurring "individual differences" and the Mendelian law pertaining to particulate (in de Vries's formulation), non-blending, discontinuous variations was not readily apparent. Kellogg iterated objections to Darwinian natural selection, most of which focused on problems of variation: the problem of the size of variations, i.e., whether "continuous variations" could provide the raw material for selection or whether "discontinuous variations" played a role; the problem of the swamping of unusual variations by interbreeding; the problem of producing a sufficient number of similar variants at a given time to alleviate the swamping problem; the problem that many variations that distinguished species showed no utility (Kellogg, 1908, Ch. 3) Darwin had focused much attention on adaptations; subsequent work had turned up many instances of traits whose adaptive significance was not apparent. In discussing alternatives to Darwinism, Kellogg suggested that a number of people had come to the view: "Why may not variation be the actual determinant factor in species-forming, in descent?" (Kellogg, 1908, p. 34).

Although de Vries rediscovered Mendel's law of segregation, he did not see it as applicable to the problem of the origin of species. Kellogg prominently featured de Vries's mutation theory of 1901–1903 as one of the alternatives to Darwinian selection that did in fact make the origin of certain types of variations and the origin of new species coincide. De Vries proposed that progressive mutations arose in organisms and were of such a magnitude that they gave rise to new, true-breeding species in a single generation. According to de Vries, progressive mutations did not segregate, nor were they necessarily adaptive. Selection served merely to eliminate the most harmful; thus, the numerous seemingly non-adaptive variants could be explained. He based his theory primarily on his findings from experiments with the

Oenothera, the evening primrose. The mutations appearing in the evening primrose were due to its being in a mutating period, de Vries believed. Similarly, de Vries thought, other organisms would go through such periods and thus sufficient mutations would be produced. (de Vries, 1909–10).

If de Vries had been right that a new species can arise with the formation of a single mutation in a hereditary unit, then the fields of genetics and evolutionary studies would not be separate. If the study of how variations arise had been, as Kellogg had suggested, the actual determinant factor in forming new species, then the study of mutations and their inheritance would have been the key to the origin of new species. No synthesis of Darwinism and Mendelism would have been required; no separation of genetic studies and evolutionary studies would have occurred. But de Vries's theory turned out ot be incorrect; the evening primrose was not an appropriate model organism since it has unusual chromosomal mechanisms. An understanding of the origin of species did require more than the study of genetics alone.

The Field of Genetics

William Bateson is rightly called the founder of genetics, for he pursued Mendelian inheritance and its relation to the discontinuous variations that he had long advocated as the key to evolutionary change. Mendelism emerged as a separate field in 1900 with the rediscovery of the law of segregation. A field may be characterized as having the following elements: a central problem that characterizes the subject matter of the field; a set of facts related to the problem; concepts, laws and theories that aid in solving the central problem; and techniques and methods for studying the phenomena or analyzing the concept. (Sometimes fields may also be characterized as having goals providing expectations as to how the problems are to be solved; however, the goals are often very difficult to identify in the historical record.)

The emergence of the field of Mendelian genetics in 1900 was marked by both continuity with the past and change from that past. Some of the elements of the new field had been present in biology prior to 1900; others came into being with the emergence. The problem of heredity was an old problem: what are the patterns of resemblance of offspring and parents and how are they to be explained? It had puzzled people for centuries. Moreover, the qualitative phenomena that made up the domain of the new field had been known. Offspring sometimes resemble grandparents rather than parents, thus exhibiting "reversions." The techniques of hybridizations or aritificial breeding – the main experimental technique of the new field – had a long history. Similarly, the appeal to material units as explanatory factors in heredity was not new.

However, even though these elements had existed separately, they had not been related to one another as they were once genetics emerged was a separate field. The problem of heredity existed but how as it to be solved? Many types of reversions had been noted, but little quantitative analysis and careful steps of crossings had been carried out. Hybridization had been primarily used for practical purposes of artificial breeding. Material units of heredity had been postulated, but no way of investigating then experimentally had been found.

These separate elements came together in a new way with the discovery of quantitative ratios in types of characters in hybrid crosses. Where a yellow pea was crossed with a green pea, all the hybrids were yellow; but when the hybrids were crossed, the result was 3 yellow to 1 green. Such data was explained by postulating a unit, variously called a "pangen," a "factor," and "allelomorph," later "gene," that occurred in pairs which segregated in the formation of sex cells that combined randomly to produce the ratios. Investigations of the generality of these hybridization ratios occupied Bateson and others in the period immediately following 1900.

The new field of genetics that thus came into being may be characterized in the following way. Its central problem was the explanation of patterns of inheritance of characteristics. The technique of artificial breeding was used to investigate the characteristics. The laws were Mendel's laws of segregation and independent assortment, and the gene was the primary theoretical concept of the new field.

Mendelian traits were seen by Bateson and others as representing "discontinuous" variations. These were contrasted with the smaller scale differences that graded into each other called "continuous variations." The biometricians were seen as the proponents of Darwinism. They advocated evolutionary change by means of small "individual differences," as Darwin had called them, or "continuous variations," as they came to be called (Provine, 1971). It took a number of years of work within Mendelism before quantitative variations were explained as due to multiple, interacting Mendelian factors; the two types of variation were thereby shown not to be fundamentally different. Thus, the early disputes between Mendelians and Darwinians were resolved as the result of extension of the scope of Mendelism.

The studies of mutation carried out by T. H. Morgan and his students working on the fruit fly *Drosophila* provided the most extensive modifications of Mendelism after 1900 and provided much of the knowledge on which Dobzhansky was able to draw by 1937. Mutations found in *Drosophila* were of a much smaller scale than de Vries's mutations; they served well as raw material for a more gradual process of evolutionary change. Modifying genes in which one gene modified the effect of another allowed even smaller scale changes. The fact that one gene can affect more than one character provided Dobzhansky with the ability to explain seemingly non-adaptive traits – they might be caused by genes that produced other traits that were adpative (Dobzhansky, 1937, p. 29). The discovery of genes that produce sterility helped to alleviate the concern that the Mendelian differences were not sufficient to account for species differences (Dobzhansky, 1937, p. 264). The extensive development of the chromosome theory of Mendelian heredity, linking Mendelian genes to chromosomes, provided key information for Dobzhansky to use in explaining larger scale differences between varieties and species. The Morgan school's development of this theory and the subsequent work by cytologists in the 1920s and '30s added substantial new information. In short, subsequent developments of Mendelism in the period from 1900 to 1937 were crucial to the utilization of Mendelism and cytological studies in the evolutionary synthesis.

Similar stories can be told for the stages of development of mathematical population genetics as well as the experimental and field studies of populations (Provine, 1971; Mayr and Provine, 1980; Mayr, 1982). We will not dwell on the details here except to point out that the stages of development by 1937 were also crucial to the possibility of a synthesis occurring.

The Synthetic Theory versus Interfield Theories

In a previous paper, Nancy Maull and I discussed a type of theory that we called "interfield theories" (Darden and Maull, 1977). These theories relate two fields by providing relations among components of the two fields. For example, the chromosome theory of Mendelian heredity was an interfield theory that related the fields of genetics and cytology by postulating that genes were parts of chromosomes. Interfield theories may postulate other relations besides part-whole ones: they may propose a structure-function relation or specify the physical nature of an entity or process. Interfield theories function to solve problems that arise within a field but cannot be solved with the techniques and concepts available in that field. The interfield theory serves as a bridge between two previously separate fields that may have been working on the same problem from different perspectives. Predictions may be made for both fields on the basis of the other; thus, the relation is reciprocal. By postulating a physical relation among entities or processes in two fields, interfield theories thus provide a kind of unity of science.

We failed in that paper to stress the importance of the stage of development of a field in determining to what it can be linked. However, in our examples we discussed genetics in an early stage being linked to cytology, then genetics at a much later stage being linked to biochemistry to produce the developments in the operon theory of gene regulation. Thus, implicitly we were aware of the changing nature of fields and the effect of such changes on what kinds of interfield connections are possible.

The synthetic theory differs somewhat from the previously studied interfield theories since it is a multi-field theory. It postulates a process by which new species are formed. This process depends on changes at multiple levels. An understanding of those changes is provided by multiple fields. The stage of development of those fields was important in providing adequate information so that the synthetic theory could be formulated.

The problem of the origin of species did not arise within one of the fields and thus initiate a search for information in another field to solve it. Instead, it was an old problem that had numerous proposed solutions, some involving different fields and levels of organization than those integrated by the synthetic theory.

The synthetic theory not only related knowledge already present in existing fields, it also postulated the need for a new area of study at a higher level of organization, namely at the species level. Thus, the theory did not merely build bridges among existing fields. In addition to integrating knowledge from other fields, it added a new component with new research problems – the study of isolating mechanisms in the formation of new species. The solution of an old problem may thus not only require relating existing knowledge but also necessitate the emergence of a new area of study.

A causal relation exists among the mechanisms studied by the separate fields of the synthetic theory. Gene mutation and recombination provide the raw material for population changes, which, in turn, are necessary for the formation of new species. Because of these hierarchical, causal relations among the fields, the relations among the fields are not reciprocal as in the interfield theory cases. In Dobzhansky's formulation of the theory, knowledge about gene mutations provided information about population changes but knowledge about selection processes did not provide new knowledge for the geneticist about mutations. Mendelian genetics proceeded relatively independently of population genetics, but population genetics was dependent on new findings of mutational processes in genetics. Knowledge was merely taken from genetics to be used in the postulation of a new process; the genes themselves were not connected to any new entities as they were when they were connected to chromosomes or to DNA. Similarly, population processes, such as selection and migration, can continue whether or not isolation has occurred. The asymmetry of the causal relations among the fields thus makes their relations less reciprocal than in previously-studied interfield theories.

122

Since genetics can proceed without taking into account the populational level, it retains more relative independence than do the other fields in the synthesis. Dobzhansky was wrestling with the question of the relations of fields within the synthesis when he said:

> It should be reiterated that genetics as a discipline is not synonymous with the evolution theory, nor is the evolution theory synonymous with any subdivision of genetics. Nevertheless, it remains true that genetics has so profound a bearing on the problem of the mechanisms of evolution that any evolution theory which disregards the established genetic principles is faulty at its source. (Dobzhansky, 1937, p. 8)

Genetics was clearly a key component of the theory, but it remained relatively independent. It was not affected by the higher levels in the way that it affected them, in Dobzhansky's formulation of the theory. (From a more contemporary viewpoint, it might be argued that higher levels do influence the lower. For example, selection may influence the evolution of gene mutation mechanisms. However, instances of influence of selection on the evolution of genetic processes were not part of the original synthetic theory proposed by Dobzhansky; our slim knowledge of the evolution of genetic mechanisms is due to post-1937 developments.)

Conclusion

The synthetic theory is a multi-field, multi-level theory that postulates a well-integrated causal process combining mechanisms from the different fields to solve the old problem of the origin of species. It established the study of isolating mechanisms as an important new area of study while relating the already developed fields of Mendelian genetics, mathematical population genetics, and experimental and field studies of populations. The stage of development of these fields was crucial to their role in the evolutionary synthesis. The new synthetic theory that was formed has provided the basis for evolutionary studies from the time it was proposed to the present. It has been one of the truly remarkable syntheses of twentieth century science.

Acknowledgement

I gratefully acknowledge the support of the General Research Board of the University of Maryland for the research and writing of this paper. I wish to thank John Beatty and William Bechtel for comments that improved the final version of this paper.

References

Darden, Lindley (1976). Reasoning in scientific change: Charles Darwin, Hugo de Vries, and the discovery of segregation. *Studies in the History and Philosophy of Science, 7,* 127–169.

Darden, Lindley (1977). William Bateson and the promise of mendelism. *Journal of the History of Biology,* **10,** 87–106.

Darden, Lindley (1980). Theory construction in genetics. In Thomas Nickles, (ed.) *Scientific Discovery, Case Studies.* Dordrecht, Holland: Reidel. Pages 151–170.

Darden, Lindley and Maull, Nancy (1977). Interfield theories. *Philosophy of Science,* **44,** 43–64.

Darwin, Charles (1859). *On the origin of species, A facsimile of the first edition.* Cambridge, Mass.: Harvard University Press, 1966.

Darwin, Charles (1868). *The variation of plants and animals under domestication.* 2 Vols. New York: Organge Judd and Co.

Dobzhansky, Theodosius (1937). *Genetics and the origin of species.* New York: Columbia University Press.

Galton, Francis (1871). Experiments in pangenesis. *Proceedings of the Royal Society (Biology),* **19,** 393–404.

Kellogg, Vernon (1908). *Darwinism today.* New York: Henry Holt.

Kottler, Malcolm (1979). Hugo de Vries and the rediscovery of Mendel's laws. *Annals of Science,* **36,** 517–538.

Mayr, Ernst (1982). *The growth of biological thought.* Cambridge, Mass.: Harvard University Press.

Mayr, Ernst and Provine, William (eds.) (1980). *The evolutionary synthesis.* Cambridge, Mass.: Harvard University Press.

Provine, William (1971). *The origin of theoretical population genetics.* Chicago: University of Chicago Press.

Vries, Hugo de (1900). The law of segregation of hybrids. Transl. from the German and reprinted in Stern, C. and Sherwood, E. (ed.), *The origin of genetics: A Mendel sourcebook.* San Francisco: W. H. Freeman, 1966. Pages 107–117.

Vries, Hugo de (1909–10). *The mutation theory.* 2 Vols. Transl. J. B. Farmer and A. D. Darbishire. New York: Kraus Reprint Co., 1969.

Vorzinner, Peter (1970). *Charles Darwin, The years of controversy.* Philadelphia: Temple University Press.

The Synthesis and the Synthetic Theory

JOHN BEATTY

*Department of Ecology and Behavioral Biology, University of
Minnesota, Minneapolis, Minnesota 55455, U.S.A.*

Commentators on the evolutionary synthesis frequently identify that event with the formulation of the "synthetic theory." According to the simplest such characterization of the synthesis, Mendelian genetic theory and Darwinian evolutionary theory – once considered irreconcilable – were eventually reconciled in the theory of population genetics, which is the core of the synthetic theory. That reconciliation itself constituted the much-heralded synthesis.[1]

The story can be made more complicated by bringing more theories to bear – i.e., by construing the synthesis as an integration of more theories than just Mendelian genetic theory and Darwinian evolutionary theory. Lindley Darden, for one, characterizes the synthesis in terms of the integration of theories from a wide range of fields, effectively identifying the synthesis with the formulation of one grand "multi-field theory" (Darden in the preceding chapter).

---

[1]   The story of the reconciliation of Darwinian evolutionary theory and Mendelian genetic theory has been recounted in greatest detail by William Provine (1971). The story goes something like this. According to Darwin's theory, *if* small variations arise which even slightly increase the fitness of their possessors, and *if* those variations are inherited, then the same variations will increase in frequency in a species over the course of generations. The applicability of the theory rests on the assumption that such small variations actually arise and are passed down without too much modification from parents to offspring. But at the turn of the century, there was no well accepted theory of inheritance which accounted for the origin of these small variations, and which ensured their preservation from generation to generation. Mendel's laws, rediscovered in 1900, were thought at first to pertain only to the inheritance of large variations, and hence were not considered complementary to Darwinian evolutionary theory.

The synthesis, to which the name "synthetic theory of evolution" refers, was in large part a matter of reconciling Mendel's and Darwin's theories – invoking Mendel's laws in order to explain how the natural selection of small variations produces gradual, systematic, evolutionary change. (At present, there is less emphasis on the gradualness of evolutionary change. Mendel's laws are now used to link various rates of evolutionary change to natural selection.) The theory of population genetics based on the Hardy–Weinberg principle – the core of the synthetic theory – can be derived straightforwardly from Mendel's laws. That theory is discussed more fully later in this chapter.

Bechtel, W (ed), Integrating Scientific Disciplines. ISBN 90-247-3242-5.
© *1986, Martinus Nijhoff Publishers, Dordrecht. Printed in The Netherlands.*

There is much of merit in Darden's analysis. While not exactly historically intricate, it provides what promises to be an insightful framework for further historical analyses of the synthesis.

It's principal deficiency, it seems to me, is that it construes the synthesis as a purely *theoretical* affair – no more than the formulation of a new theory. That is especially unfortunate inasmuch as Darden chooses to discuss the contributions of Theodosius Dobzhansky as exemplifying the synthesis as she construes it. I think it is questionable whether Dobzhansky contributed anything new, theoretically speaking, to the synthesis. He contributed a lot, but no new theory.

In order to accomodate Dobzhansky's contributions, we need to broaden our conception of the synthesis. There was more to the synthesis than theory. There was observation as well. In what follows, I hope to be able to explain why this is more than just a nitpicking point. I do not just mean that the synthesis consisted of the formulation of a new theory plus the gathering of observational evidence for that theory. That really *would* be nitpicking.

Among Dobzhansky's greatest contributions to the evolutionary synthesis were his experimental and field studies of laboratory and natural populations of the fruitfly *Drosophila* – e.g., as reported in his long series of articles "Genetics of Natural Populations I–LXIII" (reprinted in Lewontin et al., 1981) and in the various editions of *Genetics and the Origin of Species* (Dobzhansky, 1937, 1941, 1951). Darden actually lists "experimental and field studies" among the fields from which the multi-field, synthetic theory emerged. She follows Ernst Mayr in this regard.[2] Both presumably have Dobzhansky's contributions (among others') in mind. But neither Darden nor Mayr is very specific about how those experimental and field studies contributed to the formulation of the synthetic *theory* – i.e., as opposed to just serving as *evidence* for that theory. Darden only mentions experimental and field studies in passing anyway.

It may very well be the case that experimental and field studies *did* contribute to the formulation of the synthetic theory (vs. just serving as evidence for the theory). At least, that is a topic worth exploring. What concerns me here, though, is another very different and very important role of experimental and field studies in the synthesis. Again, the sort of contribution I have in mind is not purely evidential.

To see what more of a role those studies played, we need first to consider the

---

[2]   Mayr has painted a more comprehensive picture of the synthesis than anyone else so far (Mayr 1980, 1982). His analysis is currently very influential, its credibility deriving not only from its inherent persuasiveness, but also from the fact that, as an important contributor to the synthesis, Mayr is acknowledged to have a special insight into the whole affair.

synthetic theory in more detail, and then the experimental and field studies in question. The core of the synthetic theory is pretty much just the theory of population genetics, developed by Sergei Chetverikov (1926), J. B. S. Haldane (1924–1932), R. A. Fisher (1930), and Sewall Wright (1931), and based on the so-called "Hardy–Weinberg equilibrium principle." The Hardy–Weinberg equilibrium principle describes the conditions under which evolutionary equilibrium is maintained – that is, the conditions under which changes in gene and genotype frequencies do not occur. Specifically, in an infinitely large population in which there is no selection, mutation, or migration, the gene and genotype frequencies of the population will remain stable. Plug in nonzero values for the variables associated with natural selection, mutation, and migration, and you can calculate the evolutionary consequences. Thus, for instance, one can calculate the evolutionary consequences of moderate mutation together with heavy selection, or moderate mutation together with moderate selection and heavy migration, or heavy mutation together with heavy selection and moderate migration, etc. Assigning these variables different values results in a number of different "modes" of evolutionary change.

The theory can also (fortunately) be extended to finite populations by introducing population size as a variable. One can then predict the probability that gene and genotype frequencies will fluctuate to a certain extent as a result of change alone – that is, even in the absence of selection, mutation, and migration. The smaller the population size, the larger the probability of large chance fluctuations. This mode of evolutionary change is called evolution by random drift.[3]

So much for the synthetic theory. There was, however, more to the synthis than the synthetic theory. The components of the evolutionary synthesis consisted of more than theories. They consisted of observations as well – experimental and field observations of laboratory and natural populations.

To speak of the relationship of experimental and field observations to theory in terms of a "synthesis" may sound peculiar. To the extent that it does, that is probably because we are accustomed to thinking of observations only as *evidence* for theories. We certainly do not refer to every confirmation of a theory as a "synthesis" of theory and observation. To do so would just be to use the term "synthesis" in a confusing way. I think it is quite legitimate, however, to talk about

---

[3] Population genetic theory is considered by virtually everyone to be the core of the synthetic theory. There is, however, considerable disagreement as to how much *more* there is to the synthetic theory over and above population genetic theory. One's position with regard to this issue depends on how many more theories, besides Mendelian genetic theory and Darwinian evolutionary theory, one considers to have been incorporated into the synthetic theory.

128

a synthesis of theory and observation in the context of the evolutionary synthesis, and especially with regard to Dobzhansky's contributions to the synthesis. As, for instance, Richard Lewontin expresses the contribution of Dobzhansky's experimental and field studies to the synthesis, "The successful melding of the theory of gene frequency change with the known facts of genetic variation ... marks the first real *synthesis* in biology of a complex mathematical theory with a large body of observation and experiment" (Lewontin 1982, p. 99; my emphasis).

But what does it mean to talk about a "synthesis" of population genetic theory together with experimental and field studies of laboratory and natural populations? What is Lewontin talking about? As I already suggested, the primary role of those observations in the evolutionary synthesis was not simply to serve as evidence for the theory of population genetics. Evolutionary biologists did not go into the field, and into their labs, in order to test the synthetic theory as just described. They had two other ends in mind. First, they wanted to measure the values of the variables of the theory in particular cases. Second, they wanted to extrapolate from the values of the variables in particular cases to the overall relative weights of the variables, or, to put it somewhat differently, they wanted to extrapolate from the values of the variables in particular cases to the overall relative importances of the various possible modes of evolutionary change (see also Provine 1978). Let's take up these two aims in order.

Population genetic theory describes numerous *possible* modes of evolutionary change – evolution by natural selection, by random drift, by combinations of natural selection and random drift, etc. – but it says nothing about whether any or all of those modes *actually* ever occur. Experimental and field studies are required to determine the values of the variables of the theory for each particular trait in each particular taxon, and hence the particular mode of evolution in each particular case. Thus, for instance, one of Dobzhansky's main contributions to the synthesis was his measurement of selection coefficients associated with morphologically different chromosome types in populations of *Drosophila pseudoobscura*. He originally believed that selection played no role in determining the frequencies of these variations. He believed instead that the frequencies of the variations fluctuated entirely randomly. He later gave up the random drift account when he found considerable selection pressures influencing the chromosome frequencies.

On the basis of these measurements of the values of the variables of population genetic theory in specific cases, some evolutionary biologists extrapolate to the overall relative importances of the variables – the overall relative importances of the various possible modes of evolutionary change. On the basis of case-by-case measurements, in other words, biologists seek answers to questions like the following. Is selection more important, in general, than random drift? Or is random drift more important, in general, than selection? How important, in general, is migration relative to selection and random drift?

Thus, for example, Dobzhansky's selectionist reinterpretation of chromosomal variations in *Drosophila pseudoobscura*, together with a number of other selectionist reinterpretations of patterns of nature previously attributed to random drift, occasioned a significant shift in his thinking about the overall relative importances of selection and random drift. He had earlier believed that random drift was, in general, very significant, but later believed that it was, in general, much less significant than natural selection.[4]

These questions about the relative importances of the various modes of evolution are questions of great weight among evolutionary biologists, who have haggled over the relative importances of random drift, mutation, migration, and selection for quite some time.

It is remarkable how many modern evolutionary biologists have been of the opinion that the work of most pressing importance, following the general acceptance of population genetic theory, is that of empirically determining the values of the variables of the theory for specific cases, and from that determining the relative importances of the various modes of evolution. Timofeef–Ressovsky's 1940 assessment of the work-to-be-done is representative of the first aim, and has been echoed often since then. As Timofeef explained the situation, the theory developed by Chetverikov, Haldane, Fisher, and Wright,

> show[s] us the relative efficacy of various evolutionary factors under the different conditions possible within the populations (Wright 1932). It does not, however, tell us anything about the real conditions in nature, or the actual empirical values of the coefficients of mutation, selection, or isolation. It is the task of the immediate future to discover the order of magnitude of the coefficients in free-living populations of different plants and animals; this should form the aim and content of an empirical population genetics (Buzzati–Traverso, Jucci, and Timofeef–Ressovsky 1938). (Timofeef–Ressovsky 1940, p. 104).[5]

---

[4]   This claim – i.e., the claim that Dobzhansky later considered selection more important that random drift – stands in need of qualification. Evolutionary biologists are not only concerned with the relative importance of selection, but also with the relative importances of the various *forms* of selection, like selection in favor of heterozygotes vs. selection in favor of homozygotes. Dobzhansky did not consider all forms of selection more important than random drift. He only considered selection in favor of heterozygotes important. Selection in favor of heterozygotes was more important, he believed, than random drift or any other form of selection (see Beatty forthcoming and Provine forthcoming for discussions of the development of Dobzhansky's views in this regard).

[5]   A. J. Cain expressed the same position much more recently, "The researches of R. A. Fisher, J. B. S. Haldane, and Sewall Wright laid the foundations of the mathematics of population genetics.... What they did was to provide a mathematical theory covering all possible contingencies, from which →

The second aim – that of determining the relative importances of the various modes of evolution – has also been frequently expressed. Dobzhansky, for instance, put the point very clearly, very early:

> Since evolution as a biogenic process obviously involves an interaction of all the above agents [of evolutionary change], the problem of the relative importance of the different agents unavoidably presents itself. For years this problem has been the subject of discussion. (Dobzhansky 1937, p. 186).

Dobzhansky himself contributed to the evolutionary synthesis in both of these respects, by measuring the values of the variables of population genetic theory in specific cases, and by extrapolating to the relative importances of the various modes of evolution. A lot of people ask me, pointedly, whether Dobzhansky really contributed anything to the synthesis. These people have heard him referred to, along with the likes of Fisher, Wright, and Haldane, as one of the major architexts of the synthesis, but they are suspicious of the magnitude of his contribution. These people sometimes suggest that his only contribution was the popularization of the ideas of the others: it was his very readable *Genetics and the Origin of Species* that first made population genetic theory intelligible to most biologists. I think, though, that the reason that many people overlook Dobzhansky's contribution is that they tend to think of contributions to scientific syntheses as *theoretical* contributions, whereas, in order to appreciate Dobzhansky's contribution, we must consider the possibility that observations as well as theories may count among the components of a scientific synthesis.

Of course, one might allow that experimental and field studies of laboratory and natural populations provided more than just evidence for population genetic theory, but deny that the additional contribution constituted any significant *part* of the evolutionary synthesis. The experimental and field findings might, in other words, be considered *additional to* rather than *an addition to* (i.e., constitutive of) the synthesis. At least when evolutionary biologists characterize the synthesis, though, they include the additional contribution of experimental and field studies. For example, they include in their characterization of the synthesis the relative importances attributed to the various modes of evolution, based on studies of particular cases. They disagree, to be sure, as to which mode of evolutionary change is or was accorded most importance in the synthesis. Mayr, for instance, characterizes the synthesis as a vindication of the singular importance of natural

→

quantitative predictions of both deterministic and stochastic processes could be made. That was a great gain, but it does not tell us what of all these possibilities are acutally exemplified in the wild – what, in short, are relevant to the actual process of evolution.'' (Cain 1979, pp. 599–600.)

selection (e.g., Mayr 1980, 1982). Stephen Orzack views it as a more pluralistic affair (Orzack 1981). Stephen Gould believes that it started-out a pluralistic affair, but has become increasingly one-sided in favor of selection (Gould 1980, 1981, 1982, 1983). More important than the disagreement, though (at least for our purposes), is the common acknowledgement that experimental and field illustrations of the various modes of evolution allowed by population genetic theory, and extrapolations to the relative importances of those various modes, were part and parcel of the evolutionary synthesis. Experimental and field studies of evolving populations contributed significantly to the synthesis in this regard, and truly deserve, for this reason, to be counted among the areas of biology integrated into the synthesis.

Along the same lines, it is useful to keep in mind that the core of the synthetic theory, as described here, was well formulated by the early thirties, and yet the synthesis is commonly said to have taken place during the thirties and forties (e.g., Mayr 1980, 1982). There are, to be sure, a *number* of good reasons for considering the synthesis to have been a protracted affair (e.g., because there were more theories integrated into the synthetic theory than just Mendelian genetic theory and Darwinian evolutionary theory – see especially Mayr 1980, 1982). But *one* reason, it seems to me, is that the synthesis consisted of more than just theoretical treatment of the various modes of evolution. It consisted also of illustrations of the occurrences of the various modes. Such illustrations were chronicled by experimentalists and field naturalists like Dobzhansky, who, during the thirties and forties and thereafter, sought to understand nature in terms of the theory of the early thirties.

Before concluding, I would like to take up a possible objection to my analysis of the synthesis. That is, it may have struck you already that attempts to determine the relative weights of the variables of the synthetic theory might be construed as a form of theory testing. The theories in question would be theories about the relative importance of the various modes of evolution – for example, theories to the effect that random drift is more important than natural selection, theories to the effect that selection is more important than random drift, etc. These theories would then be positively or negatively supported by measurements of the variables of population genetic theory in natural selection would lend greater support to the first sort of theory than to the second.

Does this fact – the fact that measurements of the values of the variables of population genetic theory can be formally construed as evidence for or against some theory – undermine the view that the relation of observation to theory in the evolutionary synthesis was something other than evidential? No. Note that the theories being tested in this case are different from the synthetic theory – i.e.,

132

different from the theory of population genetics based on the Hardy–Weinburg equilibrium principle. Evolutionary biologists measuring the values of the variables of population genetic theory in particular cases, even when they're doing it for the purpose of extrapolating to the overall weights of those variables, cannot be construed as testing the synthetic theory. They are testing *separate* theories about the relative weights of the variables of the synthetic theory. The synthetic theory is silent with regard to the relative significance of the various possible modes of evolution it describes. Any determination of the relative overall significance of the various possible modes of evolution cannot, then, be considered a test of *that* theory.

There were indeed "theories" developed concerning the relative weights of the variables of the synthetic theory. Wright's so-called "shifting balance theory" according to which random drift was of considerable significance, was one such theory (e.g., Wright 1931, 1932, 1940). Similarly, Fisher had reasons, contrary to Wright's, for believing that natural selection was of utmost significance (Fisher 1930). But, again, it is important not to confuse these theories with the theoretical core of the evolutionary synthesis – the theory of population genetics based on the Hardy–Weinburg equilibrium principle. Wright and Fisher were officially in agreement with regard to the latter theory (see Provine forthcoming on the Wright–Fisher dispute). Fisher and Wright both sought to bring observations to bear on their theories about the relative weights of the variables of the synthetic theory, but not on the synthetic theory itself. *Within the context of the evolutionary synthesis*, the relation of observation to theory was not primarily evidential. The synthetic theory prompted observations that were by no means intended to test *that* theory.

In conclusion, I want to suggest that we take seriously Darden, Mayr, Lewontin, and others when they include, among the areas of biology integrated into the synthesis, not only Mendelian genetic theory and Darwinian evolutionary theory, but also experimental and field studies of laboratory and natural populations. In particular, let's take seriously the idea that not only theories, but also observations, were important components of the synthesis. I offer this suggestion in the same spirit that Darden and Nancy Maull suggested almost a decade ago that there were many more interesting modes of integration in science than the notion of "theory reduction" alone captured (Darden and Maull 1977).

One of the consequences of taking seriously the contribution of experimental and field studies in the evolutionary synthesis is that it forces us to think afresh about the relations between theoretical and observational pursuits. Theory-biased historians and philosophers of science have expressed so little interest in experimental and field work that we seem to have either forgotten, or overlooked the fact that experimental and observational investigations serve more than the purely subservient role of providing confirmation for theories. They play a creative role as well. As Thomas Kuhn once described that role, contrasting it with the role of observation as evidence,

> Probably the rarest and most profound sort of genius in physical science is that displayed by men who, like Newton, Lavoisier, or Einstein, enunciate a whole new theory that brings potential order to a vast number of natural phenomena. Yet radical reformulations of this sort are extremely rare, largely because the state of science very seldom provides occasion for them. Moreover, they are not the only truly essential and creative events in the development of scientific knowledge. The new order provided by a revolutionary new theory in the natural sciences is always overwhelmingly a *potential* order. Much work and skill, together with occasional genius, are requried to make it *actual*. And actual it must be made, for only through the process of actualization can occasions for new theoretical reformulations be discovered. The bulk of scientific practice is thus a complex and consuming mopping-up operation that consolidates the ground made available by the most recent theoretical breakthrough and that provides essential preparation for the breakthrough to follow. In such mopping-up operations, measurement has its overwhelmingly most common scientific function. (Kuhn 1961, in Kuhn 1971, p. 188).[6]

Similarly, the *actual* synthesis (vs. just the potential synthesis) involved more than just the formulation of a theory, and more too than the confirmation of that theory. It also involved the creative contributions of experimentalists and field naturalists like Dobzhansky, who discovered much more than one can discover from the theory of population genetics alone about how evolution *actually* occurs.

---

[6]   Kuhn is not completely alone in emphasizing the non-evidential roles of observation relative to theory. Ian Hacking and Bas van Fraassen have also recently stressed the non-evidential, measurement role of observation (Hacking 1983, van Fraassen 1980). Their analyses are, however, of limited use in discussing the evolutionary synthesis, since they attend only to measurements of the values of theoretical *constants*, while the non-evidential, measurement role of observation in the synthesis had more to do with the measurement of the values of theoretical *variables* (those variables, of course, taking different values in different cases).

For reasons that I am not prepared to treat adequately here, I think Hacking's and van Fraassen's emphasis on the measurement of constants vs. the measurement of variables has to do with the fact that they are both more concerned with the physical sciences than with the biological sciences. In the physical sciences, knowledge about particular cases is not regarded as a means of advancing science. The way to advance science in the physical sciences is to discover generally applicable laws, and likewise, to determine the values of generally applicable constants. Of course, the measurement of variables in the physical sciences is important for *testing* theories, but not for advancing science in any other, non-evidential way.

In biology, on the other hand, and especially in evolutionary biology and ecology, one very definitely advances science by discovering the facts about specific cases. Each case is special. Remember Timofeef–Ressovsky's call for evolutionary biologists to go out and measure the variables of population genetic theory in specific cases. That sort of knowledge is really additional to theory in biology. That sort of knowledge is, moreover, an addition of a very cherished sort.

Acknowledgement

I would like to thank William Provine and William Bechtel for very helpful discussions concerning the issues treated here.

References

Beatty, J. (forthcoming). Dobzhansky and drift: Facts, values, and chance in evolutionary biology. In L. Kruger (ed.), *The probabilistic revolution*. Cambridge: MIT Press, Bradford Books.

Buzzati-Traverso; A., Jucci, C.; and Timofeef-Ressovsky, N. W. (1938). Genetics of populations. *Ricerca Scientifica,* **9,** (1).

Cain, A. J. (1979). Introduction to general discussion. *Proceedings of the Royal Society of London,* **B205,** 599–604.

Chetverikov, S. S. (1926). On certain aspects of the evolutionary process from the standpoint of modern genetics. M. Barkers (transl.). *American Philosophical Society Proceedings,* **105,** (1961), 167–195.

Darden, L. and Maull, N. (1977). Interfield theories. *Philosophy of Science,* **44,** 43–64.

Dobzhansky, T. (1937). *Genetics and the origin of species (1st ed.). New York: Columbia University Press.*

Dobzhansky, T. (1941). Genetics and the origin of species (2nd ed.). New York: Columbia University Press.

Dobzhansky, T. (1951). *Genetics and the origin of species* (3rd ed.). New York: Columbia University Press.

Fisher, R. A. (1930). *The genetical theory of natural selection.* Oxford: Oxford University Press.

Gould, S. J. (1980). G. G. Simpson, paleontology, and the modern synthesis. In E. Mayr and W. B. Provine (eds.), *The evolutionary synthesis.* Cambridge: Harvard University Press.

Gould, S. J. (1981). But not Wright enough: reply to Orzack. *Paleobiology,* **7,** 131–134.

Gould, S. J. (1982). Introduction. In Columbia Classics in Evolution Series reprint of the first edition of T. Dobzhansky's *Genetics and the origin of species* (1937). New York: Columbia University Press.

Gould, S. J. (1983). The hardening of the synthesis. In M. Grene (ed.), *Dimensions of Darwinism.* Cambridge: Cambridge University Press.

Hacking, I. (1983). *Representing and intervening.* Cambridge: Cambridge University Press.

Haldane, J. B. S. (1924–1932). A mathematical theory of natural and artificial selection (9 parts). *Transactions and Proceedings of the Cambridge Philosophical Society.*

Kuhn, T. S. (1961). The function of measurement in modern physical science. Reprinted in Kuhn, *The essential tension.* Chicago: University of Chicago Press, 1971.

Lewontin, R. C. (1982). Theoretical population genetics in the evolutionary synthesis. In E. Mayr and W. B. Provine (eds.), *The evolutionary synthesis.* Cambridge: Harvard University Press.

Lewontin, R. C.; Moore, J. A.; Provine, W. B.; and Wallace, B. (eds.). (1981). *Dobzhansky's genetics of natural populations I–XLIII.* New York: Columbia University Press.

Mayr, E. (1980). Prologue: Some thoughts on the history of the evolutionary synthesis. In Mayr, E. and W. B. Provine (eds.), *The evolutionary synthesis.* Cambridge: Harvard University Press.

Mayr, E. (1882). *The growth of biological thought.* Cambridge: Harvard University Press.

Orzack, S. (1981). The evolutionary synthesis is partly wright. *Paleobiology,* **7,** 128–131.

Provine, W. B. (1971). *The origins of theoretical population genetics.* Chicago: University of Chicago Press.

Provine, W. B. (1978). The role of mathematical population genetics in the evolutionary synthesis of the 1930s and 1940s. *Studies in the History of Biology, 2*, 167–192.

Provine, W. B. (forthcoming). *Sewall Wright: Geneticist and evolutionist.* Chicago: University of Chicago Press.

Timofeef–Ressovsky, N. W. (1940). Mutations and geographical variation. In J. S. Huxley (ed.), *The new synthesis.* Oxford: Oxford University Press.

van Fraassen, B. (1980). *The scientific image.* Oxford: Oxford University Press.

Wright, S. (1931). Evolution in Mendelian populations. *Genetics, 16*, 97–159.

Wright, S. (1932). The roles of mutation, inbreeding, crossbreeding, and selection in evolution. *Proceedings of the Sixth International Congress of Genetics, 1*, 356–366.

Wright, S. (1940). The statistical consequences of Mendelian heredity in relation to speciation. In J. S. Huxley (ed.), *The new systematics.* Oxford: Oxford University Press.

Editor's Commentary

There is little question today that the evolutionary synthesis was a major event in the history of many disciplines and that Dobzhansky played a critical role in this event. Beyond that, however, there is a fair amount of disagreement about just what the synthesis involved and what it has meant to the disciplines that participated in it. One thing is clear – it did not produce a new discipline. Evolutionary biology continues to exist as an "interdisciplinary research cluster" in which a variety of disciplines (including genetics, population genetics, systematics, paleontology, zoology, and botany) continue to interact. In the Introduction I have described some of the institutional features of this interdisciplinary research cluster. What is often claimed for the synthesis is that it produced a new theory, "the synthetic theory of evolution." It is far from clear, though, that there is one well-articulated theory that was the product of the synthesis. Gould (1983) speaks of the current textbook versions of evolutionary theory, with its focus on gradualism and on natural selection as the principal force of evolution as "the hardened synthesis," and argues that during the period of the evolutionary synthesis there was considerably greater pluralism. Mayr (1982) on the other hand, seems to take this theory as the principal product of the synthesis, while Stebbins and Ayala (1985) treat it as a framework that can be altered and amplified.

Even if we cannot agree on the precise character of the product that resulted from the evolutionary synthesis, it is clear that considerable integration of disciplines did occur during the synthesis. This integration was both social and conceptual in nature. The social character of the integration is partly revealed by the degree to which scientists from the various fields came to communicate and interact with each other. Mayr (1980), for example, contrasts the degree of discord at conferences of paleontologists and geneticists in the 1920s and early 1930s with the general harmony at an international conference held in Princeton in 1947. What made this communication possible was the development of a conceptual basis for integration. The interesting question is: what did this conceptual basis consist in? For Mayr

138

(1982), it consists in the fact that what he takes to be the two main opposing traditions prior to the synthesis, that of the naturalists and the geneticists, both learned central concepts from each other. In particular, he points to the fact that geneticists learned, amongst other things, what he refers to as "populational thinking" (which recognizes the fact that species are not a natural kind but that there is variability within species) from the naturalists, while the naturalists learned the new ways in which geneticists characterized genetic material.

The papers in this section offer two alternatives to Mayr's account of the conceptual integration achieved in the synthesis, at least in the work of Dobzhansky. The difference between Darden's and Beatty's views (and also between their views and Mayr's) begins with their conception of the disciplines that were integrated through Dobzhansky's work. Darden differentiates disciplines (fields) in terms of the level at which inquiry was pursued. Thus, she distinguishes the chromosomal level that was the focus of Mendelian genetics and cytology, the population level which was the focus of mathematical population genetics and experimental and field studies of populations, and the species level, where one would study the isolating mechanisms involved in the formation of new species. She represents Dobzhansky's role as postulating the need for independent consideration of the species level and establishing the relation that should hold between the three levels in an explanation of the origin of new species. Beatty distinguishes instead among scientific pursuits, arguing that there is great variety in the nature of the pursuits that is concealed by Darden's focusing on fields and levels of inquiry. Beatty focuses primarily on the different pursuits of theory development, experimental studies and field studies, and argues that Dobzhansky's main contribution was to integrate these.

In light of this difference in the way Darden and Beatty differentiate the disciplines that were integrated, it is not surprising that they point to different features in Dobzhansky's work as constituting his main contribution to integrating disciplines. Darden presents Dobzhansky as creating an interfield theory, although one different in nature from the interfield theories she and Maull have previously considered. One significant difference she notes in this case is that one does not find one field turning to another field for assistance, but rather two fields initially competing to solve the same problem. Some of this competitive aspect of interdisciplinary exchange was also found in the background to the development of biochemistry. Such competitive interactions have a sociological component which Darden does not take up, but which deserves further consideration. The general phenomenon of conflict between fields Darden highlights in this case may be a rather common feature of the historical backdrop to cross-disciplinary exchanges that invites further examination.

A second difference Darden notes between this case and the ones she has examined previously is that in this case neither contributing field was able to solve the problem. Rather, Dobzhansky's theory involved the postulation of an additional set of mechanisms serving to isolate species. These became the object of study in a new field. (In this paper Beatty takes issue with the claim that this contribution can be credited to Dobzhansky, but does allow that it was an important theoretical development that emerged from various contributors to the synthesis.) The overall account of speciation Dobzhansky proposed incorported mechanisms studied in each field, but was not itself contained in any. In giving rise to a new field of research, this case is again similar to that of biochemistry. But it is different in that the recognition of a new field served primarily to permit an integration of mechanisms from separate fields instead of bringing them all into one field. This is part of the reason that the synthetic theory remained an interfield one (rather than becoming the core of a fairly autonomous new discipline).

Finally, the synthesis postulated a temporal ordering between the processes studied in each field: "gene mutations precede populational changes which precede the development of isolating mechanisms". This temporal ordering of the process provides an ordering amongst the fields included in the synthesis: genetics accounts for the material included in populational changes, but is unaffected by theories about populational changes. Darden has related that for Dobzhansky (although not for Mayr) a similar relation is supposed to hold between studies of population changes and studies of speciation. This again distinguishes this case from the biochemistry case, where such an ordering was intially expected (with the postulation of linear pathways), but where the actual pattern allowed for feedback processes that undercut this temporal ordering.

While admitting that there are significant differences between the synthetic theory and the cases of interfield theories studied by Darden and Maull previously, it may be unwise to make too much of the differences. As with the previously discussed interfield theories, the background knowledge accumulated by the 1930s suggested linkages between genetic processes and selectional processes (as is evident in the work on population genetics of Fisher, Haldane, and Wright); both fields had a common interest in explaining the origin of species; and yet the problem remained unsolved. What Dobzhansky offered, in Darden's analysis, is a theory that bridged the two fields and the new field at a new level. This theory made clear how the mechanisms of the separate fields interface with each other, so providing a coherent framework for developing explanations.

The kind of conceptual integration discussed by Beatty is quite different, but equally important. From Beatty's perspective, Dobzhansky's interfield contribution did not involve so much the development of a new theoretical framework, but the use of results from two different research approaches (experimental and field

140

studies) to constrain the development of evolutionary theory. The underlying theory, as Beatty presents the case, is the Hardy–Weinberg law, which predates the synthesis by several decades. The Hardy–Weinberg law, when embelished to allow for selection, limited population size, and similar factors, gives a generalized perspective of what happens to frequencies of genes in a population under a variety of conditions and, as such, provided the basis for the theorizing of Fisher, Haldane and Wright. Their theorizing involved setting different constraints on the basic Hardy–Weinberg principles (e.g., small population size versus large population size) to show how populations could change under different possible conditions. What was required to make this the basis of an empirical theory of the evolution of life forms was a way of determining the values of various variables and background conditions. From this, one could then determine which modes of evolution are most important. Beatty contends that Dobzhansky's contribution was to integrate the use of both experimental and field observational studies in the task of specifying the values of the variables and the boundary conditions under which the Hardy–Weinberg law should be applied. Hence, his focus is not on the integration of fields as defined by Darden, but rather of research approaches.

The question arises as to the degree to which the analyses of Darden and Beatty are incommensurable with one another. In the discussion at the conference, where earlier versions of these papers were presented, Darden indicated a way one might bring them together. She noted that to integrate fields of research as she distinguished them it was necessary to integrate components of the fields. Thus, integrating the kinds of pursuits Beatty has distinguished might be one part of the task of integrating fields. Beatty offered one interesting example of this by pointing to Dobzhansky's experiments relating a theory in the field of evolution to Morgan's theories in the field of Mendelian genetics. Dobzhansky tried to determine whether the kind of variability ones sees in the laboratory existed in nature. But Beatty's approach is not merely a component of Darden's, since the kind of integration he sees Dobzhansky producing could occur within fields as Darden construes them, not just between them. What Beatty is focusing on is the question, too often neglected, of how theoretical and empirical investigations are to be integrated. He brings out a point not often considered – that empirical studies serve not just to test theories, but also to fix important parameters within them. One thing that is difficult to determine when such integrating work is accomplished by one researcher is how he or she made decisions as to how to relate empirical and theoretical models. All we have is the final record of the integration that was accomplished. Determining how such decisions get made is a relevant issue since there is a great deal of flexibility in deciding how to use empirical studies to determine parameters. The insights stemming from the laboratory studies of Latour and Wolgar (1979) and of Knorr

(1981) may be useful in helping us understand how scientists go about deciding how to relate empirical data to theoretical models and for further investigating the kind of integration which Beatty has highlighted.

These papers of Darden and Beatty can be viewed as two more contributions to the literature on Dobzhansky and the synthesis. However, in this context they serve a greater purpose, since they point to two important modes of conceptual integration that are involved in bridging scientific disciplines.

References

Gould, Steven J. (1983). The hardening of the modern synthesis. In M. Grene, (ed.), *Dimensions of Darwinism*. Cambridge: Cambridge University Press.

Knorr, Karin D. (1981). *The manufacture of knowledge. Toward a constructivist and contextual theory of science*. Oxford: Pergamon.

Latour, Bruno and Wolgar, Steve (1979). *Laboratory life. The social construction of scientific facts*. Beverly Hills: Sage Publications.

Mayr, Ernst (1980). Prologue: Some thoughts on the history of the evolutionary synthesis. In E. Mayr and W. B. Provine (eds.), *The evolutionary synthesis: Perspectives on the unification of biology*. Cambridge: Harvard University Press.

Mayr, Ernst (1982). *The growth of biological thought*. Cambridge: Harvard University Press.

Stebbins, G. Ledyard and Ayala, Francisco J. (1985). The evolution of Darwinism. *Scientific American, **253**, 72–82.

PART III
Incorparating Development Biology into the Evolutionary Synthesis

Introduction

MARJORIE GRENE
*Department of Philosophy, University of California, Davis, California
95616, U.S.A.*

Conventionally and in very introductory presentations, the Evolutionary Synthesis is said to have been initiated with the work of Fisher, Haldane and Wright in the early thirties (Fisher, 1930; Wright, 1932; and Haldane, 1932). It is certainly the case that in the first decades of this century there appeared to be a very fundamental conflict, even a contradiction, between Darwinian natural history, the home of evolutionary speculation, and Mendelian genetics, a rapidly developing experimental science. Darwinians thought of variations so small that continuous evolution of adaptive features could result from them; Mendelians thought of their variations, 'mutations', of whatever magnitude, as *dis*continuous, segregating and, even when expressed in phenotypic 'blends', continuing their separate ways in a kind of patchwork 'germplasm'. Bateson's journal, *Genetics,* was said to have been initiated precisely to oppose Darwinism (at least Waddington said so). And then, lo and behold, Fisher, Haldane, and Wright, in different mathematical modes, showed that genetic discontinuities could be handled so as to produce nice continuous curves – the seeming contradiction was no contradiction at all. Experimental genetics, in particular, population genetics, became the underpinning of Darwinian theory, and at the same time and by the same conceptual move, Darwinian theory became the theoretical explanation of Mendel's laws. Moreover, even if Mayr and Provine begin their story in 1936, there is no doubt that this triumphant reconciliation, initiated in 1930–1932, was given its classic and foundational expresssion in Dobzhansky's *Genetics and the Origin of Species* in 1937 (based rather on Wright's than on Fisher's or Haldane's versions of population genetic algebra).

The synthesis, moreover, continued to develop. Bruce Wallace was one of Dobzhansky's leading students and collaborators, and as one of the co-editors of *Dobzhansky's Genetics of Natural Populations* (Lewontin, Moore, Provine, and Wallace, 1981), is not only an authority on the experimental methods and results associated with the synthesis; he is himself one of its major representatives. Many

Bechtel, W (ed), Integrating Scientific Disciplines. ISBN 90-247-3242-5.
© *1986, Martinus Nijhoff Publishers, Dordrecht. Printed in The Netherlands.*

of the classics of the synthesis appeared only in the forties, and Stebbins' *magnum opus* only in 1950. As a long-term spectator of these developments, I myself would date the history of the evolutionary synthesis from 1930 to 1959: the date of the centennial of the *Origin* and the celebrations that issue in the volumes edited by Sol Tax (true, there were a few still small voices of dissent in these otherwise triumphant tomes, but just a few) (Tax, 1960).

In that period, moreover, which marks both the initiation of the synthesis and the transition to what Gould calls its 'hardened' form, the synthesis was celebrated not only as 'reconciling' Darwinian evolution and Mendelian genetics, but as bringing to a 'synthesis' many other fields as well (Gould, 1983, Grene, 1981). Systematists had resisted the synthesis, finding natural selection too speculative as an explanation for their data (e.g., Robson and Richards, 1936); Mayr brought systematics squarely within its scope (Mayr, 1942). Indeed, from his point of view, that *is* the major synthesis (Mayr and Provine, 1980). Paleontologists, too, had proved recalcitrant; the fossil record is too gappy for those smooth curves: after all, Darwin himself had introduced paleontology as a *difficulty* for his theory. Yet, G. G. Simpson proved him wrong: paleontology in Simpson's person welcomed the synthesis with open arms (Simpson, 1943, 1953). Stebbins in 1950 safely incorporated botany (Stebbins, 1950). So the synthesis was represented as 'synthesizing' many biological fields. Indeed, development itself, the topic addressed in these papers, was said to be included. The work usually mentioned was one that Wallace here twice refers to: Schmalhausen's *Factors of Evolution*, published in 1949 translation from the Russian, with an introduction by Dobzhansky himself (Schmalhausen, 1949). Yet, as Wallace's argument shows (I'd bet on development for the mother-in-law, not the fecund and lovely daughter!), embryology certainly did not major work in the development of the synthesis.

Latterly, however – and this is the reason for the present discussion – there have been suggestions, some fairly strident, others more restrained, to the effect that embryology could, and should, play a more positive role in evolutionary theory than it has so far done (i.e., Goodwin, Holder, and Wylie, 1983). Three arguments to that effect, one by a biologist and two by philosophers, are presented here. (Both philosophers, it should be added, work closely with biological colleagues.) Stuart Kauffman is concerned to construct a theory of the constraints imposed by physical and topological laws on the development of genetic systems. So far as I understand what I know of his work, he is not, I believe, as some more radical evolutionists are now doing, offering an alternative to the synthesis, but rather a complement to it. As Wallace rightly points out, no one was ever (or for long) strictly a "beanbag geneticist', and the complexifications of the genome consequent on the development of biochemical and molecular genetics do not as such 'destroy' the synthesis. But if evolution is 'differential gene frequency' within strict limits of what possible genomic structures allow, those structures may themselves be said, and will be said by those who spend their lives studying them, to add a further causal dimension to the theory itself.

Wimsatt, starting more explicitly from a traditional developmental 'law', might be said to be presenting a still more radical challenge to the synthesis; on the other hand, when one takes into account all the consequences he wants to draw from his developmental lock, it may turn out that the explanatory core of the synthesis will rest unchallenged. Burian takes his start from recent genetic studies of genomic reorganization, and argues that, because the same processes of reorganization figure in the germ line and somatic cells, the recent developmental studies of genomic reorganization in somatic cells may be equally informative to evolutionary theorists, whose prime interest is in the germ cells. This approach would seem to pose no substantial threat to the synthesis but only to open the way for developmental biologists and evolutionary geneticists to work together on a phenomenon of common interest. In summary, these three papers provide material, if not for actual interdisciplinary investigation, at least for the conceptual analysis of the possibilities of such work,

References

Dobzhansky, Theodosius. (1937). *Genetics and the origin of species*. First edition. New York: Columbia University Press.

Fisher, R. A. (1930). *The genetical theory of natural selection*. Oxford: Clarendon Press.

Goodwin, Brian; Holder, Nigel; and Wylie, C. C. (1983). *Development and evolution*. Cambridge: Cambridge University Press.

Gould, Stephen, J. (1983). The hardening of the synthesis. In M. Grene (Ed.), *Dimensions of Darwinism*. Cambridge: Cambridge University Press.

Grene, Marjorie (1981). Changing concepts of Darwinism in evolution. *The Monist, 64*, 197–213.

Haldane, J. B. S. (1932). *The causes of evolution*. London: Longmans, Green.

Lewontin, R. C.; Moore, J. A.; Provine, W. B.; and Wallace, B. (Eds.) (1981). *Dobzhansky's genetics of natural populations I–XLIII*. New York: Columbia University Press.

Mayr, Ernst (1942). *Systematics and the origin of species*. New York: Columbia University Press.

Mayr, Ernst and Provine, William (Eds.)(1980). *The evolutionary synthesis*. Cambridge: Harvard University Press.

Robson, G. C. and Richards, O. W. (1936). *The variation of animals in nature*. London: Longmans, Green.

Schmalhausen, I. I. (1949). *Factors of evolution*. Philadelphia: Blakiston.

Simpson, George G. (1943). *Tempo and mode in evolution*. New York: Columbia University Press.

Simpson, George G. (1953). *The major features of evolution*. New York: Columbia University Press.

Stebbins, G. Ledyard (1950). *Variation and evolution in plants*. New York: Columbia University Press.

Tax, Sol (Ed.) (1960). *Evolution after Darwin. I. The evolution of life: Its origin, history and future.* Chicago: University of Chicago Press.
Wright, Sewall (1932). The roles of mutation, inbreeding, crossbreeding, and selection in evolution. *Proceedings of the Sixth International Congress of Genetics,* **1,** 356–366.

Can Embryologists Contribute to an Understanding of Evolutionary Mechanisms?

BRUCE WALLACE

Department of Biology, Virginia Polytechnic Institute and State University, Blacksburg, VA 24061, U.S.A.

The issue for this symposium is whether developmental biology requires a rethinking of the Modern Synthesis. I shall argue that it does not.

I will begin by discussing the challenge to the evolutionary synthesis posed by Richard Goldschmidt and his "hopeful monster". Here was a claim that new species arise in the guise of single mutant individuals, coupled with a denial that mutations with small effects (often, and I belief, erroneously called "small mutations") and changing gene frequencies have any evolutionary consequence beyond the differentiation of local populations. Although there are no references to Goldschmidt in the *Genetics of Natural Populations* series of publications (Lewontin, *et al.*, 1981), Dobzhansky did mention him in the field, in his laboratory, in the classroom, and in other publications of that era. Because Goldschmidt had identified the chromosomal inversions of *Drosophila pseudoobscura* as possible subliminal forerunners of a genomic reorganization that might eventually lead to instant speciation, much of Dobzhansky's efforts during the 1940s were spent showing that the occurrence and fate of inversions in natural (and, later, in laboratory) populations were adequately explained by virtue of the differing gene contents of chromosomes of various gene arrangements and of various geographic localities. There was, in Dobzhansky's view, no need to imagine that the points of chromosome breakage which gave rise to the inversions were accompanied by evolutionary important "position effects." Throughout this period of intellectual sparring, I might add, Dobzhansky, even in private, expressed only the highest regard for Richard Goldschmidt, the person.

In the pages that follow, I shall further argue that problems concerned with the orderly development of the individual are unrelated to those of the evolution of organisms through time just as, for example, the means by which individuals store memories of past experiences are unrelated to the obelisks, victory arches, heroic monuments, and scrolls by which ancient civilizations recorded their histories for posterity. In this argument, I shall join (at the risk of defending "entrenched

Bechtel, W (ed), Integrating Scientific Disciplines. ISBN 90-247-3242-5.
© *1986, Martinus Nijhoff Publishers, Dordrecht. Printed in The Netherlands.*

150

orthodoxy'') Dobzhansky and Boesiger (1983) and Mayr (1984). Evolution consists of changing, within the collective genetic resources of populations, the frequencies of those genetic programs that lead to the successful development and survival of individuals under prevailing environmental conditions. This claim, many will recognize, is merely a verbose paraphrase of the terser statement: evolution is a change in gene frequency. This statement, of course, is also controversial today (Mayr, 1984).

In their history of the Modern Synthesis, Mayr and Provine (1980) asked representatives of various biological disciplines as well as those of many nations to comment on the relationship of their discipline or their nation to what has become known as the Modern Synthesis – neo-Darwinism. Among these persons was Viktor Hamburger who related the contribution of embryology to the Modern Synthesis. His account of Waddington's (1953) ''genetic assimilation'' is largely responsible for the thrust of this article. It would appear that Hamburger (perhaps most embryologists) even now has little understanding of the genetic basis of Waddington's observations and, by inference, of neo-Darwinism. Evolutionary geneticists have the responsibility for explaining the origins and subsequent fates of the genetic programs which determine developmental programs; embryologists, on the contrary, need not explain how somatic development might affect the evolution of these developmental programs. Except for achieving success in reproduction, they do not.

Figure 1 illustrates Hamburger's view of genetic assimilation as presented in Mayr and Provine (1980). Within a given environment (I) the modal phenotype of a population or organisms coincides with the optimal phenotype; because of limitations imposed upon diagrammatic representations of complex situations, only two facets (*i* and *j*) of the total phenotype can be represented. The phenotypic variation of the population about the mode is ascribed by Hamburger to non-heritable (environmental) variation.

Following a change in the environment (from I to II), the modal and optimal phenotypes no longer coincide. Nevertheless, the population continues to exist because some of the non-heritable variants exhibit, by chance, the new optimal phenotype. Eventually, according to Hamburger, a newly arisen mutation, (or sequence of such mutations) whill shift the modal phenotype so that it coincides once more with the optimal one. That shift, presumably, represents ''genetic assimilation.'' Missing from this account is the very heart of neo-Darwinism; the above account of genetic assimilation is an example of what Mayr refers to as ''typological thinking.''

The element which is lacking in Figure 1 is the understanding that the phenotypic variation displayed by the population in environment I is the result of both

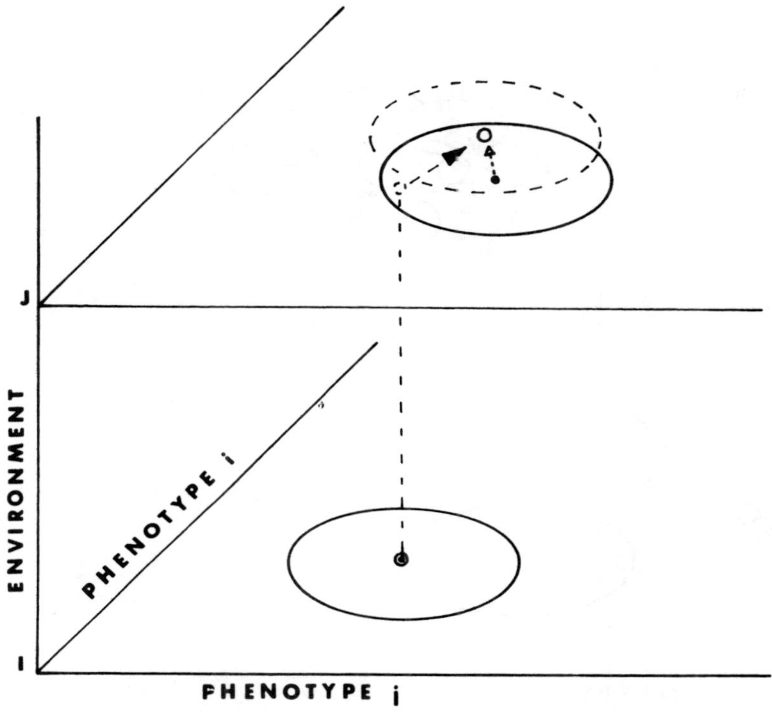

Figure 1. A diagrammatic representation of Hamburger's (Mayr and Provine, 1980) account of genetic assimilation. In the lower place (Environment I), various facets (*i* and *j*) of the modal phenotype () coincide with that optimal for the environment (O); variation about this optimum is said to be non-heritable. In the altered environment (J), the modal phenotype no longer coincides with that which is optimal under the changed conditions; the optimum is achieved only by some rare, non-heritable variants. Mutation is then said to cause the modal phenotype to move so that eventually it once more coincides with the optimal one.

environmental and genetic variation. Thus, when environment II is substituted for environment I (Figure 2), each of the various genotypes that were included within the previous phenotypic variants now react somewhat differently to the new environment. Each genotype may have its own modular phenotype and its own newly acquired range of phenotypic variability. One of these modes may fall near the new optimal phenotype; that genotype will tend to increase in frequency. Furthermore, since genotypes in a mendelian population are not fixed but, rather, appear and disappear under recombination, newly formed geotypes whose phenotypic modes also lie near the optimal phenotype, will also increase over time,

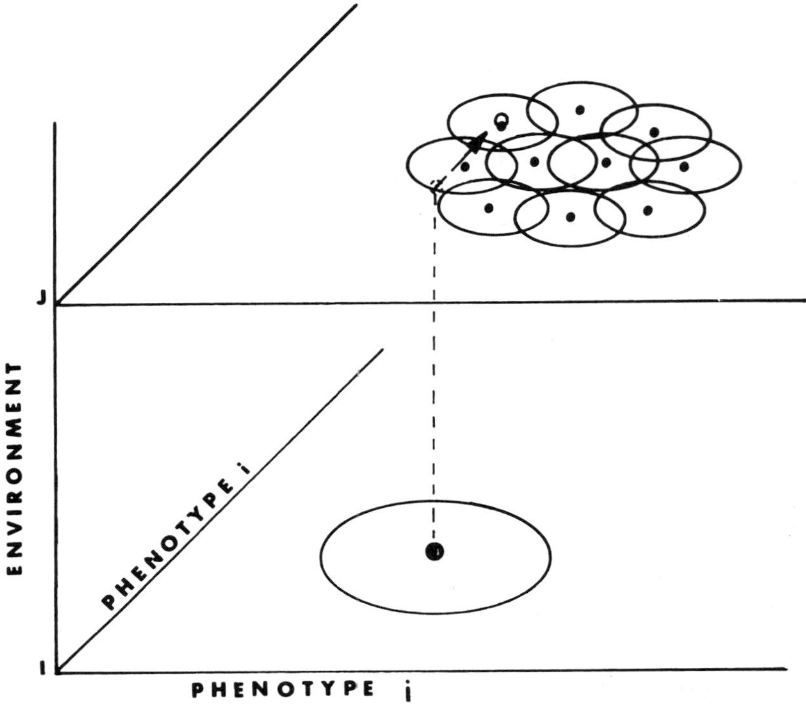

Figure 2. A diagram illustrating the neo-Darwinian view of adaptation to an altered environment. The variation around the (super-imposed) modal and optimal phenotypes is recognized as being both genetic and environmental in origin. In the altered environment, gene-environment interactions cause the many different genotypes to take on novel modal phenotypes (), one or more of which may lie near the new optimum (o). The proportion of the latter genotypes, as well as that of others which may arise by recombination, increases; this increase being natural selection. Mutation is relegated to a minor role in the population's adaptation to new environmental demands.

while continually approaching that optimum. The outcome that results from the interplay of recombination and natural selection will be a population whose modal and optimal phenotypes coincide once more but one possessing phenotypic variants whose deviations from the mode are once again caused by both genetic and environmental factors.

The environmental shocks employed by Waddington (1953) in his experiments on genetic assimilation resulted in low frequencies of abnormal phenotypes which subsequently responded to artificial selection. Hamburger refers to these as non-heritable variants; on the contrary, what Waddington's experiments demonstrate is that phenocopies and other abnormal phenotypes that result from environmental

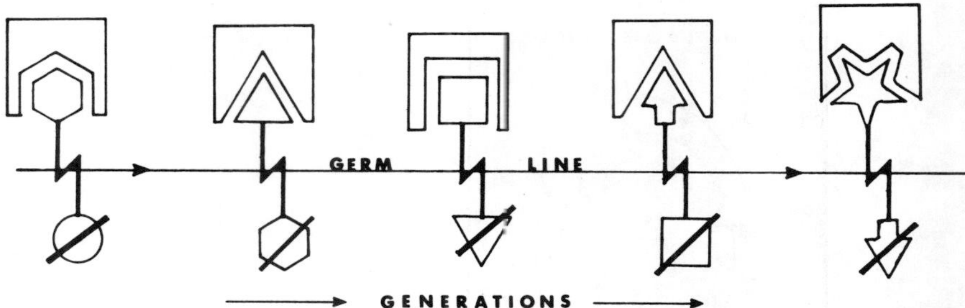

Figure 3. The role of the environment in guiding the evolution of the phenotype. At each time level, the environment assumes a different shape, thus excluding the previously adapted phenotype while placing a selective premium on one that is compatible with the new demands. The one thread that continues unbroken through the passage of time is the germ line.

stress can have a *genetic* basis – hence, they respond to artificial (or, in Figure 2, to natural) selection.

Figure 3 illustrates in an extremely diagrammatic way both the cause and the course of evolution through time. At every time level, the environment imposes demands on the organism that must be met if the organism is to survive (exist); these demands change with time. The series of evolving forms, each fitting, in turn, the demands of the current environment, is shown in the Figure; for emphasis, the preceding form – now a misfit – is shown with an obliterating line. In every instance, had the acceptable form not arisen in this diagrammatic lineage, the line would have terminated. The need to fit the environment, in other words, is shown as being severe; neither a hexagon nor a square in this representation, could fit a triangular demand although two different forms, a triangle and an open arrow, were adequate.

One thread continues unbroken throughout Figure 3: the germ line. If one were to trace the germ line from the right to the left, that is, one would easily find that it is continuous. If, on the other hand, one were to trace the germ line from left to right, one would discover this continuity only with the greatest difficulty because one would have entered a made of unbelievable complexity. Many of the paths which one might follow in tracing the germ line from left to right would lead to individuals who, prior to reproduction, lost their lives for seemingly irrelevant reasons – drownings, falls, and accidents of all sorts. Some lines would lead with numerous others as coherent clusters only to terminate; these would reflect entire populations that were wiped out in the past by events of cataclysmic magnitude or, on the contrary, of seemingly trivial dimensions. To escape from such a cul de sac in attempting to trace the one continuous line of descent, one would need to retrace one's steps back through considerable time. Despite the difficulty in identifying in

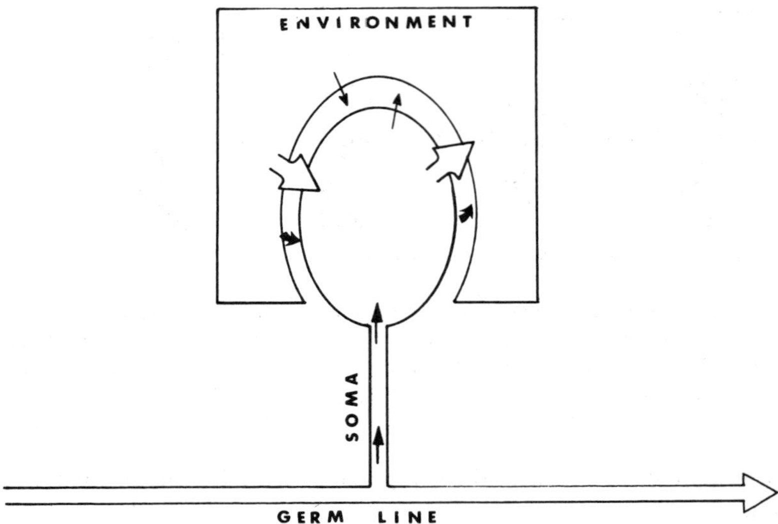

Figure 4. An enlarged view of one generation in Figure 3 showing the inter-relations between environment, phenotype (soma), and germ line. The open arrows connecting environment and soma represent adaptive interactions, the heavy curved arrows represent non-adaptive ones, and the linear ones represent trivial interactions. The arrows from the germ line to the soma emphasize that the program for normal development (including the concomitant interactions) exists in the germ line.

advance which line of descent will lead to the future, at least one line must – a line that is seen rather clearly in retrospect.

The environment is not a passive agent – a mere mould – in shaping evolutionary change; it does not make demands on a take-it-or-leave-it basis. Figure 4 illustrates at higher magnification a time level such as those illustrated in Figure 3. The germ line that has given rise to the individual is shown running across the bottom. The individual itself (soma) arises as a cluster of cells that eventually form the tissues, organs, and organ systems which are the individual. As a functional being, the individual interacts with its environment in several ways: adaptive, non-adaptive, and trivial.

To a considerable extent, individuals of many species "improve" their immediate environment by burrowing, building nests and other shelters, or by accumulating a variety of protective items for warmth or camouflage. Some individuals, as we learned from the three little pigs, are more successful than others.

The impingement of the environment on the individual very often evokes adaptive somatic responses. The growth of scarcely any individual is so deterministic that it fails to respond to environmental signals. As a whole, these adaptive responses are the "modifications" of Schmalhausen (1949); such modifications would include, for example, environmentally induced polymorphisms (Hazel and West, 1979; Gilbert, 1980) or more prosaic adaptive responses such as the formation of callouses on workers' hands.

Individuals also interact with their environment in non-adaptive ways. They excrete toxic waste substances (one such case limits the alcohol content of wine) or use up non-renewable or non-replenishable resources. Non-adaptive responses to environmental stimuli have been called "morphoses" by Schmalhausen (1949). Morphoses can be regarded as reactions that lack an evolutionary history; in contrast, reactions to commonly encountered environmental stimuli have been tempered by generations of selection so that they now appear as adaptive modifications.

Trivial environment-organism interactions, in both directions, occur but may be dismissed. Even though they may be individually identifiable, they fall in the category of those negligible quantities such as the impetus given to the earth by a departing rocket (Lewontin, 1974).

The crucial feature to note in Figure 4 is that the germ line is independent of the soma; this is the situation which applies to most higher animals. Thus, the germ line brings in from the left those genetic programs and sub-programs that lead to the development of the soma vis-a-vis the environment and the reciprocal interactions between the organism and its environment. These programs continue to the right only if they have succeeded in leading to viable, fertile individuals. Those programs that fail this requirement terminate on the spot as far as this thread in the web of descent is concerned. Whether or not the genetic program continues to the right determines whether the frequency of this program in the web of descent changes in going from the left side to the right side of the figure.

Figure 5 emphasizes that higher plants possess hereditary options that are not available to most higher animals. Germ cells in plants – as the position of flowers on all branches of a bush or tree suggests – arise from somatic cells. In theory, a plant has the opportunity to test its original genetic program in relation to the prevailing environment (only reciprocally adaptive interactions are indicated in Figure 5), to hit upon an improved program in one branch of the plant, and then to incorporate this newly improved genetic program within the germ cells that arise within the reproductive structures of that branch. This additional opportunity that is available to plants (and to some lower animals, as well) has been stressed by Buss (1983).

The opportunity possessed by plants that is illustrated in Figure 5 does not guarantee evolutionary success to these organisms. An example of the type of problem accompanying the ability to make intra-somatic alterations of genetic

156

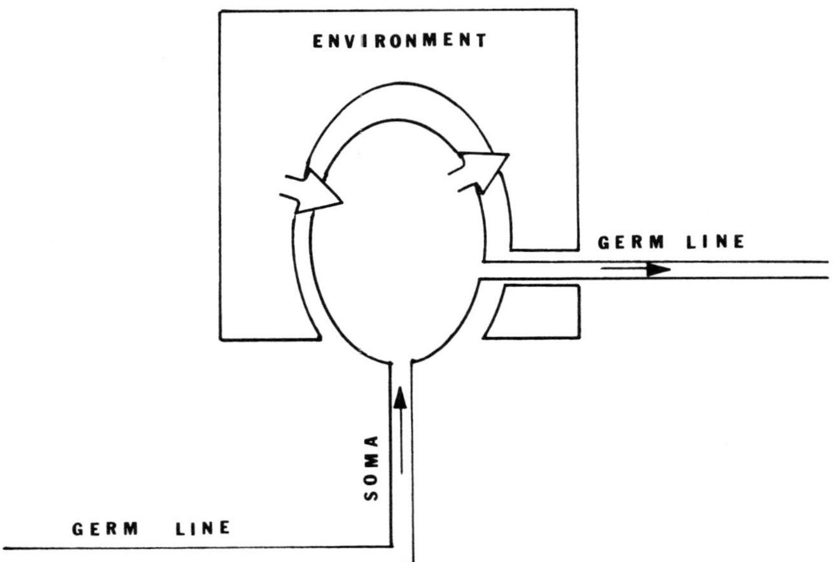

Figure 5. A diagram depicting an option that is available to plants but which is not available to most higher animals: Germ cells arise from somatic cells; therefore, adaptive genetic changes arising in the soma can, in theory, be incorporated into the germ line to the benefit of future generations.

programs is presented in Figure 6: selective forces that involve cells within a single individual need not correspond to populational ones that operate between individuals which compete with one another for survival, and for the opportunity to transmit their genetic programs to their offspring.

Growth of the sort illustrated by the May Apple (*Podophyllum peltatum*) may, however, offer an escape from the difficulty posed in Figure 6. The rhizome of the May Apple (Figure 7) grows continuously at one end while decaying at the other. The living segment that bears the leaves (and the fruit if sexual reproduction occurs) is only several years old. Somatic mutations – that is, changes in the genetic program that occur within the May Apple's somatic cells – are, in this case, tested at the individual level in the arena of both intra- and inter-specific competition; the successful innovations are then available for incorporation into the individual's germ cells.

In summarizing the discussion that has been emphasized in Figures 1 to 7, the following points can be made:

1. The embryologist's view of the Modern Synthesis as illustrated by Viktor Hamburger's analysis of Waddington's genetic assimilation reveals a lack of

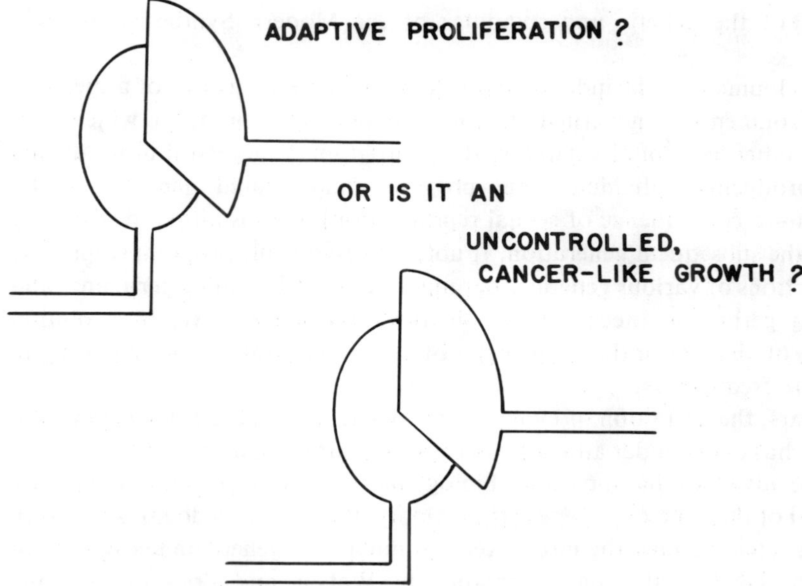

Figure 6. A diagram emphasizing schematically the difficulty that accompanies a reliance on somatic mutations for phylogenetic change (evolution): Intra-individual selective forces may run counter to inter-individual ones.

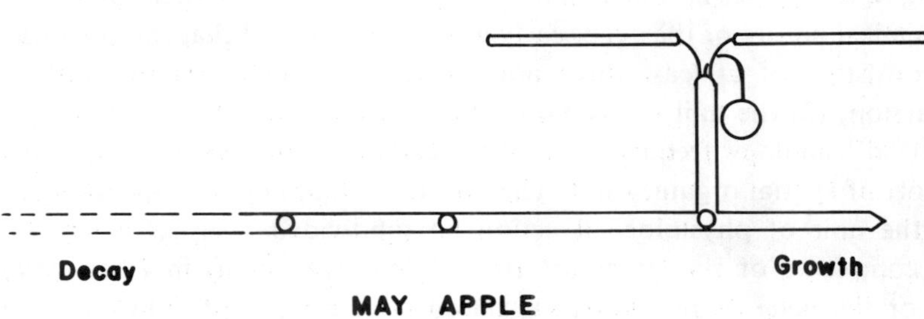

Figure 7. A diagrammatic representation of a May Apple showing its mode of growth. The rhizome grows at one end and decays at the other; at any moment, the rhizome consists of cells whose ages span several years (small nodules reveal the location of flowing shoots of past years). Adaptive chnges of somatic origin could be easily incorporated into a plant possessing this growth habit; these changes in turn could subsequently be transmitted through the germ line.

158

understanding of the genetic underpinnings of the Modern Synthesis, of Neo-Darwinism.

2. The development of the individual (or the *soma*) within any one of a specified range of environments is governed by the developmental program which that individual has inherited from its parents. If the program is successful in producing an adult, reproducing individual, that program (and related ones arising by recombination – a consequence of sexual reproduction) is transmitted to offspring who make up the subsequent generation. If not, the responsible program stops. The relative proportions of various genetic programs among the incoming germ lines and the continuing germ lines need not be identical: the lowest level of evolution consists, then, of changes in the frequencies of genetic programs – or, stated more simply, of gene frequencies.

In recent years, the definition of evolution as "a change in gene frequency within a population" has come under attack from several quarters. One of these criticisms arises from the advances that molecular biology has made in elucidating the nature of the gene and of the control of gene action. Oddly, the criticisms do not stem from the era when protein seemed the most likely candidate for genetic material only to be displaced by DNA, following the creation by Watson and Crick of the now-accepted DNA-model. Rather, the bulk of the recent criticism seems to stem from data showing that the DNA at a particular chromosomal region can be subdivided into (1) subregions concerned with initiating, promoting, modulating, and attenuating transcription and (2) the structural gene itself, the region of DNA that actually specifies the sequence of amino acids in the corresponding polypeptide chain. This complexity is frequently contrasted with the early (non-elaborated) concept of "a bead on a string."

In responding to this criticism, one might emphasize that, despite their ignorance of the exact chemical nature of the gene, early geneticists realized that *the gene* was a composite consisting of at least three non-identical concepts: (1) the unit of physiological action, (2) the unit of recombination, and (3) the unit of mutation. Despite the "bead" analogy frequently used in textbooks, the last two concepts require some sort of spatial organization. The unit of recombination, for example, required that the unit of physiological action be subdivided – either genes are clustered into complexes of similar action (the "white eye" locus in *Drosophila melanogaster*) or the gene of physiological action is composed of subgenes (the "scute" locus in *D. melanogaster*). Even the unit of mutation, despite the lack of a clear understanding of the gene's physical nature, called for separate mutational "sites" within a region previously known simply as the "gene." The number of such sites was, on occasion, estimated by dividing forward by reverse mutation rates.

Despite their ignorance of the physical structure of the gene, early geneticists were

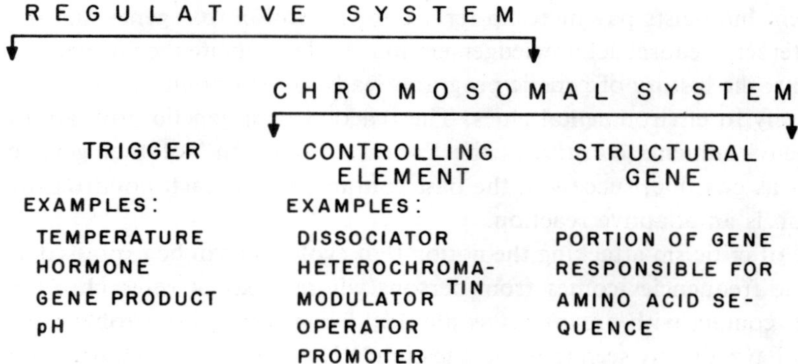

Figure 8. A diagram that emphasizes the dual nature of gene control systems. One part (the structural gene and the adjacent DNA which serves to promote, attenuate, and modulate transcription) follows the classical transmission rules of genetics; the other extends from the environment to the controlling elements of the structural gene by ways of hormones, physiologically active substances, or products synthesized by genes at other loci (see the open arrows of Figure 4). (After Wallace, 1963)

well aware that under the term "gene" lurked a number of phenomena awaiting explanation. The explanations are now arriving, but the complexities that are being explained are not at all new; some were recognized in the early 1900s.

Figure 8 is intended to serve two purposes: First, it illustrates some of the complexities now known to exist in the regulation or control of gene action; second, it illustrates that these complexities have been recognized for a quarter-century at least and, therefore, "recent" discoveries are not all that crucial in evaluating the adequacy of the phrase "gene frequency."

The relationship between two quasi-independent concepts of gene action are illustrated in Figure 8, a diagram based on one to be found in Wallace (1963). To the right, lies the structural gene, the stretch of DNA that specifies the sequence of amino acids in the polypeptide chain whose synthesis is controlled at that gene locus. "Upstream" of this structural gene and, hence, tightly linked to it by the vast majority of genetic tests designed to "locate" genes, is the controlling region of that locus. The structural gene and its associated control region constitute the gene of physiological action, the Mendelian gene of old.

To the left, terminating in the control region of the structional gene, is the entire regulatory system. Although the regulatory system terminates in a Mendelizing unit, its opposite end may lie outside the individual; it may lie in the environment – in hours of daylight, in mean daily temperature, or in the availability of nutrients. The regulatory system is the means by which the environment is internalized within the individual. Often, biologists pay mere lip service to the notion that genes and the environment interact; a causal acknowledgement may underestimate the intimacy of these interactions: the history of genetic programs leads to their ability to read and respond adaptively to environmental clues. The reaction of a genetic program to each of many environments constitutes the "norm of reaction" of that genetic program; as far as possible, success in the past guarantees that each point on the norm of reaction is an adaptive reaction.

A second line of criticism attacking the notion that evolution can be expressed as a change in gene frequencies comes from persons whose areas of research bring them into closer contact with organisms as physical beings than with problems of inheritance. We have already seen that the question, "How does a genetic program become translated into a developmental one?" might be an exciting question for a molecular biologist studying problems of development. The inverse question, "How does a developing organism alter the genetic program carried by its originating germ cells?" is a non-question except possibly for organisms such as the May Apple (Figure 7).

The matter extends beyond developmental biologists; it includes many non-geneticists such as paleontologists, taxonomists, and morphologists. Figures 9 and 10 are intended to illustrate that persons who deal with individuals (and the appearances or morphologies of individuals) and geneticists may merely be wrestling for control of the word: *evolution.*

Figure 9 illustrates the origin and divergence of several varieties of the domestic pigeon through artificial selection; the common ancestor to these pigeon varieties is the rock dove. Because centuries-old illustrations of birds representing earlier phases of the different varieties are available, Figure 9 shows some of the evolutionary changes that have occurred. Nearly all present-day phenotypes were much less extremely developed in past centuries than they are today and, presumably, less so today than they will be sometime in the 21st century. Figure 9, then, represents the type of material that is available to most biologists who wish to discuss evolution.

The landscape shown in Figure 9 has an inverse side; that side of the landscape is shown in Figure 10. No pigeons are to be seen in this view (except for tiny feet that protrude through the frail surface); only lines of descent – germ lines – are shown. Even the germ lines have been simplified because sexual reproduction leads to a net-work, not to a simple system of branching lines as the figure depicts.

Despite the oversimplification of the lines of descent that are illustrated in Figure 10, the diagram does illustrate two points: First, individuals that were illustrated in Figure 9 as being on the line of descent from the rock dove to a present-day variety may, in reality, have lain in cul de sacs having little to do with subsequent selectional

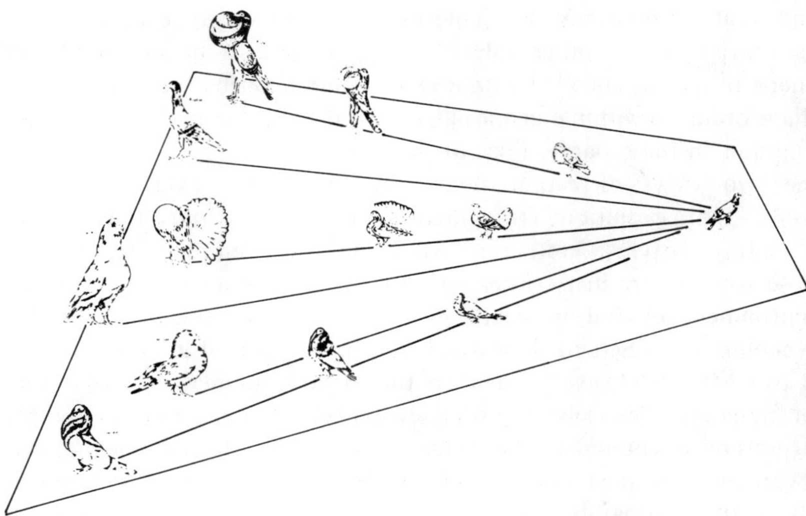

Figure 9. A schematic sketch of the evolution of varieties of the domestic pigeon that have descended from the rock dove; a number of intermediate forms from past centuries are shown in several lines of descent.

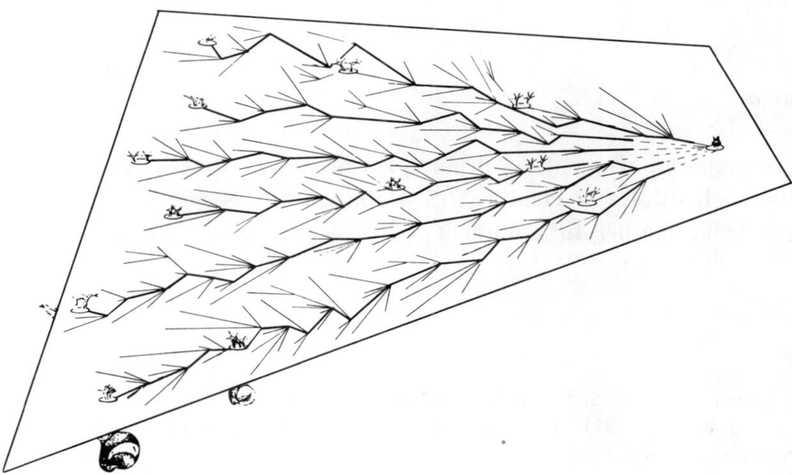

Figure 10. A diagram illustrating another view of Figure 9, a geneticist's view. Here (grossly oversimplified) is shown the passage of germ cells through time from the rock dove to modern varieties. Several "intermediate" forms are seen to lie in cul de sacs rather than on the main line of descent. Seen from this angle, evolution becomes synonymous with "change in gene frequency" where the term "gene" includes sytems such as those illustrated in Figure 8.

change. Second, that, without any first hand knowledge of the appearance of the birds hanging unseen on the other side of the landscape, a person skilled in evaluating genetic programs encoded within DNA could describe the evolutionary changes that have occurred within pigeons in terms of the changes in the frequencies of various programs in the separate lines of descent.

Perhaps the two views of evolution – one emphasizing (Mayr, personal communication) "that evolution is a matter of orchids, butterflies, warm bloodedness, mating systems, etc., etc." and the other emphasizing genetic programs whose somatic products either interact sufficiently adaptively with the prevailing environment or fail in death or sterility – are equally valid but, nevertheless, incompatible. Sets of data which can be interpreted in contradictory ways are said to be "inconclusive"; scientists then search for new data that may resolve the contradiction. The views regarding evolution may more nearly resemble the optical illusions of psychologists: the sketch, for example, which in one moment shows an old woman, in the next a beautiful young lady, but at no time both images simultaneously. If this analogy is valid, some biologists will see evolution only in terms of the vast array of morphologies and behaviors, others will see it only in terms of genetic programs and their changing frequencies, while a few (and I want to be among them) will be able to switch from one view to the other, taking the best advantage of each view.

Acknowledgement

This paper was prepared while the Author's research was supported by grant GM31687, National Institutes of Health, U.S. Public Health Services. The author wishes to acknowledge the helpful comments of Professor David West.

References

Buss, L. W. 1983. Somatic variation and evolution. *Paleobiology,* **9**, 12–16.

Dobzhansky, Th. and E. Boesiger. 1983. *Human culture: A moment in evolution.* (ed. Bruce Wallace). Columbia University Press, New York.

Gilbert, J. J. 1980. Developmental polymorphism in the rotifer *Asplanchna sieboldi. American Scientist,* **68**, 636–646.

Hazel, W. N. and D. A. West. 1979. Environmental control of pupal colour in swallowtail butterflies (Lepidoptera: Papilioninae): *Battus philenor* (L.) and *Papilo polyxenes. Fabr. Ecol. Entomol.* **4**, 393–400.

Lewontin, R. C. 1974. *The genetic basis of evolutionary change.* Columbia University Press, New York.

Lewontin, R. C., J. A. Moore, W. Provine, and B. Wallace (eds.). 1981. *Dobzhansky's genetics of natural populations I–XLIII*. Columbia University Press, New York.

Mayr, E. 1984. The triumph of evolutionary synthesis. *The Times Literary Supplement,* 2 November 1984, pp. 1261–1262.

Mayr, E. and W. Provine (eds.). 1980. *The evolutionary synthesis.* Harvard University Press, Cambridge, Mass.

Schmalhausen, I. I. 1949. *Factors of evolution.* Blakiston, Philadelphia.

Waddington, C. H. 1958. Genetic assimilation of an acquired character. *Evolution,* **7,** 118–126.

Wallace, B. 1963. The annual invitation lecture. Genetic diversity, genetic uniformity, and heterosis. *Canadian Journal of Genetics and Cytology,* **5,** 239–253.

A Framework to Think About Evolving Genetic Regulatory Systems

STUART A. KAUFFMAN

Department of Biochemistry and Biophysics, University of Pennsylvania, School of Medicine, Philadelphia, Pennsylvania 19104, U.S.A.

Introduction

The aim of this article is to suggest a framework to study the expected regulatory connections and coordinated patterns of gene expression in large genetic systems undergoing persistent mutation and selection.

It is obvious that a central problem in developmental biology concerns the large regulatory system coordinating the expression of perhaps 5,000 to 100,000 genes during ontogeny (Bishop 1974, Brown 1981). It is equally obvious, if rarely discussed, that analysis of such, in principle, complex systems confronts us with very substantial epistemological problems. We shall almost certainly not take apart in detail a system with thousands of genes, whose activities directly or indirectly regulate one another. Rather, we shall succeed in analyzing local patches of such a network, for example, the well known puffing sequence in *Drosophila* salivary gland polytene chromosomes induced by ecdysone (Ashburner 1970). From such local information we will have to build up a picture of the organization and behavior of the entire regulatory system. Almost certainly, such a picture will have to be partially statistical: using local features we will build theories about larger regulatory systems whose local features are those we have found. Inherently, this involves analysis of alternative plausible classes of larger systems with the observed local features. Characterization of these alternative classes, and prediction of new alternative testable consequences involves analysis of the statistically expected features of the classes.

Analysis of the expected properties of classes of large genetic regulatory systems are important for a second reason. Suppose we had analyzed in detail the regulatory system coupling the activities of thousands of genes in some specific isogenic strain of some higher eukaryote. Mutational events including both point mutations, and chromosomal mutations such as inversions, deletions, duplications, transpositions, translocations, and conversions (Bush *et al.* 1977, Cameron *et al.* 1979, Chaleff and

Bechtel, W (ed), Integrating Scientific Disciplines. ISBN 90-247-3242-5.
© *1986, Martinus Nijhoff Publishers, Dordrecht. Printed in The Netherlands.*

Fink 1980, Corces *et al.* 1981, Dover *et al.* 1982, Green 1980, Sherman and Helms 1978), persistently alter not only specific coding and control sequences, but move cis acting and trans acting regulatory loci to new positions in the chromosome set, altering the regulatory connections in the system, and the behavior of regulated loci. Even were we to understand in detail the construction of a specific regulatory system, we would need to develop theory characterizing the ways mutations alter the structure and behavior of the regulatory system. We will need theories telling us that if an arbitrary regulatory gene is deleted, or "rewired" to control some other genes in the system, then the coordinated expression of the set of 50,000 genes will be expected to change in such and such ways, or to such and such an extent. In short, we need to develop a body of theory about the expected similarities and differences in the structure and behavior of large genetic regulatory systems as mutation alters components of the system.

Further, it is necessary to develop theory about how selection can act on genetic regulatory systems with thousands of genes. How precisely can coordinated patterns of gene expression be maintained in any cell type of an organism? How precisely can pathways of differentiation between cell types be maintained? How does the precision which can be achieved by selection relate to the size and complexity of the genomic system upon which selection is working? Perhaps high precision can be maintained in sufficiently small genetic systems, as in phage or bacteria, but in systems with many thousands of components, substantial regulatory "sloppiness" must be endured.

The Expected "Wiring Diagram" of Complex Genetic Regulatory Systems

Consider an idealized chromosome set with structural genes, cis and trans acting regulatory loci, and some bounded domain on each chromosome limiting the region over which each particular cis acting locus acts. In this idealization, I do not consider independently the RNA and protein products by which one gene may act on another gene, but simply schematize the "wiring diagram" of a genetic regulatory system as a diagram showing each gene as a point, with an arrow between two genes A and B, if A directly or via its product acts on B. Thus, I draw an arrow from a cis acting gene to each of the structural and other genes in the domain which it influences, I draw an arrow from any trans acting gene to the genes its product acts upon, etc. This point and arrow picture shows the regulatory connections of the genomic system, that is, its "wiring diagram", and is a form of directed graph. Directed graphs are mathematical structures with N points, or nodes, connected by Marrows (Berge 1962).

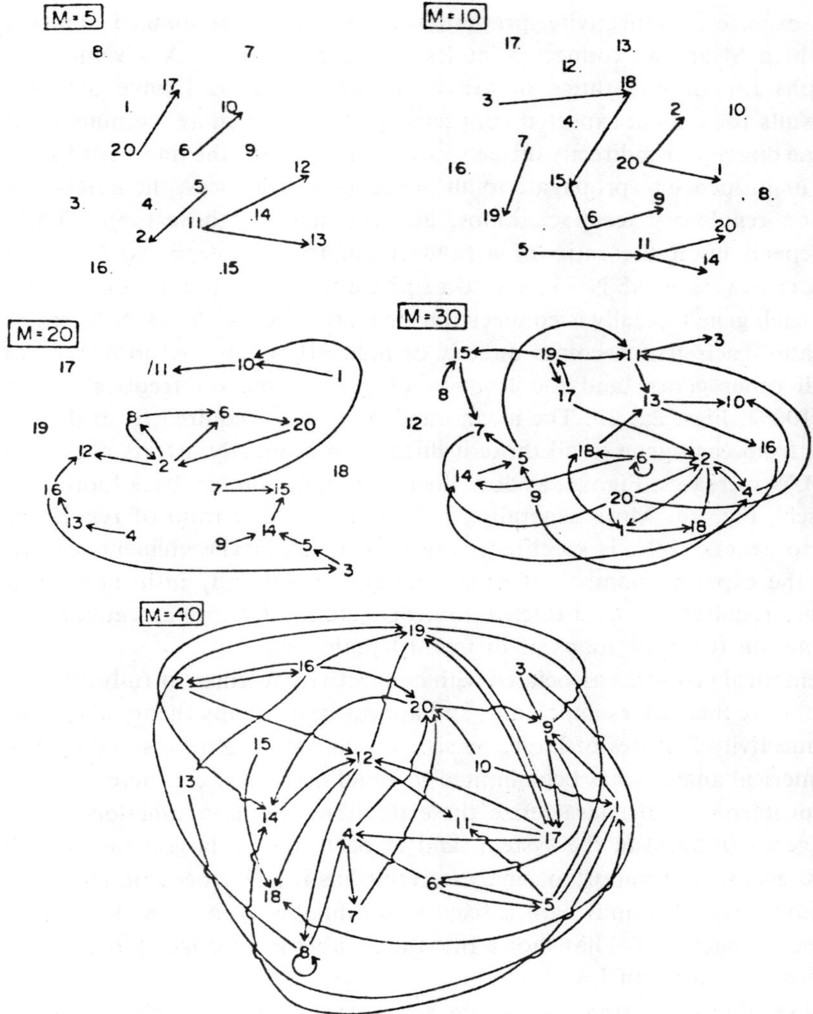

Figure 1. Random directed graphs among 20 genes as the number of connections, M, increases from 5 to 40.

Chromosomal mutations moving cis and trans acting loci "scramble" the wiring diagram by changing regulatory connections (Corces *et al.* 1981, Errede *et al.* 1980, Finnegan *et al.* 1982, Flavell 1982, Hayward *et al.* 1981, McClintock 1957, Peterson 1981, Sherman and Helms 1978). What will such persistently mutating networks look like? A first and most fundamental approach to take to this problem is to analyze the expected connectivity properties of randomly scrambled directed graphs, in which M arrows connect N nodes at random, Fig. 1. Analyzing such random graphs for varying ratios of arrows to nodes, M/N, I have obtained numerical results for critical expected connectivity features such as the number of genes any gene directly or indirectly influences via the network, the number of steps required for its influence to propagate to all the genes it influences, the number of genes lying on regulatory feedback loops, and the sizes of those loops. These properties depend upon the ratio of arrows (regulatory connections) to nodes (genes). At a critical ratio M/N = 1, a kind of phase transition occurs. Below that critical ratio each gene typically is connected to few other genes. As M/N increases above that ratio, each gene becomes directly or indirectly connected to many and eventually all other genes, and the fraction of genes lying on feedback loops increases to 100%, Figs. 2a, 2b. The mean number of steps required for influence to propagate from each gene to all those it directly or indirectly affects rises then declines as M/N increases, Fig. 2c, as does the mean minimum feedback loop from a gene to itself, Fig. 2d. More generally, if the numbers and ratio of regulatory connections to genes, M/N, is specified, very strong statistical statements can be made about the expected number of genes any gene potentially influences, how many steps are required for its influence to propagate to all those it regulates, its chance of lying on feedback loops of different lengths, etc.

The mathematical problems associated with connectivity features of fully random directed graphs are themselves important, and appear basic to any theory about the expected connectivity features of large, persistently mutating genetic systems. The available numerical analysis has been limited to small networks, e.g. here N = 200. One set of problems is to characterize these distributions as a function of the numbers of genes (modes) in the system, and in particular in large systems with 5000–100,000 genes. One important analytic result has recently been obtained for infinite random directed graphs with a fixed mean number of arrows, K, leading from each node. Cohen (1984) has shown that the number of feedback loops of any size L is *Poisson*, with mean $I = KL/L$.

Fully random directed graphs constitute a beginning point for studies of the connectivity features of mutating genetic systems, but are only a background. The actual ways chromosomal mutations "scramble" the genetic regulatory system are not fully random in an equivalent sense. For example, a major mechanism creating novel regulatory connections presumably involves duplication of a sequence and its dispersion via transposition, conversion, or other processes, to new positions in the genome (Dover *et al.* 1982, Flavell 1982). The probabilities of generating a new duplication are almost certainly not independent of the number and positions of copies of that sequence already present. For example, unequal recombination is

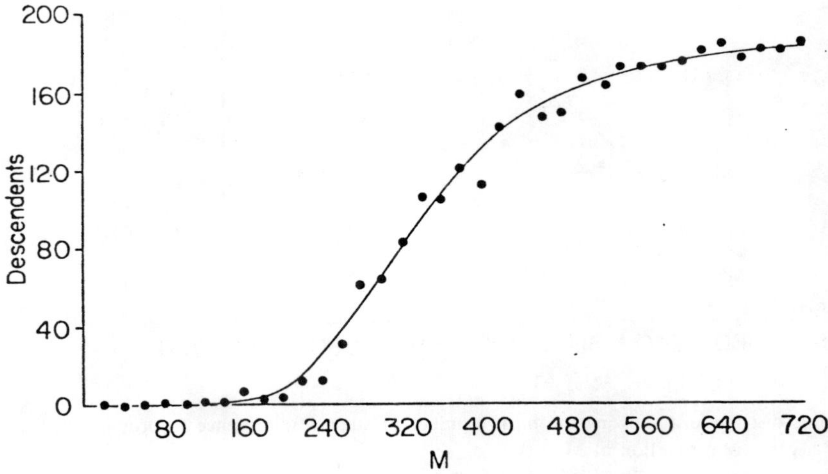

Figure 2a. Descendent distribution, showing the average number of genes each gene directly or indirectly influences as a function of the number of genes (200) and regulatory interactions, M.

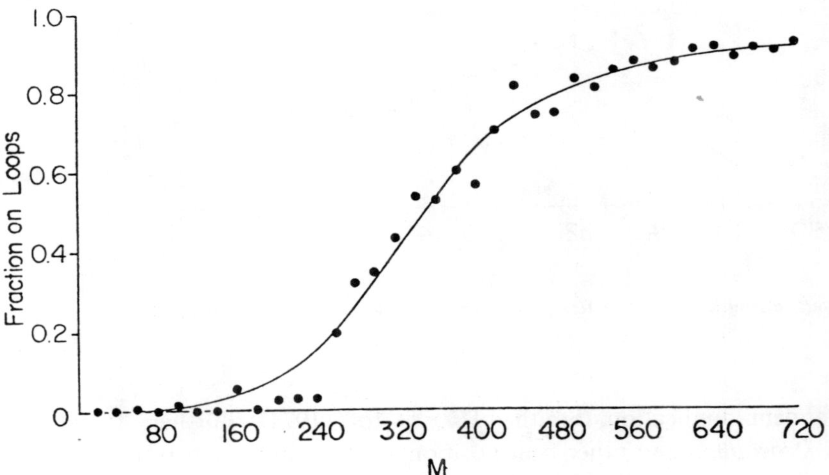

Figure 2b. Average number of genes lying on feedback loops, as a function of M.

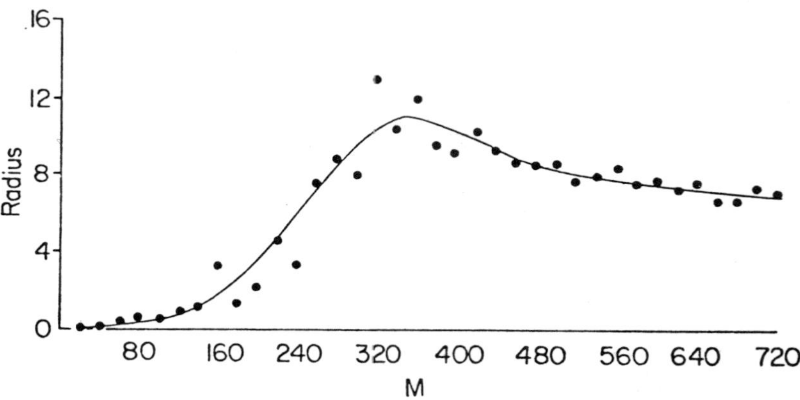

Figure 2c. Radius distribution, showing the mean number of steps for influence to propagate to all descendents of a gene, as a function of M.

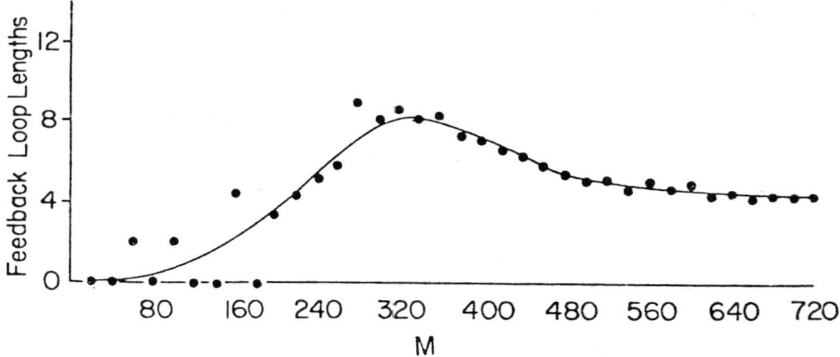

Figure 2d. Average length of shortest feedback loops genes lie on as a function of M.

aided by tandem duplication (Smith 1974, Tartoff 1974). Further, P element insertion in *Drosophila* and other transpositional events show that transpositional insertion is often biased towards "hot spots" (Green 1980, Spradling and Rubin 1981). A general approach which seems reasonable is to attempt to characterize with increasing accuracy the ways known mutational processes generating, degrading and moving putative regulatory genes and connections tune the mean connectivity of the network and bias the "scrambling" process which occurs, hence affect the statistical features of the regulatory networks generated. For example, the familiar, if unproved hypothesis that mid-repetitive DNA families are regulatory loci (Britten and Davidson 1969, 1971), implies a correlation between mid-repeat family sizes and intra-family diversity and the convergent and divergent "fan" of arrows in an evolving regulatory network. A large dispersed family of identical sequences, plus the hypothesis that each is a target for the same regulatory signal, implies that that regulatory signal affects many sites. If creation and dispersion of new copies of a sequence depend on the numbers and locations of existing copies, then, for different plausible hypotheses about those dependencies, what is the expected evolutionary time course of generation of different families like? How do the relative rates of creation and degradation of putative regulatory genes affect mean connectivity? What do the resulting network wiring diagrams look like?

Two aspects of this line of theory development should be emphasized. First, it would appear important. We have at present, essentially no idea what complex genetic regulatory systems subject to persistent mutations in regulatory connections might be expected to look like. Any insight is better than our current complete ignorance. Second, points where theory can come into testable contact with potential data must be considered critically, but with a certain sympathy. Not all aspects of theory, particularly in new directions, can be expected to be immediately testable, even if the theory lays groundwork for later understanding and experimental approaches. A sophisticated theory of the expected "wiring diagram" of a complex genetic regulatory system incorporating all the constraints known about the ways mutations alter regulatory connections would predict features, among others, such as those noted above, e.g. on average, how many genes any gene might influence, how many steps are required for such influence to propagate, numbers of genes on feedback loops of any specific length, variances of these distributions, etc. Because we are far from being able to assess such "wiring diagram" features, immediate testing of these inferences may not be possible. On the other hand, related predictions about distributions of tandem and dispersed repetitive sequences, sizes of repetitive families, similarities of families, etc. (Ohta 1983) are more immediately testable.

Expected Coordinated Gene Expression in Genetic Regulatory Systems

How shall we think about the coordinated expression of 5000–100,000 genes underlying ontogeny? How can we approach analysis of the behavior of such complex systems based on the partial knowledge we can attain? One way is to construct classes of large model genetic regulatory systems incorporating known features, and assess the expected behavior of such classes of regulatory systems. This involves studying networks of model genetic systems with perhaps 10,000 or even 100,000 gene components.

An approach which appears useful is to model a single gene as an on-off device. This idealization allows study of large systems with, e.g. 10,000 genes, coupled into arbitrarily complex networks regulating one another's activities (Kauffman 1974). Consider the class of all genetic systems with N on-off (binary) genes, each of which is regulated directly by K genes (or their products) from among the N. Since each gene is modeled as an on-off device, its behavior as a function of the on or off values of the K genes controlling it, is given by a "Boolean" switching rule. For example, a gene might switch on at the next moment, only if all K of the gene controlling it were active the moment before, or if any were inactive, etc. Any such model genetic regulatory system is a finite automaton (Minsky 1967). Any combination of activity and inactivity of the N genes is a legitimate "state" of the genomic system, hence there are 2^N states. At each moment, the system follows its rules, and passes from one combination of gene activities, to a next combination of gene activities. Over time, the genomic system traces out a pathway through the 2^N states of gene activities, and eventually settles into and cycles through a recurrent subset, called an attractor state cycle. If released from a different initial state, the system might flow to this first attractor, or to a different attractor state cycle. Thus, some collection of states flows into each attractor state cycle, each attractor state cycle is one asymptotic and recurrent pattern of gene expression, and the set of distinct attractor state cycles constitutes the repertoire of distinct recurrent patterns of gene expression in the model genomic regulatory system, Fig. 3.

The preliminary interpretation I suggest is to consider each such distinct attractor state cycle as a distinct cell type in the repertoire of the genomic system. This identification seems natural in this kind of first order model; it says the recurrent patterns of gene expression correspond to cell types, while transient sequences of gene expression leading to these recurrent patterns are to be classified in terms of which cell type each transient leads to.

Identification of state cycle attractors of model genomic systems with cell types allows comparison of these model genomes to many known or testable characteristics of genuine regulatory systems: (1) Numbers of cell types in an organism; (2) similarity of gene expression patterns in different cell types of one organism; (3) the existence of a core of genes ubiquitously active in all cell types of one organism; (4) the stability of a cell type to reversals in activity of any single gene; (5) the number of cell types into which any cell type can differentiate by such signals to any specific gene; (6) the existence and range of abnormal inductions; (7) the number of genes which typically alter their expression when any arbitrary gene is deleted from the regulatory system. (8) Similarly, the number of genes which typically alter their expression when any arbitrary gene is rendered constitutively active. (9) The distribution around these means; (10) The generation of novel cell

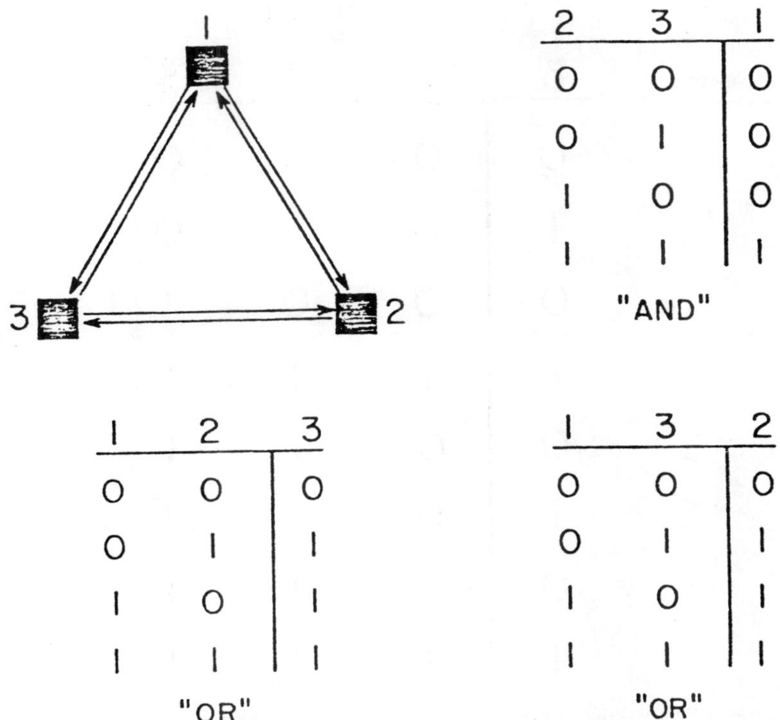

Figure 3a. Three genes, each regulated by the other two, according to the "OR" and "AND" Boolean functions.

types by regulatory mutations, e.g. how many novel cell types does a single regulatory mutant typically generate? What is the distribution about this mean? How different are the novel cell types from the "parental" cell types in terms of patterns of gene expression? (11) How many cell types in the normal repertoire of the genomic system are directly accessible from any initial "zygotic" cell type along pathways of differentiation? Can the "organism" use all its potential cell types, or are some normally unreachable?

These properties, of course, differ in different classes of model genetic regulatory systems (Walker and Gelfand 1979, Sherlock 1979a, b, Fogelman Soulie *et al.* 1982, Wolfram 1983, Kauffman 1969, 1984a). Interestingly, the class which incorporates simple, but ubiquitous features of known regulatory systems exhibits many behaviors similar to those known or expected in contemporary higher cells. Two of the most fundamental properties of genetic regulatory systems are (1) the numbers of genes which directly affect or are affected by, any particular gene; (2) the class

174

$\overline{T}$			$\overline{T+1}$		
1	2	3	1	2	3
[illegible]	[illegible]	0	0	0	0
[illegible]	[illegible]	1	0	1	0
[illegible]	[illegible]	0	0	0	1
[illegible]	[illegible]	1	1	1	1
[illegible]	[illegible]	0	0	1	1
[illegible]	[illegible]	1	0	1	1
[illegible]	[illegible]	0	0	1	1
[illegible]	[illegible]	1	1	1	1

Figure 3b. All $2^3 = 8$ states of binary activity of the three genes at time T, and the next state, at time T + 1, into which each state transforms.

of control rules describing the activity of a regulated gene as a function of the activities of those regulating it. The first property is just the local connectivity. How many genes, typically, does any one (kind of) cis acting gene influence directly? How many genes does any trans acting gene impinge upon directly? How many genes directly regulate the activity of any specific (kind of) cis acting, transacting or structural gene? While the concept of "direct influence" is somewhat imprecise at this preliminary stage, it suffices for a main purpose, which is to investigate the role of high and low connectivity per se in classes of regulatory systems. In bacteri and phage, typically each gene is directly affected by rather few other genes, 0 to perhaps 5 (Hershey 1971, Vogel 1971, Kauffman 1974). The connectivity in higher eukaryotes is unknown, but could be much higher.

The actual control rules found in bacteria and phage almost all fall into an interesting and special class in the binary (Boolean) idealization, which I call canalizing. This is exemplified by the lac operator in *E. coli*, which is bound by repressor in the absence of allolactose, but allolactose itself binds the repressor

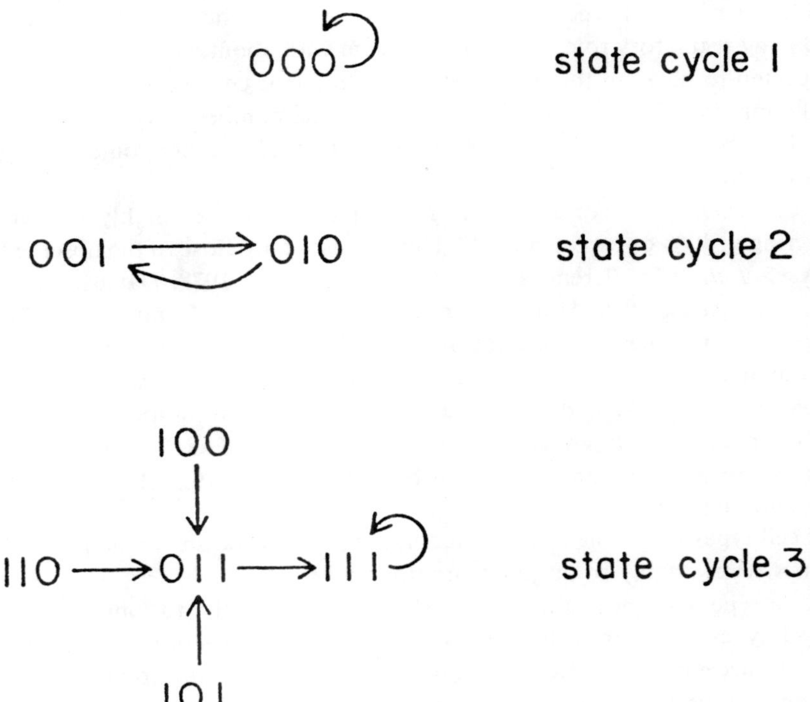

Figure 3c. The sequences of state transitions from 3b, showing three state cycles, of lengths 1, 2, and 1. The first and third state cycles are steady states. The second is an oscillation.

modifying its conformation and pulling it off the operator. This "Not If" Boolean function has the property that if allolactose is present, the operator is free regardless of the presence or absence of repressor (Zubay and Chambers 1971). Almost all known regulated loci appear to have homologous behavior; there is at least one value (active or inactive, present or absent) of at least one of the molecular variables regulating the gene which alone suffices to guarantee that the regulated gene assumes one state of activity (either active or inactive) regardless of the activities of the remaining regulatory variables (Hershey 1971, Vogel 1971, Kauffman 1974).

Modal binary genetic regulatory systems in which ech gene is directly influenced by few other genes, and is governed by one of the possible canalizing Boolean functions, typically mirror expected or known behaviors of eukaryotic cells:

1. The predicted number of cell types in an organism rises less than linearly with respect to the number of genes, as appears to be observed (Kauffman 1969). In fact,

the expected number of cell types is about a square root of the number of genes capable of playing regulatory roles; a genomic system with about 50,000 such genes would be expected to have on the order of several hundred cell types. This is not a trivial prediction; in other classes of model systems the number of cell types rise much faster than the numbers of genes. A system with 50,000 genes could readily have billions of cell types.

2. The patterns of gene expression in different cell types are highly similar, typically differing in 10% or less of the loci. This is reasonably close to the observed diferences (Axel *et al.* 1976, Berendes 1966, Chikaraishi *et al.* 1978, Davidson and Britten 1979, Ernst *et al.* 1979, Hough *et al.* 1975, Kleene and Humphries 1977).

3. A large core of genes is ubiquitously active in all model cell types in the organism. Similarly, a large core of genes is ubiquitously expressed in the heterogeneous nuclear RNA, and to a lesser extent in the cytoplastic messenger populations of many cell types of one higher eukaryote (Axel *et al.* 1976, Chikaraishi *et al.* 1978, Davidson and Britten 1979, Ernst *et al.* 1979, Hough *et al.* 1975, Kleene and Humphries 1977).

4. Model cell types are stable to most fluctuations which transiently activate or inactivate genes. Presumably so are contemporary cells.

5. Any cell type can be induced to differentiate directly to only a few neighboring cell types, which may themselves differentiate to a few other neighbors. This implies the necessary existence of branching pathways of differentiation in ontogeny, a ubiquitous feature of all metazoan ontogeny.

6. The limited range of neighbors to any cell type limits the range of abnormal differentiation pathways open to each, hence simultaneously limits the range of abnormal induction transitions. This property also offers an account of the "poised" properties of tissues in which many non-physiological or abnormal inductive agents induce the same differentiation step – e.g. early ectoderm to neurectoderm.

7. The number of genes typically altering their activity patterns when an arbitrary gene is deleted, or rendered constitutively active is modest, typically less than 10–15%. Activity errors do not ramify extensively throughout the system. Known alterations in protein synthetic patterns to single mutants or viral transformations appear to fall in the range (Strand and August 1977, 1978, Liebermann *et al.* 1980).

The catalogue of parallels between this simple class of model genomic systems and contemporary eukaryotic cells is longer, but the main point to stress is that an extremely primitive class of model systems utilizing only a few known features of contemporary genetic regulatory systems, typically exhibits many behavioral features close to those of higher organisms. The issue is not whether this primitive model is adequate, but that it encourages the hope that incorporating more refined knowledge about the actual properties of contemporary mutating genomic systems might lead to a refined class of models whose statistically typical behavioral and structural features are closely predictive of actual evolving systems. Whether contemporary cells and organisms will have properties predictable from the typical properties of such a refined class of genomic models alone, depends upon the effects of selection, discussed next.

The Interaction of Selection and the Self Organizing Properties of Mutating Genetic Regulatory Systems

While selection is undoubtedly deeply involved in the organization and dynamical behavior of contemporary genetic regulatory systems, it may be possible and conceptually useful to try to partially isolate selective forces from what the genomic system would be like in the absence of selection. Maintenance of a genetic regulatory system involves at least generation, dispersion and degradation of cis and trans acting loci; alterations of the control rules governing specific loci; and maintenance, generation, and degradation of structural genes. Movement of cis and trans acting sequences sometimes interrupts other regulatory or structural genes, leading to their loss of function. Maintenance of those functions requires selection. Similarly, relative birth and death rates of regulatory loci and structural genes controls the mean regulatory connectivity of the network, and is probably under selective control. However, granted those selective forces are operating, we can consider the "wiring diagram" organization of the regulatory system and ask what the statistically expected features of the network would look like, in the absence of further selection, under the drive of the mutational events persistenly rearranging control interactions in the system. In the absence of selection, such genetic networks would be expected to exhibit the statistically typical features of class explored by, hence defined by, the scrambling process itself. Those features are the proper null hypothesis about the properties of the genomic systems we should expect to find. Moreover, in the absence of further selection, the properties found in any genomic system under the drive of mutation will relax back toward those class typical properties which therefore constitute a fundamental constraint which selection must overcome if it "needs to".

A basic question is this: to what extent and under what conditions can selection overcome the statistically typical properties of the class which mutation is exploring? Perhaps selection cannot easily avoid those properties, and they shall turn out to be ubiquitous. In that case, many features of genetic regulatory systems

178

in contemporary organisms would be predictable based on the typical properties of the proper class of systems.

A preliminary approach to this question asks whether selection is able to maintain some arbitrary optimal genetic "wiring diagram" in the face of mutations scrambling that diagram. An example of such a wiring diagram might be the familiar hierarchical batteries of the Britten Davidson models (1969, 1971), in which some particular wiring diagram is optimal and deviants are less fit. A simplest measure of fitness is the number of "good" connections in any network, i.e. the number of connections which are identical to those in the arbitrary optimal network. Fitness might be linearly proportional to deviations from optimal, or nonlinear in a variety of ways:

$$(1) \qquad W_i = \left(\frac{G_i}{T}\right)^\alpha$$

Here, W_i is the relative fitness of the i_{th} mutating network, T is the total number of regulatory connections in a network, G_i is the number of "good" connections in the i_{th} network, and measures the linearity or non-linearity with which relative fitness decreases for networks deviating from the arbitrary optimal network, whose $G_i = T$.

In a simplest model, all "organisms" are haploid, hence all variance in fitness is heritable (Ewans 1979). I assume discrete generations and a fixed mutation rate altering genetic regulatory connections in the wiring diagram by randomly reassigning the head or tail of a randomly chosen connection arrow to one of the N genes in the network. Simulation and approximate analytic results for this simple fitness model, both for the linear (additive) fitness rule, $\alpha = 1$, and for strongly nonlinear rules $1 < \alpha \leqslant 10$, [show that the selective system has a single stationary state, reflecting a stable mean fitness in the population (Fig.4; for further development, see Kauffman 1984b). For a sufficiently small genetic wiring diagram, selection can maintain a population clustered near any arbitrary optimal genetic wiring diagram. As the number of genes and regulatory connections, T, or the mutation rate increases, however, selection becomes too weak to hold the population near any arbitrary optimal network and the single be stattainable stationary state shifts toward the mean, unselected properties of the class (Kauffman 1984b). Thus, for a sufficiently large network, even if the population is initially clustered near the optimal network, it will come to resemble typical randomly scrambled networks.

As light modification of the fitness rule leads to a result of potential interest. Let

$$(2) \qquad W_i = (1\text{-}b) \left(\frac{G_i}{T}\right)^\alpha + \beta$$

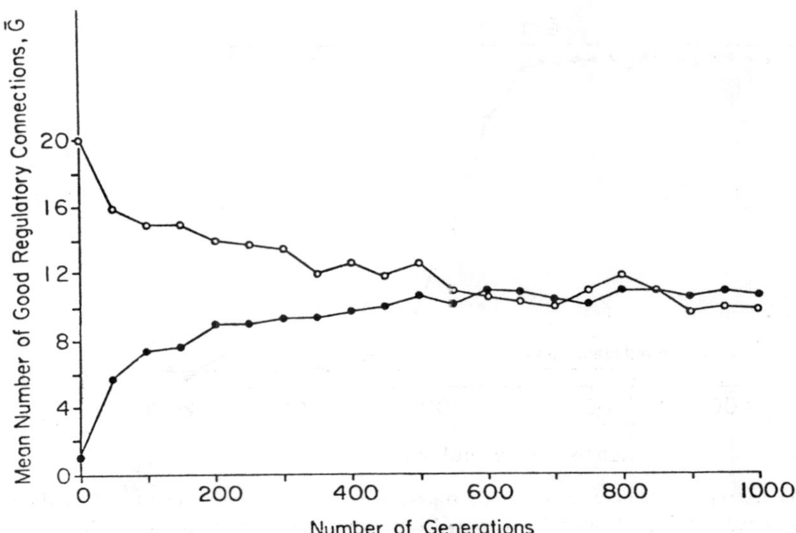

Figure 4. Population selection for 1000 generations after initiation at $G = T = 20$, and $G = 1$. G, mean number of good connections per organism. Mutation rate per arrow end $\mu^* = .005$. $\alpha = 1$, $\beta = 0$.

where b, $0 \leqslant \beta \leqslant 1$ is a residual basal fitness in the absence of any "good" connections. This has analogies to the residual relative fitness in a haploid multiplicative model with Lloci, each with two allelesof fitness 1 and (1-S), where the least fit genotype is $(1-S)^L > 0$.]

For $b > O$, and α sufficiently large, this fitness model has two alternative stable stationary states, with low and high mean numbers of good connections in population members, and fluctuation driven transitions between the low and high states (Fig. 5; for further development, see Kauffman 1984b). Unlike the simpler fitness model, (1), with a single best attainable stationary state, in this more complex mode, the final position of the population under selection is dependent on its history. In particular, if an evolving population of genetic networks were to start with small optimal structures and gradually increase the numbers of regulatory connections, the population could remain near optimal even when considerably more complex. However, if the population were to begin as substantially suboptimal large networks, it would flow to the lower stationary state and be very unlikely to undergo the transition to the near optimal stationary state (Kauffman 1984b). Thus, in the simplest model, $W_i = (G_i/T\alpha)$, maintenance of precision in a complex network is not aided by gradual evolution from a simpler optimal one, while in the more complex model, (2), the strategy can work.

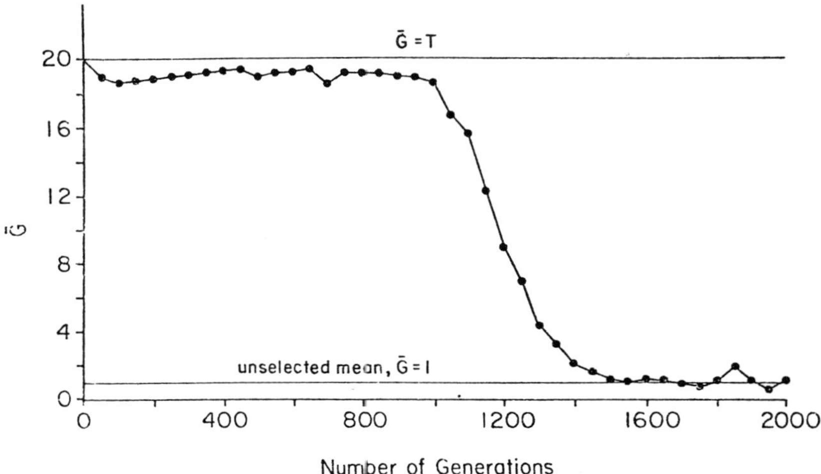

Figure 5. Selection system 2) exhibits two stationary states, one near the optimal state $\bar{G} = T$, and one near the fully unselected state, $\bar{G} = 1$, with fluctuation driven transitions between them. $T = 20$, $\alpha = 10$, $\beta = .5$. The mutation rate μ^* per arrow end $= .005$.

Help maintaining near optimal organization in (2), however, is limited. For other parameters of the selection system fixed, as the numbers of regulatory connections, T, increase, the near optimal stationary state becomes less stable, then vanishes. A "complexity catastrophy" occurs. Below a critical complexity, the population remains near optimal. Above that critical complexity, it falls abruptly to the single remaining lower fitness stationary state. This behavior is reminiscent of the "error catastrophy" found by Eigen and Schuster using selection to maintain nucleotide strings in the face of mutation, as the number of nucleotides, i.e. string length, increases beyond a critical lengh (Eigen and Schuster 1979).

A general implication of both the simpler and more complex fitness model, is that a sufficiently complex genetic regulatory wiring diagram will approach arbitrarily close to the typical organizational properties of the unselected system. Thus, for sufficiently complex genomic systems, predictions from the typical properties to be expected in the absence of further selection to those actually found in organisms would be reasonably accurate. The complexity required to approach typical depends upon the detailed assumptions. For the case of additive fitness in a related simple doploid model with T loci, each of two alleles, and fitness 1, $(1-1/2S/T)$, $(1-S/T)$ for the homozygote, heterozygote, and homozygote at each locus (Ewens 1979), it can be shown that for fixed mutation rates, the number of "bad" connections increases proportionally to T^2, Appendix II. With a mutation rate of 10^{-6}, and 30,000 regulatory connections, about 25% would be expected to be "bad". Thus, for example, with additive fitness, it would be impossible to maintain the rigid hierarchical batteries in large Britten Davidson models; networks will look characterizably more scrambled. This level of "bad" connections suggests that characterization of the proper null hypotheses about network organization may in fact be predictively useful.

The basic issue, then, is to analyze the interaction of reasonable selection rules and the underlying class of genetic regulatory systems explored by mutations. The most important problems arise, not in considering wiring diagrams, but in studying the capacity of selection to modify the coordinated patterns of gene expression underlying cell types and differentiation between cell types. For example, utilizing the canalizing Boolean models of genetic regulatory systems, or better models incorporating more known features, we need to develop theory about the extent to which selection can maintain precise coordinated patterns of gene expression in cell types, and precise differences between cell types. This should have a bearing on understanding the diversity in tissue patterns of enzyme expression in sibling species. Thus, Dickinson has found high variability in tissue patterns among a number of enzymes, tissues and sibling Picture Wing *Drosophila* species (Dickinson 1980a, b). Each enzyme is expressed in a common subset of tissues in all the species, but "twinkles" on and off in the remaining tissues in complex ways. Does this represent microhabitat selection or regulatory "sloppiness"? If regulatory precision cannot be achieved for all genes for all cell types, then what precision can be achieved? How is it influenced by the complexity of the genomic system? By linearity or non-linearity in selection rules? By mutation rate? By nonuniform fitness contributions across the genes, in which particular subsets of genes and their coordinate expression are more critical than other subsets? By the numbers and distances between alternative "optimal" systems and their distances from the typical properties of the class of mutating systems itself? Perhaps most important, how are these features influenced by the particular styles and capacities for adaptive selection of coordinate patterns of gene expression, etc., in different reasonable classes of genetic regulatory system? Are there formulable senses, perhaps analogous to ideas like exon shuffling (Gilbert 1978, Darnell 1978), in which evolution can have "learned" to use regulatory systems more adept at adapting? Can selection alter the class of regulatory systems mutation is exploring such that it incorporates heuristic rules about how to adapt (Campbell 1982)? In short, we need to develop theories about the capacities of complex mutating regulatory systems to adapt to various definable requirements, and perhaps, adapt to adapting.

182

Acknowledgments

This work was partially funded under ACS CD-30, ACS CD-149, NIH GM22341 and NSF PCM 8309206. The author is particularly grateful to Drs. W. Ewens and R. Burian for fruitful discussions.

References

Ashburner, M. (1970). Puffing patterns in *Drosophila melanogaster* and related species. *In* W. Beermann (ed.), *Developmental studies on giant chromosomes. Results and problems in cell differentiation*, Vol. 4, p. 101–151. Berlin: Springer-Verlag.

Axel, R., Feigelson, P., and Schultz, G. (1976). Aanalysis of the complexity and diversity of mRNA from chicken liver and oviduct. *Cell*, **7**, 247–254.

Berendes, H. D. (1966). Gene activities in the malphigian tubules of *Drosophila hydei* at different developmental stages. *Journal Experimental Zoology*, **162**, 209–218.

Berge, C. (1962). *The theory of graphs and its applications.* Methusena, London.

Bishop, J. O. (1974). The gene numbers game. *Cell*, **2**, 81–86.

Britten, R. J. and Davidson, E. H. (1969). Gene regulation for higher cells: a theory. *Science*, **165**, 349–357.

Britten, R. J. and Davidson, E. H. (1971). Repetitive and non-repetitive DNA sequences and a speculation on the origins of evolutionary novelty. *Quarterly Review of Biology*, **46**, 111–137.

Brown, D. D. (1981). Gene expression in eukaryotes. *Science*, **211**, 667–674.

Bush, G. L.; Case, S. M.; Wilson, A. C.; and Patton, J. L. (1977). Rapid speciation and chromosomal evolution in mammals. *Proceedings of the National Academy of Sciences, U.S.A.*, **74**, 3942–3947.

Cameron, J. R.; Loh, E. Y.; and Davis, R. W. (1979). Evidence for transposition of dispersed repititive DNA families in yeast. *Cell*, **16**, 739–751.

Campbell, J. H. (1982). *In* R. Milkman (ed.), *Perspectives on evolution.* Sunderland, Mass.: Sinauer Assoc., Inc.

Chaleff, D. T. and Fink, G. R. (1980). Genetic events associated with an insertion mutation in yeast. *Cell*, **21**, 227–237.

Chikaraishi, D. M., Deeb, S. S., and Sueoka, N. (1978). Sequence complexity of nuclear RNAs in adult rat tissues. *Cell*, **13**, 111–120.

Cohen, J. E. and Newman, C. M. (1984). Manuscript in preparation.

Corces, V., Pellicer, A.; Axel, R.; and Meselson, M. (1981). Integration, transcription and control of a *Drosophila* heat shock gene in mouse cells. *Proceedings of the Ntional Academy of Sciences, U.S.A.*, **78**, 7038–7042.

Drnell, J. E. (1978). Implications of RNA–RNA splicing in evolution of eukaryotic cells. *Science*, **202**, 1257–1260.

Davidson, E. H. and Britten, R. J. (1979). Regulation of gene expression: possible role of repetitive sequences. *Science*, **204**, 1052–1059.

Dickinson, W. J. (1980a). Evolution of patterns of gene expression in Hawaiian picture-winged *Drosophila*. *Journal of Molecular Biology*, **16**, 73–94.

Dickinson, W. J. (1980b). Complex cis-acting regulatory genes demonstrated in *Drosophila* hybrids. *Developmental Genetics*, **1**, 229–240.

Dover, G., Brown, S., Coen, E., Dallas, J., Strachan, T., and Trick, M. (1982). The dynamics of genome evolution and species differentiation. p. 343–372. *In* G. A. Dover and R. B. Flavell (eds.), *Genome evolution*. New York: Academic Press.

Eigen, M. and Schuster, P. (1979). *The hypercycle, A principle of natural self-organization*. Berlin: Springer-Verlag.

Ernst, S.; Britten, R. J.; and Davidson, E. H. (1979). Distinct single-copy sequence sets in sea urchin nuclear RNAs. *Proceedings of the National Academy of Sciences, U.S.A.*, **76**, 2209–2212.

Errede, B.; Cardillo, T. S.; Sherman, F.; Dubois, E.; Deschamps, J.; and Wiane, J.-M. (1980). Mating signals control expression of mutations resulting from insertion of a transposable repetitive element adjacent to diverse yeast genes. *Cell*, **22**, 427–436.

Ewens, W. J. (1979). *Mathematical population genetics*. Berlin: Springer-Verlag.

Finnegan, D. J.; Will, B. H.; Bayev, A. A.; Bowcock,; and Brown, L. (1982). Transposable DNA sequences in eukaryotes. p. 29–41. *In* G. A. Dover and R. B. Flavell (eds.), *Genome evolution*. New York: Academic Press.

Flavell, R. (1982). Sequence amplificxation, deletion and rearrangement: major sources of variation during species diversity. p. 301–324. *In* G. A. Dover and Flavell, R. B. (eds.). *Genome evolution*. New York: Academic Press.

Fogelman Soulie, F.; Chacc, E. Goles; and Weisbuch, G. (1982). Specific roles of the different Boolean mappings in random networks. *Bulletin of Mathematical Biology*, **44**, 715–730.

Gilbert, W. (1978). Why genes in pieces? *Nature*, **271**, 501–

Green, M. M. (1980). Transposable elements in *Drosophila* and other diptera. *Annual Review of Genetics*, **14**, 109–120.

Hayward, W. S.; Neel, B. G.; and Astrin, S. M. (1981). Activation of a cellular *onc* gene by promoter insertion in ALV-induced lymphoid leukosis. *Nature*, **290**, 475–480.

Hershey, A. D. (ed.) (1971). *The Bacteriophage Lambda*. Cold Spring Harbor Laboratory.

Hough, B. R.; Smith, M. J.; Britten, R. J.; and Davidson, E. H. (1975). Sequence complexity of heterogeneous nuclear RNA in sea urchin embyros. *Cell*, **5**, 291–299.

Kauffman, S. A. (1969). Metabolic stability and epigenesis in randomly constructed genetic nets. *Journal of Theoretical Biology*, **22**, 437–467.

Kauffman, S. A. (1974). The large scale structure and dynamics of gene control circuits: an ensemble approach. *Journal of Theoretical Biology*, **44**, 167–182.

Kauffman, S. A. (1984a). Emergent properties in random complex automata. In press, *Physica* D.

Kauffman, S. A. (1984b). Selective adaptation and its limits in automata and evolution. To appear in *Proceedings of the workshop on dynamical behavior of automata: Theory and applications*. Université Scientifique et Médicale de Grenoble.

Kleene, K. C. and Humphries, T. (1977). Similarity of hRNA sequences in blastula and pluteus stage sea urchin embryos. *Cell*, **12**, 143–155.

Liebermann, D.; Hoffman-Liebermann, B.; and Sachs, L. (1980). Molecular dissection of differentiation in normal and leukemic myeloblasts, separately programmed pathways of gene expression. *Developmental Biology*, **79**, 46–63.

McClintock, B. (1957). Controlling elements and the gene. *Cold Spring Harbor Symposium in Quantitative Biology*, **21**, 197–216.

Minsky, M. L. (1967). *Computation: Finite and infinite machines*. Englewood Cliffs, New Jersey: Prentice-Hall.

Ohta, T. (1983). On the evolution of multigene families. *Theoretical Population Biology*, **23**, 216–240.

Peterson, P. A. (1981). Diverse expression of controlling element components in maize: test of a model. *Cold Spring Harbor Symposium in Quantitative Biology*, **45**, 447–456.

Sherlock, R. A. (1979a). Analysis of the behavior of Kauffman binary networks. I. State space description and the distribution of limit cycle lengths. *Bulletin of Mathematical Biology,* **41**, 687–705.

Sherlock, R. A. (1979b). Analysis of the behavior of Kauffman binary networks. II. The state cycle fraction for networks of different connectivities. *Bulletin of Mathematical Biology,* **41**, 707–724.

Sherman, F. and Helms, C. (1978). A chromosomal translocation causing overproduction of Iso-2-cytochrome *c* in yeast. *Genetics,* **88**, 689–707.

Smith, G. P. (1974). Unequal crossover and the evolution of multigene families. *Proceedings of the Cold Spring Harbor Symposium in Quantitative Biology,* **38**, 507–513.

Spradling, A. C. and Rubin, G. M. (1981). *Drosophila* genome organization: conserved and dynamic aspects. *Annual Review of Genetics,* **15**, 219–264.

Strand, M. and August, J. T. (1977). Polypeptides of cells transformed by RNA or DNA tumor viruses. *Proceedings of the National Academy of Sciences U.S.A.,* **74**, 2729–2733.

Strand, M. and August, J. T. (1978). Polypeptide maps of cells infected with murine type C leukemia or sarcoma oncovirus. *Cell,* **13**, 399–408.

Tartof, K. (1974). Unequal mitotic sister chromatid exchange and disproportionate replication as mechanisms regulating ribosomal RNA gene redundancy. *Proceedings of the Cold Spring Harbor Symposium of Quantitative Biology,* **37**, 491–500.

Vogel, H. J. (ed.) (1976). *Metabolic pathways,* Vol. 5. New York: Academic Press.

Walker, C. C. and Gelfand, A. E. (1979). A system theoretic approach to the management of complex organizations: management by exception, priority, and input span in a class of fixed-structure models. *Behavioral Science,* **24**, 112–120.

Wolfram, S. (1983). Statistical mechanics of cellular automata. *Review of Modern Physics,* **55**, 601–621.

Zubay, G. and Chambers, D. A. (1971). *In* H. J. Vogel (ed.), *Metabolic pathways.* New York: Academic Press.

Developmental Constraints, Generative Entrenchment, and the Innate-Acquired Distinction

WILLIAM C. WIMSATT
Philosophy and Evolutionary Biology, The University of Chicago, Chicago, Illinois 60637, U.S.A.

Introduction: With Mixed Feelings and Apologies to an Old (and Dishonored) Friend

The innate-acquired distinction has long been a point of connection between evolutionary questions and developmental questions. It is also one of our oldest conceptual tools, dating back at least to the time of Plato. Further, it hs been broadly applied not just in development biology, animal ethology, and evolutionary theory, but also in epistemology, metaphysics, ethics, philosophy of science, cognitive psychology, linguistics, neurophysiology and in theories about the nature and evolution of human sociality, culture, and morality. As befits any concept which has cast such a broad and deep shadow, it has at some times and places been highly honored and at others most deeply dishonored. It has been viewed as an agent of progress and of stagnation, of clarity and of confusion, and of liberation and of bondage.

For all of these reasons, any attempt at analyzing this distinction must be interdisciplinary, tentative, and humble. It must be interdisciplinary because there are so many disciplines which use or abuse it. I will draw primarily on philosophy, philosophy of science, evolutionary biology, developmental biology, ethology, and the psychology of human problem solving for my analysis, and to a lesser extent on perspectives from linguistics and artificial intelligence. It must be tentative because there are so many sources of possibly relevant information that some are certain to be missed, and so many conflicting claims that some must be rejected. It must be humble because, for all of its checkered past, the distinction has permitted many insights which only the foolish would ignore. In the analysis which follows, I argue that a juxtaposition of elements from diverse disciplines permits a new view of the phenomena which have motivated that distinction in a way which is highly productive, though it will be necessarily reconstructive, and therefore, almost certainly contentious.

Bechtel, W (ed), Integrating Scientific Disciplines. ISBN 90-247-3242-5.
© *1986, Martinus Nijhoff Publishers, Dordrecht. Printed in The Netherlands.*

186

In this paper I will focus on the relevance of this analysis for guiding the development of constructive linkages between developmental biology and evolutionary theorizing.

The analysis I propose captures more of the phenomena commonly linked with the innate-acquired distinction than any prior analysis, and explains their relevance in a new way, but necessitates giving up at least two claims which have been dear to advocates of that distinction, and other assertions which depend on these claims. *The first claim* (or presupposition, since it is seldom claimed explicitly), *is that what is innate must be internal to the object in question.* This has been a presupposition of making this distinction from the earliest time, and is the basis for claims that innate features are in some sense independent of or unmodifiable through experience. I will argue that under appropriate conditions, information which is embodied in the environment, and enters the organism as experience, must be regarded as innate, or something close to it, and that as a result the environment can in most even moderately complex cases play a major role in the expression of the trait.

The *second claim* is a more recent accretion dating from the rise of genetics in the early 20th century. This *is the reanalysis of the classical association between the innate and the hereditary into the more modern claim that to be innate is to be genetic.* Although more recent, this claim is by now hardly less firmly entrenched than the first. While I will argue that there is a (narrow) sense in which it is correct to say that to be innate is to be "coded in a genetic program", this is a misleading and unduly narrow way of pointing to the generative role of innate elements, and leads to an incorrect identification of the innate with the (biologically) genetic. In particular, it does not follow, on the account I propose, either that if something is coded in a genetic program, it is "genetically determined" or that the information which determines it is entirely in the genome. From this it follows that not all things which are biologically genetic are innate, and not all things which are innate are biologically genetic.

Since these claims have been abused (e.g., by proponents of genetic determinism for cultural traits and for I.Q.), and since these represent conceptual confusions, I must recommend radical surgery on the innate-acquired distinction. With some sorrow (and a sense that I do too little honor to historical insights and thereby join modern eliminativists with whom I have little sympathy), I suggest that what follows be regarded as an argument for the elimination of the old distinction, not because its continued use would carry with it too many of the old associations, and provide further support for old ideas which have been both conceptually stultifying and too readily productive of morally reprehensible conclusions. I will argue that a new distinction in terms of generative entrenchment both preserves what was good in the

Table 1. Philosophical claims about innateness.

P1 Innate knowledge is prior to experience in one of the following senses:
    a. It exists prior in time to any experience.
    b. It is a precondition for experience.
P2 Innate knwoledge is independent of experience in that:
    a. Its origin or justification is independent of experience.
    b. It is invariant across different experiential histories.
P3 Innate knowledge arises as an effect of or is "triggered" by experience.
P4 Innate knowledge is knowledge of general truths.
P5 Innate knowledge is universal in the sense that it is common to all members of the class or species.
P6 Innate knowledge plays a generative role in producing other knowledge.
P7 Innate knowledge is analytic, necessary, or a priori.

Table 2. Ethological claims about innateness.

E1 Innate behavior for a given species is universal among normal members of that species in their normal environment either
    a. because the behavior has a genetic base
    b. because the behavior is 'canalized' or homeostatically regulated in development.
E2 Innate behavior appears early in development, before it could have been learned.
E3 Innate behavior is relatively resistant to evolutionary change.
E4 "Innate learning mechanisms" account for critical periods in learning or the ability to learning rapidly or on one trial.
E5 Phylogenetic parallels between behavioral and morphological traits is evidence for the innateness of the behavioral traits.
E6 Innate information is said to be "phylogenetically acquired" and hereditarily transmitted; acquired information is said to be "ontogenetically acquired" through learning.
E7 Stereotypy of behavior is evidence for its innateness.
E8 Relatively major malfunctions occur if innate features fail to appear or develop.
E9 If a trait show simple (e.g., Mendelian) inheritance patterns, it is innate.
E10 If a trait can be modified through selection, it is innate.
NOTE: The last two claims have been added since the rise of genetics and the modern synthetic theory of evolution, presumably because they are criteria for a trait's having a genetic basis.

old distinction and blocks associations which we should avoid. Under the new analysis proposed here, neither of the above claims or errors are defensible. For this reason, I will as much as possible avoid speaking of traits as "innate", except where talking about past analyses. Where I am talking about my analysis, and it is cumbersome to avoid the term "innate", I will use it in quotation marks.

Philosophical and Ethological Claims about Innateness

Since the innate-acquired distinction has had a long legacy in both philosophy and

188

ethology, I have summarized in Tables 1 and 2 the chief claims of these traditions. (The philosophical claims come from a variety of sources that I cannot individually cite here; the ethological claims are drawn primarily from Lorenz, 1965, and Mayr, 1974.) These claims often appear in different combinations and there are many close relations between different claims which I cannot develop at this point. There are also many close connections between the claims of the philosophical tradition about knowledge and those of the ethological tradition about behavior that can be detected by comparing the two tables. For example, claims P1a and P2b were carried across more or less unchanged, and inerpreted as comments about development. P2a has at least an analogue in E6, which suggests different sources for innate and acquired information. The criterion of universality, P5, was split into two, reflecting the influence of evolutionary taxonomy: Universality within a species was accepted as a criterion (E1) while presence of a trait in phylogenetically related species was accepted as a less central and indirect criterion (but still important because it indicated a genetic basis for the trait).

Some of the differences have a staight-forward explanation in the fact that ethologists eshewed, for the most part, talk of mental life and in particular knowledge that required linguistic representation. Thus, P4 and P7 appeared to the ethologists to require too rich a mental life to attribute to animals. One other difference is somewhat surprising. Most ethologists ignored any generative role for what was counted as innate. This is perhaps because innateness was seen as bearing primarily on the origins or causes of knowledge or behavior, rather than as fundamentally connected with the effects of having that knowledge or exhibiting that behavior. By contrast, I think that the generative role is the most powerful fulcrum in analyzing the relations between what have been called innate and acquired elements of behavior and knowledge (see below).

Debates Within Ethology

Those ethologists who have defended the use of the innate-acquired distinction (Lorenz and Tinbergen are probably the senior statesmen of this group) have also argued strongly for an evolutionary analysis of the significance of behavior. They have argued that the analysis of behavioral traits required the study of the organism in its natural environments, or in at least minimal experimental preturbations of these environments, and have argued that the analysis of the functions of these traits is an essential part of this study. They have also tended to make use of extensive phylogenetic cross-comparisons. They have tended to stay away from the exclusive study of organisms raised under laboratory conditions, except under special circumstances where the aim is to study the effects of deprivations of conditions which are found in the natural environment.

By contrast, another group of students of animal behavior, found primarily in the United States (many of whom were influenced by T. C. Schnierla) worked primarily with laboratory animals under experimental paradigms which were at least partially influenced by the behaviorist tradition. Probably for this reason (as good environmentalists), they tended to find the innate-acquired distinction to be of little use, and argued for its abolition. Although I have little sympathy with their program, I believe that they rised a number of serious issues worth discussing here. In what follows, I will draw primarily on the critique of D. S. Lehrman (1970), one of the most articulate and influential of these critics. I will use my discussion of his criticism as a basis for developing my proposed replacement for the innate-acquired distinction.

Lehrman argues that there are two senses of innate, not one, and that equivocations between these senses are responsible for numerous confusions. His first sense, which I will call the "genetic sense", treats a trait as innate if it is genetic or heritable (E9 and E10). The second sense, which I will call the "developmental sense," refers variously to the other criteria listed in Table 2, but especially E1b and E2. Lehrman goes on to argue that the developmental sense is confused and adopts the genetic sense. I will follow the opposite strategy and argue that the developmental sense captures the heart of the innate-acquired distinction.

Having opted for the genetic sense of innateness, Lehrman argues against the European ethologists that innate and acquired are not strict opposites, since what was acquired referred to the phenotype, and what was innate referred to the genotype. Everything in the phenotype, Lehrman argued, was partially due to the influence of experience, and thus was at least partially acquired. At best, the distinction as applied to phenotypic traits had to be treated as a matter of degree. The analogue to innateness that I will defend, generative entrenchment, is also a degree property.

A second objection Lehrman offers is to Lorenz's (1965) speaking of innate traits as being "blueprinted" in the genome. The blueprint metaphor suggested a 1–1 correspondence between genes or genetic elements and those behaviors that were innate, again making the innate-acquired distinction a dichotomous variable. Lehrman objected that the "blueprint model" of gene action was outdated and led to conceptual errors. He offered another metaphor for the relation between genes and behaviors, according to which the genome is a *program* which produces the phenotype. There are very few 1–1 mappings between components of a program and its outputs, just as there are few neat 1–1 mappings between genes and characters. (See Wimsatt, 1976.) The program model is far superior. It has been employed by

190

Mayr (1974), who offers the best positive defense of the importance of the innate-acquired distinction in the literature. I will also employ the "program" metaphor.

I agree with Lehrman on another point – that a proper way to study the origin of behavior is through developmental processes. However, I take issue with Lehrman's use of this as the exclusive approach and rejection of evolutionary and cross-species comparisons. I will argue that phylogenetic comparisons are also a powerful source of information about the character of developmental programs. One of the reasons Lehrman was led to reject these tools was his overly broad definition of experience. Lehrman draws his definition from Schnierla's "generalized" concept of experience:

> "Schnierla ... has used the concept of 'experience' to mean all kinds of stimulative effects from the environment, ranging from stimulus-involved biochemical and biological processes (having effects on the developing nervous system) to what we would ordinarily call conditioning and learning." (Lehrman, 1970, p. 21.)

With this definition, everything not in the genotype (or at least in the zygote at fertilization) is at least partially a product of experience. One problem with this definition of the innate-acquired distinction is that nothing in the phenotype is innate. The complementary problem with the nativism of, for example, Fodor, is that no (or almost no) information is acquired. I think that it is safe to say that a good definitional strategy for an important distinction is that no definition should be such that one or the other of the two dichotomous terms has no cases to which it applies. Both of these characterizations violate this condition.

A more specific problem with Lehrman's characterization of experience is that early input through sensory channels in a developing nervous system has an ambiguous status. Its deprivation often appears not just to deprive the organism of some information which it has to learn later, but of a capacity for acquiring experience through that sensory channel. (See, e.g., the series of papers on sensory deprivation collected in McCleary (1972)). Loss of a capacity, rather than simple loss of some information which could be acquired later has been a basis for claiming the innateness of the capacity. Early experience may be required for the development of this capacity, but as such, it performs a function more like food than like information. Schnierla's or Lehrman's concept of 'experience' would make no distinction between recognizing a possible meal, and eating it! The nativist would claim that in depriving the organism of this kind of experience, the system is rendered malfunctional, and is not normal, indicating an innate need for that experience. (One might say that early experience is often neither simply food, nor thought, but food for thought!) Related claims are made for deprivation of information or the right kind of experience during critical periods for learning. The idea that an organism has periods when it learns certain kinds of information or tasks easily and after which it can learn them only with great difficulty or not at all has been an embarrassment for naive empiricists, and is the basis for Mayr's (1974) distinction between closed and open programs, his substitute for the innate-acquired distinction.

The conflict between Lehrman and the European ethologists reflects and extreme polarization of positions. Lehrman accuses the Europeans of ignoring developmental processes. To the degree this is justified, it is due to the fact that the Europeans focused on behavior of species in their natural environments, which makes study of developmental processes and their experimental manipulations difficult at best and often impossible. The Europeans have charged the (primarily American) students of animal behavior with paying too little attention to evolution and the natural function of behavioral adaptations. This is due to the focus on the experimental study of the ontogeny of behavior in the laboratory. There has been some rapprochement since 1970, so that the best modern ethologists draw heavily on results from both areas and often combine both kinds of studies in their work. One advantage of the analysis I will propose is that it makes studies of both kinds equally relevant (and complementary) to the study of animal behavior, and the phenomena which the innate-acquired distinction was designed to capture.

Lehrman's strategy was largely to discount the innate-acquired distinction. He suggests (1970, p. 28) that since selection selects only for results, it does not matter whether a trait is innate or acquired. He points out that in some species, bird song is innate, while in others it is largely acquired. However, the distinction is in fact very important for the analysis of evolutionary and ecological dimensions of behavior. Mayr, for example, in defending the closed/open distinction as a replacement for the innate-acquired distinction, describes a number of general conditions in which it is advantageous for a trait to be coded for by a closed (or open) program and in which the other kind of coding would not work. Mayr uses telling and insightful examples and, for example, is able to predict for some of the cases Lehrman discusses whether a particular bird song should be innate or acquired. If the innate-acquired distinction did not exist, it or something like it would have to be invented by evolutionary biologists to talk about the evolution of the adaptive design of the phenotype in response to the structure of information in the environment.

From this discussion of Lehrman, some important desiderata for my substitute for the innate-acquired distinction have been established. While I have indicated why we cannot follow Lehrman in doing away with the distinction altogether, I have agreed with him that the distinction ought ot be a matter of degree. What would be desirable, and what I will provide in my analysis, is an indication of why certain

192

features of the innate-acquired distinction have been grouped together in terms of this distinction (neither the ethologists nor Mayr have offered such an explanation).

The Developmental Lock as a Model of Development

The analysis that I shall substitute for the innate-acquired distinction is based upon a model which I first designed for another purpose – to explain the phenomena characterized by von Baer's Law(s), which can be interpreted as providing a basis for connecting developmental biology and evolutionary biology. The fundamental phenomena underlying von Baer's Law(s) is the parallel between phylogenetic patterns and patterns of development. These regularities turn out to have a fundamental connection with the innate-acquired distinction.

The developmental lock, which is a model for a developing system, is partially based on H. A. Simon's work on the heuristics of problem solving. Simon (1962) compares two kinds of problems, one of which can be factored into sub-problems that can be solved independently (or at least where good first approximations can be achieved working independently), while the other must be solved as a whole. The first kind of problem Simon speaks of as "nearly decomposable." He represents these two kinds of problems in terms of two different kinds of combination locks. The following is a slight adaptation of his analogy. Both locks are cylindrical combination locks having 10 wheels, each with 10 positions. Only one of the 10^{10} possible combinations opens the lock. The first, or "complex" lock is represented in figure 1a. It includes no clues for partial solutions so that the expected number of trials before getting the correct combination is one half of the total, or 5×10^9. In the second, or "simple" lock (represented in figure 1b), we hear a "click" when each wheel is put into its right position, so that the expected number of trials is 5 per wheel, or 50 for the lock. The advantage of the simple lock over the complex is the ratio of the expected number of trials for the two locks, or 10^8.[1]

I will now describe a model derived by combining features of Simon's "simple" and "complex" locks, which I will call the developmental lock. It is a simple lock if worked from left to right but a complex lock if worked from right to left. (See

---

[1] For Simon, the simple lock represents a nearly decomposable problem, whereas the complex lock represents a problem that cannot be factored into subproblems. The computational advantages of a decomposable problem is a prime factor in the decision to opt for reductionist problem solving strategies (these are strategies that assume the system is nearly decomposable). In my (1980, pp. 230–235) I describe nine reductionistic problem-solving heuristics, most of which can be seen as techniques for redescribing or reanalyzing a system to facilitate this kind of problem-solving approach.

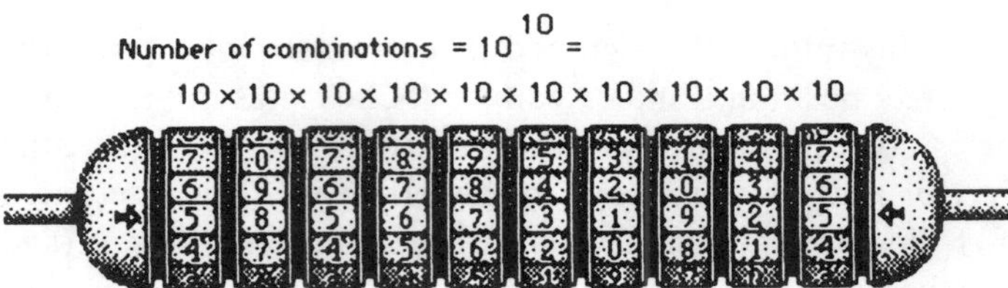

Figure 1a: Simon's "complex lock": 10 wheels with 10 positions per wheel. In the "complex" lock, the correct combination is only discoverable as a complete solution. (No clues are given for partial solutions.) Expected number of trials = 5×10^9

Figure 1b: Simon's "simple lock": Just as above, but a faint "click" is heard when each wheel is turned to its correct position, allowing partial solutions.
Expected number of trials = 50.
(The advantage of near decomposeability in problem-solutions is the ratio of expected number of trisl for the two locks = $5 \times 10^9/50 = 10^8$.)

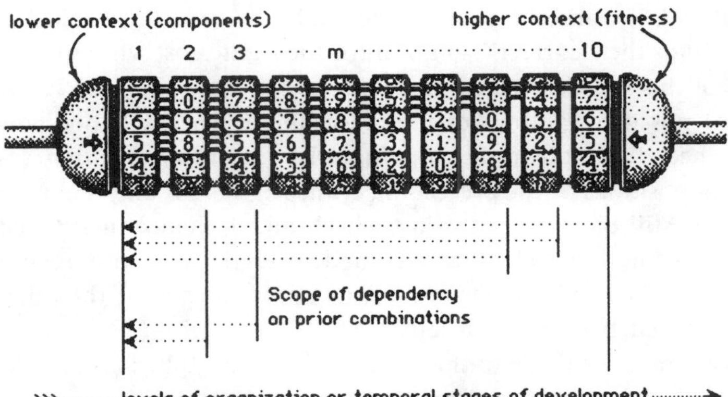

Figure 1c: Developmental Lock: (simple if worked from left to right, but complex if worked from right to left). Suppose a "click" is emitted by each wheel when it is in its "right" position, but what position is right is a function of the actual positions (whether right or wrong) of any wheels to the left of it, so that a change in position of any wheel randomly resets the combination of all wheels to the right of it.

figure 1c.) Suppose that in this third lock, each had its correct position (as in the first two locks) and the correct position is indicated by a "Click" (as in the second lock), but what position is correct for a given wheel is determined by the actual positions (*whether correct or not*) of any wheels to the left of it. Thus resetting a wheel will randomly reset the combinations for any wheels to the right of it (and earlier clicks from these wheels will no longer indicate correct solutions).

If this lock is worked from left to right, the solution for the first wheel is a simple problem (with an expected number of 5 trials before solution) since there are no wheels to the left of it. Given that this position is now set, and will be left undisturbed, the solution to the next wheel is also a simple problem: although its solution is a function of the position of the first wheel, that position has already been set. If each wheel is solved before one proceeds to the next right-most wheel, the solution to this lock is a simple decomposable problem, with an expected number of 50 trials for the whole lock. The lock is simple because once a wheel is solved (working from left to right) we never need to return to it in order to get "compound" solution. In problems whose structure is like this lock, there is a necessary order in which the sub-problems must be solved (e.g., because the solution to later sub-problems requires data from the solution of earlier sub-problems), but if this order is followed, this complex problem is decomposable into independently solvable sub-problems.

If the developmental lock is worked from the other end, however, it is similar to the complex lock. One finds the "correct" solution for the right-most wheel in five trials but the chance that this represents a correct solution for the whole lock is but 1 in 10^9, the number of possible positions of the preceeding 9 wheels. If one now turns to the second wheel from the right and finds its "correct" solution (as determined by the actual positions of the preceeding 8 wheels) there is only 1 chance in 10 that this solution is still correct for the last wheel, and 1 chance in 10^8 that it is correct for the preceeding 8 wheels. One has made essentially no progress: a complex problem remains complex! Similar arguments apply to each of the other wheels when they are worked from right to left.

One more modification must be made to this lock if it is to be applied as a model for development. An *n* wheel lock as described here has but one solution out of 10 possible combinations (suppose that there were but one possible adaptive phenotype – then there would be no adaptive radiations and no diversity of organic form! Serial evolution would be possible by adding wheels to simpler locks but the evolutionary tree would have but a single branch). We must modify the lock in a way that preserves its structure but permits a branching multiplicity of adaptive solutions. The simplest way to do this is to assume that each wheel has 100 positions, 10 of which are solutions (i.e., are adaptive) for that wheel. When the position of a wheel is changed, the 10 adaptive solutions of each wheel downstream of it are randomly reassigned to the digits 0 through 99. In this case the n wheel lock has 100^n or 102^n possible combinations, 10^n of which are adaptive. We now have many alternative solutions but all the other above conclusions are unchanged (it is still true, for example, for the 10 wheel lock that one in 10^{10} of the possible combinations are solutions, and that a random resetting of a wheel m positions to the right has 1 chance in 10^{m-1} of being correct for the wheels to the right of it.

The net effect left-to-right dependence in the combinations of the wheels is that if one takes a lock of this type which has a correct solution to all of its wheels and randomly resets a wheel which is m wheels from the right, the chance that this resetting is a solution to the lock is 10^m. *Thus, as one goes from right to left, the probability that a change in a wheel does not scramble combinations to the right of it declines exponentially!*

To apply the developmental lock to evolutionary contexts, let the wheels correspond to developmental stages of an organism, with the earliest stage at the left and successively later stages further to the right. (I don't want to suppose that development is partitioned neatly into stages. It would be almost trivial to develop a "continuous" model of this simple lock which would preserve this left-to right dependence, and permit the same qualitative conclusions, but for present purposes I will continue with the discrete multi-stage model.) Suppose that expression of a mutation at a given stage of development corresponds to the random resetting of the appropriate wheel.

The idea that earlier wheels set correct combinations for later wheels corresponds to the idea that the proper development of features at later stages presupposes or depends upon the occurrence of given features at earlier stages, and that if these features are not present at earlier stages, this has a high probability of causing a malfunction at a later stage or stages. Now suppose that the probability of getting a solution which is correct for that stage is k, rather than 1/10, and that the correctness of a solution at a given stage is independent of its probability of correctness at later stages. The probability that a mutation will be correct at a stage which is m stages from the end and also correct at the m−1 later stages is then k^m. Thus, as one goes from right to left, the probability that a mutation at a stage will permit normal development later on declines exponentially!

This model can be generalized to accommodate effects such as canalization, environmental interactions, and critical periods in development. (Critical periods will be discussed in a later section.) The first two effects can be accommodated by replacing the simple parameter k with multiple ones. To model canalization, suppose that a given mutation, denoted by *, occurs at a stage which is m stages from the right, and has a probability $k*_m$ of being adaptive at that stage. Suppose that it induces probabilities $K*_{m.m-1}$, $k*_{m.m-2}$, $k*_{m.m-3}$ ... that mutation * at stage m will not lead to maladaptive changes at these stages.

If a developmental program was strongly canalized with respect to the effects of mutation * stage m, one would expect that the probabilities $k*_{m.m-1}$, $k*_{m.m-2}$, $k*_{m.m-3}$ ... would all be close to 1 (or close to the survival probabilities for the normal organism if these are substantially different than 1), rather than fairly small. Thus the simple addition of a multi-parameter model allows one to express the fundamental intuition behind the concept of canalization. This mode does not explain canalization (one needs a model like that proposed by Kauffman in this symposium for that purpose), but it does allow us to model it.

We can extend the model to accommodate environmental dependencies by making these probabilities depend upon the environment, producing an array of probabilities $k*_{m.Ei}$ for the mutation * at stage m in environment E_i, which induces effects $k*_{m.m-1,Ei}$, $k*_{m.m-2,Ei}$, $k*_{m.m-3,Ei}$ ... at subsequent stages. In general one would have to consider not only the environment E_i in which the mutation occurred at stage m, but also the environments in which the various subsequent stages develop. This would be required to discuss cases of phenotypic switching or other cases in which the effect of an adaptation or mutation is strongly dependent on the environment.

With this model we are now in a position to create an important bridge between developmental biology and evolutionary biology. By associating with each mutation which is expressed at a given stage in development a vector whose components are the probabilities that the changes are adaptive at subsequent stages, we produce models that include a great deal of detail about developmental events and which can be readily included in population genetic models. (This is true because this vector of probabilities has basically the same form and effects as life history models or population growth models with overlapping generations which model these processes in terms of a series of probabilities of an organism's surviving from one life-history stage to the next.) The fact that these parameters describe the effects of various adaptations without explaining how they work is not a limitation in this context, for the strategy of population genetic models is to do an "engineering" analysis of the operation and effects of an adaptation on fitness (see Lewontin, 1978), and then to plug the fitness estimates from such an analysis into a genetic model to see whether and under what conditions that design will be selected for. If the genetic model predicts that a given design will be selected for over the available alternatives, we have an evolutionary explanation for the existence of that design. Why this model could be important is that, to my knowledge, no population genetic models currently exist which are capable of handling detailed developmental information about phenotypes. If no such models exist, then there is no way of talking about the evolution of designs which differ in these respects, so the introduction of such models could substantially increase the scope of evolutionary explanations. This approach will be elaborated in a future paper.

I indicated above that this model was originally developed to account for von Baer's laws of development. The details of this application are presented in Wimsatt (1983); for now I will simply note the most salient results that are relevant to this discussion of the innate-acquired distinction. Von Baer's four laws of development (see Ospovat, 1976, p. 6) are often summarized under the general principle "Differentiation proceeds from the general to the particular." This principle can be given three distinct formulations (taxonomic, morphological, and functional) depending upon the interpretation of "general," but the only one we need to consider here is the one which refers to taxonomic generality. (For more detailed discussion of ontogeny and conceptual development, functional generality is also very important.) Under the taxonomic rendering, von Baer's Laws state that features which appear earlier in development tend to apply to broader taxonomic categories than those which appear later.

I shall now show how the developmental lock model explains various important features of development summarized in Table 3. We begin with two critical features of development built into the developmental lock, captured by philosophical treatments of innateness, but ignored by the ethologists: features that arise early in

Table 3. Features of Generative Entrenchment.

D1 Features expressed earlier in development have a higher probability of being required for features which will appear later.

D2 Features expressed earlier in development will, on average, have a larger number of "downstream" features dependent on them.

D3 Features expressed earlier in development are, probabilistically speaking, phylogenetically older.

D4 Features expressed earlier in development are more likely to be widely distributed taxonomically than features which are expressed later in development.

D5 Features which are deeply generatively entrenched, if they are disturbed or fail to appear, are likely to cause major developmental abnormalities.

D6 Mutations which are expressed earlier in development are more likely to have larger, more pervasive, and more deleterious effects than those expressed later.

D7 Cross-phylogenetic analyses of developmental processes can generate significant information about the structure of those processes.

D8 Environmental information may be generatively entrenched.

198

development have a higher probability of being required for features that appear later (D1) and tend to have a larger number of downstream traits depending on them (D2). Applying what we noted earlier about such locks, we see that the probabilities of mutations being compatible with overall normal development increases exponentially as one moves from right to left, that is, from later to earlier stages. As a result, we should expect that evolution would be increasingly conservative at earlier stages of development so that features expressed earlier in development are more likely to be older (D3) and more widely distributed taxonomically (D4) than features which are expressed later in development. The last is a critical claims of von Baer's law that is here given a probabilitistic interpretation (in agreement with Gould, 1977).

Of course, a feature may occur very early in development and still have little or nothing which depends on it. For this reason it is better to distinguish features not in terms of when they appear in development, but in terms of whether they are needed for other traits. I will, therefore, speak of traits being "generatively entrenched" to the degree that they have a number of later developing traits depending on them. If we reconceptualize von Baer's Law in terms of generative entrenchment rather than earliness in development, then it appear even more likely to be true (as a probabilistic generalization) since it is not undercut by examples of features which appear early in development but are not generatively entrenched.

Features D5–7 also follow directly from the developmental lock. D5 and D6 follow from the fact that earlier changes, whether due to teratological factors or mutations are likely to disrupt the expression of many additional features and so have more widespread effects. D7 results from the fact that a high degree of taxonomic stability of features earlier in development implies a high degree of generative entrenchment. Thus, in principle we can determine features of the causal structure of developmental programs from doing cross-phylogenetic comparisons, analyzing the relative stability of taxonomic features, and looking for covariances in changes of features as clues to which later features may depend upon which earlier ones. (I will withhold comment about item D8 until the next section.)

The Innate-Acquired Distinction Revisited

Comparing Table 3 with Tables 1 and 2 reveals that the features of generative entrenchment can account for nearly all of the philosophical and ethological claims about innateness. For example, D1 and D2, which characterize the entrenchment of traits, accounts for P6, the claim that innate experience has a generative role, and suggests P1b, the claim that innate features are preconditions for experience. The wider taxonomic distribution of generative entrenched traits, D4, has E1, the universality of innate behavior with a species, as a special application. Again, the fact that generatively entrenched traits should be evolutionarily conservative because of the consequences of changing them (D6) accounts for claim E3 that innate behavior is relatively resistant to evolutionary change. Indirectly, this explains claim E5 that parallels between behavioral and morphological phylogenesis yield evidence for the innateness of the behavioral (and also the morphological) traits, since if both morphological and behavioral traits are deeply entrenched and thus evolutionarily conservative, we would expect substantial phylogenetic correlations between them.

Some of the features of the earlier analyses of innateness take on a new complexion within the framework of generative entrenchment. For example, claim E2, that innate behavior appears early in development, was understood as stemming from the fact that there is little opportunity for (post-natal) experience to influence early behavior. On the analysis I propose, earliness in development is important because the *effects* of early traits are different: they are likely to play generative roles with respect to a wide variety of other behaviors or adaptations. Hence, on this analysis, Lehrman's (1970) complaint that Lorenz (1965) is ignoring the role of pre-natal experience in the ontogeny of traits present at birth is irrelevant, since a trait is classified as "innate" on the basis of its effects rather than on the basis of the source of the information which generates it.

It is striking that all of the philosophical and ethological claims (except E9 and E10) can be explained as issuing from a single property – generative entrenchment. Thus, generative entrenchment explains the relationship among these many diverse criteria, something that no analysis before has been able to do. In particular, it explains a number of criteria (e.g., E1, E3, E5, and E8) which are simply presupposed by or which look like *ad hoc* additions to other analyses.

There is, however, a major difference between the concept of innateness and that of generative entrenchment, which makes the latter a replacement for rather than an extension of the former notion. Environmental information may count as generatively entrenched (D8). This results from the fact that the concept of generative entrenchment did not distinguish between what comes from inside and what comes from outside the organism. Thus, the account of generative entrenchment rejects ethological claims E9 and E10. (Elsewhere, I will argue that E9 and E10 are actually inconsistent with the other ethological criteria, and thus would have to be rejected in any case – see Wimsatt, forthcoming.)

The analysis presented here leads to an entirely different reading of Dawkins (1976, 1978, 1982) concept of the "extended phenotype." Dawkins takes an important step in bringing environmental features into the extended phenotype. In doing this, he is taking one step further in the tradition of the European ethologists who argued that behavior could be treated as any other phenotypic feature – it could be selected for and one could study its taxonomic distribution and phylogenetic development. The problem with Dawkin's notion is that it is "gene-centered". The "gene's eye view" of the phenotype and of evolution has been a very productive heuristic for many problems and Dawkins "extended phenotype" is a major advance in some ways over prior analyses, but it is not the correct way of formulating the relationship between genes and environment. I will return to this issue below in section 9.

We should conceive of the phenotype as the expression of the unfolding developmental program which has an initially small number of generative elements. Some of the genes, together with some of their somatic environment in the zygote, interact with each other and with aspects of the outer environment to produce the developing phenotype. In this conception, it is quite clear that information acquired from the environment can have a profound effect if it is deeply generatively entrenched relative to subsequent behavior; on this analysis, if it is generatively entrenched, it is "innate". If the early experience which is withheld in a deprivation experiment has a generative role with respect to a wide range of subsequent experience in that sensory modality, its loss will produce such far-reaching consequences that it would readily be described as a loss of capacity. If in addition, there is a critical period for the acquisition of that early experience, this loss will be at least partially irreversible after the critical period has passed.

What is required to speak to environmental information as "innate"? I suggest that the following are minimal requirements:

(1)   The acquisition of that kind of information at that stage of development is deeply generatively entrenched with respect to subsequent behavior.

(2)   The developmental program is designed to receive information of that sort at that stage of development.

(3)   The information must be of a relatively specific sort.

(4)   The environment of the developing organism is a reliable source of the required information at that stage of development.

I must limit myself to one example to illustrate this idea. According to this analysis, not only is the imprinting mechanism of the greylag goose at birth "innate" (as on the standard ethological accounts), but the object of imprinting is also "innate". When the infant goose extricates itself from its shell, it imprints upon and follows any moving object. (Sound may be important, but Lorenz reports having imprinted greylag geese on himself, his co-workers, his dogs, and even a moving toilet float!) In their natural environment, however, there is a very high probability that the young goose will properly imprint on its mother and will, in short order, learn to distinguish her cries and her appearance from that of other female greylag geese nearby. The family structure and behavior of the mother greylag goose at the time of birth makes it almost a certainty that the baby geese will imprint properly (she stays close to the nest at hatching time, and imprints upon them as well, and so is readily available during the critical period). Thus the correct information (that a close moving object first detected at birth is mother) is reliably present in the environment. This is an important adaptation in colonially nesting birds, who can depend only upon their parents for food, warmth, and protection, and therefore need to be able to distinguish them from other conspecifics. In this instance, all four of these conditions for environmental input counting as "innate" are satisfied.

The inclusion of environmental input as generatively entrenched and so "innate" has significant consequences for evolutionary theorizing. Often in population biology selection is construed as operating on genes. One reason for this is that genes are assumed to be the unit of heredity and, as such, to carry the information needed to create a new phenotype. Causally, however, selection operates on phenotypes, and if environmental factors can be included as a generatively entrenched part of the program for producing the phenotype, then they become important as agents of heredity. (In fact, as numerous geneticists have pointed out, a high degree of heritability of a trait may indicate stable factors in the environment rather than genetic determination of the trait. This point renders incorrect Burian's claim in the following paper that developmental programs are not heritable because of their dependence on environmental features.) To theorize about evolutionary processses without including an important aspect – the structure and differential stability of the phenotype which is the unit of selection – ignores power and constraints which provide important resources. As Lewontin (1970) has pointed out, what is required for evolution to occur is the existence of heritable variance in fitness. Revealing, this requirement says nothing about stability of genetic structure or of phenotype, but only of fitness, which is a relationship between phenotype and environment. The major heresy of 20th century evolutionary theory is to focus on the stability of genetic structure ignoring the fact that this is *not* fundamental to the logical structure of the evolutionary process.

Generative Entrenchment and Mayr's Closed/Open Program Distinction

It will be useful to compare the analysis I am proposing here to that proposed by

202

Mayr (1974), in which he proposed replacing the innate-acquired distinction with one between "closed" and "open" genetic programs.

For Mayr, despite appearances, "open" and "closed" are not parts of the genome – both are clearly developmental programs and as such are properties of the phenotype. A closed "genetic program" is one that does not require (or permit) informational inputs from the environment to significantly affect its execution. An open program is one that does require or permit such information (Mayr, 1974, p. 651). A "critical period" is simply the temporal window during development or during a behavioral activity when the appropriate open program will receive relevant information.

Mayr exploited these notions to make clear adaptive sense of the structure of developmental programs for behavior in terms of the evolutionary and ecological contexts in which the organisms find themselves. He also has given the best discussion available of the general conditions under which it would be advantageous for a program to be closed or open, and on the basis of his analysis suggested a broad classification of types of behavior which went well beyond that provided by the traditional innate-acquired distinction.

I cannot do justice to the richness of his analysis here, but I wish to argue that the main features of his distinction – that of a "critical period" and the distinction between "closed" and "open" programs – can also be captured by the generalized version of the developmental lock model. If my claim is correct, then all of the resources of his analysis should be deployable through this model.

A critical period is a relatively delimited period during the lifetime of an organism when certain kinds of input received by the organism have important effects on its subsequent development or behavior. As Mayr says, it is a "window" in time through which environmental inputs of a delimited sort can enter. Inputs of the same sort received before or after that time in development have quite different effects, and often little or no effect at all.

This idea can be easily captured by the multiple-parameter generalized version of the developmental lock model discussed above. Suppose that we generalize from the notion of a mutation which is first expressed at a given stage, m, of development, to speak of any perturbations of the system, whether genetic or environmental in origin, acting at that stage of the life cycle. (We have in effect already made this extension to talk in the preceding section about environmental information as "innate.") To say that information can get into a "window" in the developmental program during the critical period and not at other times is to make two claims about the structure and probabilities associated with the developmental lock.

The first claim is simply that an input at stage m, of the appropriate informational sort, is generatively entrenched with respect to subsequent behavior and development. The probability that a mutation, perturbation, or environmental input acting at that stage will be adaptive at that stage is simply the probability that in the normal environment the "correct" input or inputs will be received. If the right input is received, then the subsequent probabilities are simply the normal conditional probabilities of the organism surviving through the various successive stages of its life cycle as are found in life-history or age-structure models in population genetics. If the wrong input is received, the idea of generative entrenchment requires that one or more of the probabilities for later stages will almost certainly be substantially lower (such that their net product is substantially lower in the normal environment.)

The second claim (which introduces the "frame" of the window) is that the same inputs, whether right or wrong, do not have that effect (or have a much smaller effect) on the probabilities if they are introduced at earlier or at later stages than the one in question.

This proposed explication of "critical period" ignores some of the complexities of that notion. Thus most critical periods are not "one-shot" "all-or-nothing" affairs, but a sequential series of overlapping windows with complex conditional interactions between them. None of these ideas however introduce complexities of a new kind, and all of them should be expressible in terms of conditional effects on the probability vectors associated with the generalized developmental lock model.

Similarly, an "open" program is simply a program which has at least one such window, and a closed program is one which has none – or at least none which is open for the inputs in question during the period in development being investigated. (It seems likely that there are no totally closed programs in development. The idea of a closed program must be viewed as a relative one – relative to the period of time of development under investigation, the class of inputs being investigated, and probably also to the environment and the prior state of the developing phenotype.)

With this last move, we have now captured all of the criteria of the traditional innate-acquired distinction worth capturing, at least for ethological uses of that term. Criterion (E4), which refers to critical periods or unusually rapid or "one-shot" learning now falls within the domain of the developmental lock model.

Different Strategies for Relating Development Considerations and Evolution

In this paper I have discussed the innate-acquired distinction, which has been employed in the ethological applications of evolutionary biology. I have tried to show how, when recast in terms of generative entrenchment, it can be explained and analyzed through a class of models which more generally allow encorporating

204

developmental insights into an evolutionary analysis. However, this is not the only possible approach. In the accompanying paper, Stuart Kauffman has adopted a different approach. I shall conclude with a few comments on Kauffman's approach and its relationship to my own. I regard Kauffman's approach as one of the most important and probably the most interesting and original new approach to modelling developmental constraints in the literature. (I am tempted to say that it is the only really new idea in recent literature on the topic.) I believe that his approach and mine are rarely in conflict, and far more frequently are complementary approaches to the modelling of development. I have been following his work in this area since 1969 but it has only been within the last 2 or 3 years that my views have taken more substantial shape and been elaborated sufficiently that he and I have again begun to talk seriously about developmental models. In this discussion, then, I will take his paper as given, and will draw primarily on conversations we have had about our two approaches. I hope that he would not disagree with my interpretations of these exchanges.

(1) Kauffman has argued correctly that my approach is fundamentally conservative (no pun intended!) – that I have basically modelled the neo-Darwinian (indeed, the Darwinian) abhorrence of macro-mutations. In this model, I have merely systematized the cascading effects possible for earlier developmental changes, and argued that they would be strongly selected against, producing increasing evolutionary conservatism of earlier developmental stages. It is a conservative model because developmental constraints are seen as joint products of the organization of developmental programs and the operation of selection, and the model thus basically falls within the adaptationist program. No challenge to traditional neo-Darwinism here! Kauffman argues by contrast that his generic constraints are in some sense prior to the operation of selection, and what's more, that under quite a broad set of conditions, are capable of overwhelming selection forces which happen to be going in other directs. His generic constraints are thermodynamic in character (under a generalized information-theoretic interpretation of thermodynamics) and represent a kind of physical constraint on possible (or highly probable) modes of genetic organization.

This way of describing the two approaches makes them seem to be in conflict. But this conflict is largely illusory. The illusions are of two sorts: (1) The assumption that selective forces and generic constraints will be in opposition is gratuitous: it is primarily a consequence of the fact that one of the most interesting directions for the development of Kauffman's theory lies in considering models where they are opposed. (2) The domain of phenomena to which our models apply are only partially overlapping. Each model has things to say about phenomena which the other does not address. I will address these points in turn in the next few paragraphs.

(2)   If Kauffman's generic constraints are found very early in development (as most of the ones he discusses are, since they deal largely with the differentiation of different cell types) then we would expect that they would have a high probability of becoming deeply generatively entrenched as well. They would then be doubly reinforced as developmental constraints: not only will most mutations not change them (as generic constraints), but also any mutations which did act to change them would be strongly selected against (because they are deeply generatively entrenched). In such cases, his theory and mine are obviously complementary rather than inconsistent.

(3)   The description of generic constraints makes it appear as if there are no selection forces, only entropic ones at work. But at this level of organization it may be hard to tell the difference (for real systems) between the two. It is no accident that Dawkins, for example (1976, 1982) sees the origin of selection in the relative thermodynamic stability of primitive macromolecules. Generic constraints are the consequence of a similar kind of stability one or two levels up, stability in the molar behavioral properties of genetic control networks under random mutations in the individual control elements.

(4)   Kauffman starts with an ensemble of certain types of system. Why this type? As his own earlier analysis shows (Kauffman, 1969), the parameters of this class may be products of selection. (Kauffman's use of gene control networks with an average of two connections per element is a conscious decision, based partially on the properties of the operon model, and partly on his demonstration that randomly connected networks of binary switching elements have an expected minimum cycle time for about two connections per element. As he argues, if short generation time is advantageous (as it is), then so is short cycle time. Thus, two connections per element, a defining property of the ensemble of systems he looks at, is probably a product of selection!) A somewhat different but related point is that if generative entrenchment produces sufficiently broad taxonomic generality, the system properties may look (and truly be) generic. (Consider the near universality of the genetic code, although most people argue that many characteristics of the code are products of selection, and it is nothing if not deeply generatively entrenched!)

(5)   I would argue that generative entrenchment is itself a generic feature of the design of the phenotype, because complex multiply-connected networks tend to have long causal pathways and many loops of varying lengths. One could also argue that generative entrenchment is itself deeply generatively entrenched. This is because generative efficiency (producing a lot from a little) is highly efficient, and strongly hierarchial gene control systems would have enormous evolutionary advantages over those with no significant hierarchial organization. (See Simon, 1962, 1981, and also the discussion of the importance of near-decomposeability and quasi-

206

independence in organic design in (Wimsatt, 1981, pp. 141–142.))

(6) Finally, generative entrenchment applies to some kinds of constraints which generic constraints can't touch, at least in the present form of Kauffman's model. (Thus, his analysis presently models only the behavior of gene control networks, and not the expression of structural genes or the changing consequences of their expression and interaction with changes in the control structure.) The same applies to generic constraints, which appear to be capable of explaining aspects of phenotypic behavior which can only be presupposed in the developmental lock model. Probably the most striking example of this is Kauffman's demonstration that his networks are buffered or canalized against major changes in the face of random mutations in control structure connections.

For all of these reasons, I think it is more fruitful (and more correct) to regard Kauffman's model and my own as complementary, rather than as competitors. While there might well be room for substantial argument as to whether a given constraint is due to generative entrenchment, to its generic character, to both, or to neither, the theoretical structures of our two models support no natural opposition between them.

Acknowledgments

I first hit on the basic developmental lock model in 1972 or 1973 while looking for a way of explaining Haeckel's Biogenetic Law in a hierarchical way – an idea in effect suggested by Herbert Simon in his (1962). The paper Steve Gould gave at a conference at the American Academy of Arts and Sciences in 1974, where I first presented the basic developmental lock model (a paper which became a major theme of his (1977)) led me to realize that what I had explained was not Haeckel's Law, but von Baer's Law. Since Steve there argued that von Baer's Law, unlike Haeckel's was true, I was encouraged to continue. My interest in applying this model to the innate-acquired distinction, to cognitive development, and to the elaboration of scientific theories dates from about that time, though I was not then aware of most of what is discussed in this paper, except in broadest outline. Over the years since, I have benefitted substantially from the encouragement, elaborations, scepticism and criticism of many people, particularly Stuart Altmann, Bill Bechtel, Robert Brandon, Dick Burian, Guy Bush, Bob Glassman (with whom I co-authored a paper which gave the developmental lock its first public exposure, Glassman and Wimsatt, 1984), Jim Griesemer, Joe Hanna, Stuart Kauffman, Bob MacCauley, Jane Maenschein, Joe Maxwell, Ernst Mayr, Bob Richards, Bob Richardson, Tom Roeper, Marty Sereno, Herb Simon, and students in many classes, particularly those in my 1983 and 1985 seminars on Evolution and Epistemology. This paper is most closely related to the content of the Donald Lipkind Memorial Lecture, which I was privileged to give at Chicago in the spring of 1985.

I would like to dedicate this paper to the memory of two people who have had a largely indirect influence on these ideas, an influence which undoubtedly would have been much more direct had they lived to see it. The first was Don Lipkind, who was an excellent and deeply beloved student at Chicago from 1972 to 1977, and who was in the first course I gave (with Tom Roeper) on philosophy in 1972, in which I first seriously considered the furor surrounding the innate-acquired distinction. The second was my father, William Abell Wimsatt, who as a student, teacher, and researcher in histology, reproductive physiology, and embryology at Cornell gave me paradigms of an information rich environment, an (often over-) extended phenotype, and above all, an emotionally nurturing and supportive development in, which most of the constraints were highly generative. I only wish that each of them could have lived to argue with me over this paper.

References

Chomsky, Noam. (1967). Recent contributions to the theory of innate ideas. *Synthese,* **17,** pp. 22–11.

Fodor, Jerry. (1981). The current status of the innateness controversy. In his Representations: Philosophical essays on the foundations of cognitive science. Cambridge: MIT Press–Bradford Books.

Glassman, R. B. and Wimsatt, W. C. (1984). Evolutionary advantages and limitations of early plasticity. In S. Finger and C. R. Almli (Eds.). *Early brain damage. Volume I: Research orientations and clinical observations.* New York: Academic Press, pp. 35–58.

Gould, Stephen J. (1977). *Ontogeny and phylogeny.* Cambridge: Harvard University Press.

Kauffman, Stuart A. (1969). Metabolic stability and epigenesis in randomly constructed genetic networks. *Journal for Theoretical Biology,* **22,** 437–467.

Kuo, Zing-Yang (1932). Ontogeny of embryonic behavior in *Aves,* IV. The influence of embryonic movements upon the behavior after hatching. *Journal of Comparative Psychology,* **14,** 109–122.

Lehrman, D. S. (1953). A critique of Konrad Lorenz's theory of instinctive behavior. *Quarterly Review of Biology,* **28,** 337–363.

Lehrman, D. S. (1970). Semantic and conceptual issues in the nature-nurture controversy. In L. R. Aronson et al. (Eds.), Evolution and Development of Behavior (a Festschrift for T. C. Schnierla), San Francisco: Freeman, pp. 17–52.

Lewontin, R. C. (1970): "The units of selection", *Annual Review of Ecology and Systematics,* **1,** 1–18.

Lewontin, R. C. (1974). *The genetic basis of evolutionary change.* New York: Columbia University Press.

Lewontin, R. C. (1978). "Adaptation", *Scientific American* (September)

Lorenz, Konrad Z. (1941). "Kant's categories in light of contemporary biology", reprinted in English in *Yearbook of the Society for General Systems Research,* Vol. 7, (1959).

Lorenz, Konrad Z. (1965). Evolution and modification of behavior. Chicago: University of Chicago Press.

Maynard-Smith, John. (1975). *The theory of evolution, 3rd. ed.,* London: Pelican.

Mayr, Ernst, (1974). "Behavior programs and evolutionary strategies", *American Scientist,* **62,** 650–659.

McCleary, R. A. (Ed.) (1970). *Genetic and experiential factors in perception.* Glenview: Scott-Foresman.

Ospovat, D. (1976). The influence of Karl Ernst von Baer's embryology, 1828–1859, *Journal of the History of Biology,* **9,** 1–28.

Putnam, Hilary (1967). The 'innateness hypothesis' and explanatory models in linguistics, *Synthese,* **17,** 12–22.

Simon, H. A. (1962). The architecture of complexity. *Proceedings of the American Philosophical Society.* Reprinted as chapter 7 of Simon (1981), pp. 193–229.

Simon, H. A. (1966). Thinking of computers. In R. G. Colodny (Ed.), *Mind and Cosmos,* Pittsburgh: University of Pittsburgh Press, pp. 3–20.

Simon, H. A. (1981). *The sciences of the artificial. 2nd ed.* Cambridge: MIT Press.

Sober, Elliott (Ed.) (1984). *Conceptual issues in evolutionary biology.* Cambridge: MIT Press–Bradford Books.

Wimsatt, W. C. (1976). Reductive explanation: A functional account. In R. S. Cohen, C. A. Hoker, A. C. Michalos, and J. van Evra eds., *PSA–1974, Boston Studies in the Philosophy of Science,* Vol. 32. Dordrecht: D. Reidel, pp. 671–710. Reprinted in Sober, 1984.

Wimsatt, W. C. (1980). Reductionist research strategies and their biases in the units of selection controversy. In T. Nickles (Ed.), *Scientific discovery. V. II: Case studies.* Dordrecht: D. Reidel, pp. 213–259. Reprinted in Sober, 1984.

Wimsatt, W. C. (1981). Units of selection and the structure of the multi-level genomen. In R. Giere and P. Asquith (Eds.), *PSA–1980, V. II.* East Lansing, Mich.: The Philosophy of Science Association, pp. 121–183.

Wimsatt, W. C. (1983). Von Baer's law of development. Generative entrenchment, and scientific change. Currently under revision for publication in *Philosophy of Science.* (52 page typescript).

Wimsatt, W. C. (forthcoming). The innate-acquired distinction reconsidered in ligh of developmental biology.

On Integrating the Study of Evolution and of Development

RICHARD M. BURIAN

Department of Philosophy, Virginia Polytechnic Institute and State University, Blacksburg, Va. 24061, U.S.A.

Background

In the early twentieth century, studies of heredity, development and evolution went in separate directions. Although there were never any wholly impassable barriers between the resultant disciplines, each developed its own terminology, groups of focal problems, and peculiar perspective. In short, development, evolution, and genetics evolved into fairly independent biological disciplines. The resultant theoretical lacunae and barriers to communication are of quite general interest. So are the numerous attempts to weld synthetic or integrative theories bridging these disciplines.

The difficulties of such integrative theories are enormous, largely because the problems which such theories must address are terribly difficult. Throughout the first half of our century, for example, virtually the only way in which genes could be identified was by their phenotypic effect. The vast majority of genes were therefore identified by noting the pattern of inheritance of some visible effect, usually on the adult, but occasionally on some other developmental stage. (Most of the exceptions concerned biochemical traits which could, in some instances, be identified by the presence or absence of some particular form of some enzyme or protein, as detected by biochemical tests.) Nothing in this way of identifying genes gives any real clue as to how they go about producing their particular effects nor what sort of mediating processes are or might be involved in getting from gene to phenotype.

Both the critics and defenders of Mendelian theory quickly recognized that the effects of genes are brought about by a complicated and ill-understood series of mediating steps. During the first half of the century, virtually nothing was known about the means by which these mediating processes are controlled. Many developmental biologists and geneticists with developmental interests recognized the need to cope with these problems. (I have in mind such figures as Bateson,

Bechtel, W (ed), Integrating Scientific Disciplines. ISBN 90-247-3242-5.
© *1986, Martinus Nijhoff Publishers, Dordrecht. Printed in The Netherlands.*

210

Darlington, Ephrussi, Goldschmidt, and Waddington.) A variety of figures in France and Germany, with some degree of support from such Americans as Conklin and Whitman, even insisted that transmission genetics was radically incomplete and perhaps the wrong starting point for laying bare the foundations of heredity. They held that transmission genetics was incapable of showing that nuclear genes bear primary responsibility for bringing about the end products which, as transmission geneticists had shown, are under hereditary control.[1]

There was thus a fairly long, though low level, history of complaints of the inadequacy of Morgan-style genetics as the basic theory of heredity precisely because it was unable to handle the problem or problems of development. Yet the authors of such complaints were perpetually frustrated. G. Evelyn Hutchinson and Stan Rachootin state the problem succinctly in their introduction to a recent reissue of William Bateson's *Problems of Genetics*, a book based on lectures delivered in 1907 at Yale:

> William Bateson had a gift for choosing problems that in his day were insoluble. We can with hindsight say that often they were the right problems. This is especially true of his attempts to show the connections between development and evolution; for him this was the fundamental problem of genetics (Bateson, 1979, p. viii).

Indeed, anyone who took the connections between development and evolution to be a fundamental problem of genetics during the first half of this century entered upon a tangle of problems which was, at the time, insoluble. As of that time, it was not yet possible to provide an intrinsic characterization of the units of heredity. Nor was there any detailed knowledge of the means by which those units produce signals or make the intermediate substances that initiate the series of steps yielding definite products. There was even less knowledge of the means by which those signals or the

---

[1]   The complex history of resistance to Mendelian theory among embryologists and developmental biologists has not yet been well told. A substantial factor in that resistance was the notion that in order for nuclear genes to bring about effects on cellular or organismal morphology, they must produce something that is processed and altered *in the cytoplasm. How* this processing occurs was, of course, a great mystery, but (so the argument ran) the whole system for such processing must preexist in the egg, and so must be inherited independently of the genes in the nucleus. This *second system of heredity* was thought by some to account for major features of the organism (familial differences, generic differences, and *Bauplaene* as opposed to the specific differences and decorative characters thought to be controlled by the nuclear genes); it was, in any event, a prerequisite for the proper expression of nuclear genes. This system was invisible in Mendelian experiments *because it was held constant in those experiments.* For such reasons, it was argued, Mendelian experiments did not reveal the workings of the fundamental system of heredity. Parts of this history will be well told in Jan Sapp's forthcoming book on cytoplasmic inheritance (Oxford University Press), based on Sapp (1984); I hope to tell more of it in *A Conceptual History of the Gene*, in preparation.

subsequent steps are regulated or modulated. In the absence of all of these, one could not even begin to develop analytical techniques adequate to such a task. For example, consider the long-standing "paradox" with which Mendelians and their critics wrestled in the 1930s and 1940s. Since (as experiment had shown) nearly all cells of higher organisms contained *all* of the chromosomes (and nuclear genes) of the organism, how could the differences among cells be due to the genes, and how could there be so many cell types? Was it not, rather, differences in the cells' means for making "stuffs" out of genes that were responsible for the differences between cell types? And weren't the systems for making the relevant substances located in the cytoplasm? Didn't these considerations show that nuclear genes are not the only system of heredity that must be taken account of in genetics?

Evolutionary theorists had rather less difficulty in integrating developmental considerations into their views than geneticists. Whatever the details of the developmental system, if an organism is to be successful it must go through all developmental stages up to that at which it reproduces. Thus the filter of selection must test every developmental stage. Even if, for practical reasons, one concentrated on certain critical or accessible stages in the life history of the organisms one studied, it remained clear that natural selection operated on the entire apparatus or system or program (or whatever) by means of which organisms developed from the egg to the reproductive stage. One had also to add to these considerations further ones based on what Darwin called "the mysterious laws of correlation of parts." If selection favored a particular trait at some particular developmental stage, the presence of that part or trait might well have ramifications on the constitution of the organism at prior or (especially) at subsequent stages. This point might well prove useful in explaining some of the odd, seemingly non-adaptive features of organisms tolerated by selection, for the benefits accruing from a favorable trait early in the development of an organism might far outweigh the damage done by correlated traits arising later in its life. If one started asking detailed questions, to be sure, this integration of developmental and evolutionary considerations remained quite superficial, but it was a significant part of many evolutionists' stock in trade.

First Approximations

Let us turn to the articles by Kauffman, Wallace, and Wimsatt. They start from a new situation; molecular genetics has radically altered the problem situation in genetics. We have made considerable progress in identifying the relevant genetic units. They are made up of DNA and RNA, and they can be characterized

212

intrinsically (e.g., by nucleotide sequencing). We are well along in our quest for detailed knowledge of the means by which DNA and, parasitically, RNA provide the information utilized in constructing organic matter. For certain phenotypes – most notably amino acid sequence in proteins – we can correlate gene structure (nucleotide sequence) and function (signalling amino acid sequence). We have further to go in the struggle to learn how the genes' signals and the subsequent steps are regulated and modulated; as Wimsatt points out, it is very easy to underestimate th roles of the somatic and the external environments in determining the information content of the genetic material.[2] And when we deal with phenotypes relatively far removed from the genes, we are still a very long way from understanding how immediate products are coordinated so as to yield cells, tissues, organs, organisms, and behaviors. Nevertheless, the situation has been drastically altered by the advent of molecular genetics – enough so that it is not unreasonable to attempt to construct a new genetically based theory encompassing both evolution and development. Such attempts may fail, but there is surely better hope of success in such an enterprise now than there was a few scant decades ago.

If we are to integrate a theory of development with a theory of evolution (perhaps on the basis of molecular genetics) – a matter about which Wallace is sceptical – we are immediately faced with the need to make simplifying assumptions as first approximations. It will be useful to examine the differences among our symposiasts in this connection: what sorts of simplifying assumptions do they make, what sorts do they deny? Each has a characteristic approach which, *if* successful, would have marked consequences for the degree and charcter of the synthesis between the study of evolution and that of development.

To a first approximation, Wallace denies that knowledge of development will shed (much) light on evolution because he denies the simplifying assumption that the genetic mechanisms prominent in development play an important direct role in evolution. His ground for denying this assumption has an honorable history; it is the isolation of the (contents of the) germ line from direct influence of the environment or the soma. In this respect, Wallace's view is a conservative one, belonging in the mainstream of the Mendelian tradition.

At heart, according to Wallace, evolution is statistically biased selection of germ lines. Selection operates from generation to generation among the available (but changing) pool of genotypes. (This, he claims, is the proper reading of Dobzhansky's *mot*, "evolution is change of gene frequency.") Development is the

---

[2]   The context dependence of genetic information is beautifully illustrated by the fact that there are variant genetic codes in mitochondria. The information contained in a given string of nucleotides is, therefore, different when that string is located in a nucleus than when it is located in a mitochondrion.

consequence of the interactions of an organism's genotype (or "genetic program") with the available environments. The mechanisms that govern selection do not govern development and – with one caveat – *vice versa*. The caveat is that a genetic program, if it is to remain in the evolutionary competition, must be good enough to enable the organism that it generates to survive and reproduce. (Technically, the program must yield high enough likelihood of survival and reproduction in the available environments, both absolutely and relative to its competitors.) But, *the mechanisms of development do not themselves play a significant role in evolution* beyond getting the germ lines carried by developed organisms into the next generation. And for this reason a genetically based theory of development will have little to offer to a genetically based theory of evolution, and *vice versa*.

Kauffman, Wimsatt, and I all disagree with Wallace's minimization of the importance of knowledge regarding development for our understanding of evolution. I shall offer some arguments against Wallace's stance below, but first let me provide thumbnail approximations of the positions of Kauffman and Wimsatt. Kauffman, a developmental geneticist well versed in current knowledge (such as it is) of the molecular mechanisms underlying development, hopes to provide a very abstract and general characterization of the structure of the genome. This simplifying characterization, if properly drawn, pertains equally well to evolution and development, for its draws on basic features common to the genetic systems of all complex organisms. Using this characterization, Kauffman seeks to classify "natural" or "generic" statistical states of such systems – states normally occupied in the absence of extraordinary forces or conditions. If successful, he will be able to show that certain sorts of genetic architectures and genetic programs are unlikely to arise or prevail *in virtue of the general characteristics of complex genetic systems*. While Kauffman, thus far forth, need not quarrel with Wallace's way of characterizing evolution as change in the frequencies of genetic programs, he claims to illuminate the architecture and structure of those genetic programs that are available to evolution. To this extent, his tactic is to seek appropriate simplifying assumptions about the structure and mechanisms of complex genetic systems that will allow the treatment of evolution and development within a (partially) unified framework.

Wimsatt, too seeks such a framework, but he attempts to forge it in a different way. He begins from certain phenomena of development – particularly those encapsulated in his reading of von Baer's law of development, with its emphasis on development as proceeding from the general to the particular, from the better entrenched to the less entrenched. Such phenomena form the fixed point for his model of the developmental lock, a model (or, better, class of models) that allows an abstract treatment of a great variety of cascading sequences of choices, controls,

214

and environmentally triggered interactions. Thus although Wimsatt, too, is concerned with the architecture of the developmental system, rather than beginning from the architecture of the underlying genetic system, he derives his simplifications from a general characterization of the sequencing of, and formal interrelationships among, the stages or steps taken in development. Indeed, his notion of *generative entrenchment* allows him to abstract from the details of the genetic (and other) mechanisms underlying development, while his faithfulness to the phenomena described in von Baer's law provides an anchor to supply real content to his models. Whether, and how, they connect back to evolution will be the subject of discussion below.

Wallace

Prof. Wallace's position, as I understand it, rests fundamentally on the Weismannian view that the germ line is to a very large extent unaltered by somatic and environmental influences – except that the premature death of the soma results in termination of the germ line. This position requires a number of qualifications, whose importance will be discussed below. Wallace accepts most of the qualifications to be discussed, but his assessment of their importance for the issues in hand differs from mine. For example, in sexual reproduction (and, in various special circumstances, in asexual and eucaryotic reproduction) various forms of genetic recombination and exchange of genetic material take place. In some instances, environmental cues trigger the exchange or shuffling of genetic material. One such class of environmental cues are those that lead to mating in sexual organisms. Another are the triggers of the so-called rec A system in *E. coli*.[3] This latter is, roughly, an "emergency scramble" system that is triggered when a cell is in serious trouble due to a hostile chemical environment, lack of nutrition, and so on. It leads to systematic, partly random and partly controlled scrambling of the bacterium's DNA. It is thought that this process, though it often leads directly to cell death, increases the odds of finding a way ot survive in a hostile environment. (There are alternative hypotheses, however.)

In support of Wallace's position, it is important to recognize that the environment does not have any direct control over the outcome of such recombination, or of the germ line result of matings. In thise sense, the processes just alluded to are properly viewed as endogenous to the germ line even if they are triggered by environmental

---

[3]   There is a brief discussion of this system in Campbell (1985), pp. 147 ff. A good technical review it Little and Mount (1982).

cues. So far, so good. Nonetheless, on an evolutionary scale, the clean separation of germ and somatic lines is undermined by the consequences of sexual crossing and, more especially, of various sorts of genomic reorganization and scrambling.[4] My argument to this effect will not depend on the dubious claim that there is feedback from the somatic to the germ line. Rather, I will show that the mechanisms governing genomic reorganization in development can come into play in the germ line and are likely to have significant evolutionary consequences.

I do not mean to prejudge factual questions; whether or not mechanisms like those governing DNA reorganization in development play an important role in germline evolution cannot be settled from a philosopher's armchair. But if one supposes, as Wallace does, that the study of development will shed little light on genetic evolution, one may fail to ask questions or to undertake experimental investigations of considerable importance to our understanding of evolution.

My primary objective, then, is to remove theoretical blinders by showing that the Weismannian basis of Wallace's argument does not justify rejection of all current attempts to unify the study of development and of evolution. More specifically, current biological knowledge makes it plausible that the mechanisms of DNA reorganization now being studied by developmental biologists (among others) may be of considerable evolutionary importance.

My strategy will be, first, to raise doubt about the decisive importance of the lack of feedback from the somatic to the germ line, second, to indicate some of the molecular phenomena that might support the relevance of current studies of development to the genetic theory of evolution, and, third, to hint at an account of the potential relevance of studies of development to this issues.

Wallace maintains that unsuccessful genetic programs terminate, while successful ones continue in the web of descent. But recombination and sexual crossing weaken this claim. Sexual reproduction, to take the extreme case, imposes the "cost of meiosis" (Maynard Smith, 1978; Williams, 1975); at each mating there is a stochastic replacement, on average, of one-half of each of the two programs in question.[5] Less extreme forms of recombination, replacement, and shuffling of the

---

[4]   Molecular genetics has revealed a startling number of complex mechanisms with the potential to scramble or reorganize the genome. A few of these are surveyed in articles contained in Maynard Smith (1983). Two technical volumes that go into some detail are Dover and Flavell (1982) and Nei and Koihn (1983). A readable survey of some of the salient phenomena is offered by Hunkapiller et al. (1982). I hope to provide an outsider's account of some of these complications in an article currently in progress, tentatively entitled "The Current Revolution in Molecular Genetics."

[5]   There are many mating systems in which things are not so simple – e.g., the haplodiploid system of some insects. However, the complications that such systems present do not significantly alter the argument.

216

genome work more slowly – mutation most slowly of all. But *over evolutionary time* genetic programs will very seldom remain intact and unaltered – and the precise ways in which they are altered depends on the mechanisms governing the reorganization of DNA, at least if what follows points in the right direction. Even if environmental factors do not guide germ line genomic change, but only trigger it, on an evolutionary as opposed to organismic scale there must be feedback between environmental and somatic change on the one hand and the rate and character of genomic change in the germ line on the other. This is my first point.

The second point requires a brief (but lighthanded) excursion into molecular genetics. It will suffice to show that there is often genomic reorganization during developent and that, viewed in terms of the mechanics of DNA interactions, the mechanisms that appear in development are of quite general importance.

One of the key features of the recent molecular findings is the dynamic character of the genetic material.[6] DNA turns out to behave in ways that are far more complicated than were dreamed of fifteen years ago. By virtue of a great variety of mechanical tricks, DNA can respond to a variety of stimuli by reconfiguring itself, by altering the relative position of particular segments, by assuming distinctive configurations that make it sensitive to the action of particular enzymes or protect it from their action, and so on. One such trick involves the use of sequences of nucleotides that are inverted repeats of one another; these pair together to form a characteristic stem and loop configuration that traps any intervening DNA. The entire package can then be excised as a unit and transported to alternative positions in the genome.

Many respects of the fundamental linear order that was expected in virtue of the findings of Mendelian transmission genetics are preserved by the new findings – most proteins, for example, are coded for by nucleotide sequences that read in the same order as the amino acids of the protein. But this DNA is often not contiguous. In eucaryotes (i.e., organisms with nucleated cells), for example, almost all of the DNA that codes for protein is interspersed with interruptions ("introns") which, although they are transcribed (copied onto RNA), are excised from the transcript while it is still in the nucleus. For this reason, introns yield transcripts that do not normally leave the nucleus and are not normally "translated" into (i.e., do not serve as a template for the formation of) protein. Accordingly, the linear order of the nucleotides in the DNA does not correspond exactly with the order of the amino acids in the protein(s) for which it codes – at least not until the interruptions are removed. Although introns may play a regulatory role, e.g., in mediating the transportation of the protein-coding portion of the RNA transcript within which

---

[6]   Cf. the references listed in n. 4.

they are embedded, their precise roles are a topic of continuing debate. Yet it is clear that they play an evolutionarily significant role; an understanding of their various functions in the mechanics of DNA rearrangement, transcription, and translation suggests the possibility of weak (but evolutionary important) connections between the dynamic character of DNA in somatic cells and evolutionary changes in germ cells.

An extended example suffices to enforce this point. It concerns the transformation of the genetic material that codes for antibody proteins in mammals. The fertilized egg of a mammal (e.g., a human being) does not contain a separate segment of DNA coding for each of the aproximately one million immune proteins that will be present in the adult organism. Rather, it contains a far more modest number of segments of DNA organized into an extremely complex multigene family.[7] Very roughly, during the course of development some sort of signal triggers a systematic rearrangement of the DNA belonging to this multigene family in the precursors of those cells that are fated to become antibody manufacturing cells. Depending on the class of antibody proteins, there are four or five relevant "domains" (functionally distinct regions) in an antibody protein. Corresponding to each domain there are multiple, variant segments of DNA present in the gene family. Each precursor cell goes through a phase in which its DNA is reorganized in a partly controlled, partly stochastic way. The result of this rearrangement is that a new sequence of DNA is constructed in each precursor cell such that that cell and its descendents have a properly assembled and activated gene for manufacturing an antibody. This means that in the resulting cell lineage each cell has a distinctive DNA molecule that is properly organized so that (1) one representative sub-gene is present for each domain of the antibody protein, (2) these subunits are placed along the DNA molecule in the correct linear order, (3) they are associated with an appropriate "start" signal such that, in appropriate circumstances, transcription begins at the site of the newly assembled gene, and (4) they are separated by introns, spacers, and joining segments in such a way that transcription and tanslation, and, ultimately, the folding and functioning of the protein will proceed properly. In different cell lineages, there are different DNA molecules coding for antibody proteins, each constructed by a similar procedure from the components of the original gene family. Furthermore, to ensure the manufacture of a sufficient variety of antibody molecules, the DNA coding for those critical regions of the antibody protein that are involved in recognizing and attaching to antigens goes through a limited and controlled period of (apparently) random mutation. This entire process ensures the diversity of antibody molecules – and also provides as clear a

[7]   Hunkapiller et al. (1982) provide a clear exposition of the structure of the antibody gene families.

218

demonstration as one could seek of the fact that *the genetic information contained in the adult was not present in the fertilized egg. The systematic rearrangement of DNA within controlled regions and in controlled ways significantly alters the information content of the genetic material during the course of development.*

This example may be extended into an evolutionary context. When this is done, one can see that the mechanisms for reorganizing and rearranging DNA sometimes operate in germ lines. When they do, they have a systematic effect on the combinations of genetic units explored during the course of evolution. Close analysis of the functional domains of the antibody proteins (and the corresponding DNA) show that *those functional domains are evolutionarily homologous.* That is, all four or five major pieces of DNA entering into the manufacture of antibody proteins are ultimately derived from a common ancestor. This is common in a large number of the proteins that have been examined closely; they are stitched together by combining segments that were once functionally independent. And the DNA that codes for those proteins is also stitched together out of coding segments separated by introns in such a way that each of the coding segments corresponds to a functional domain of the current protein.

Although the evidence is a little bit shaky, this construction appears to reflect the fact that whole segments of DNA, coding for protein domains, have been relocated within the genome. Such segments, then, are evolutionary units. Thus it must often be true that the DNA segments coding for separate functionl domains in a contemporary protein were originally independent of one another (except in those instances when they are modified duplicates of each other) and were joined together, over evolutionary time, by some sort of adaptive process. The fact that introns characteristically separate functional domains in a number of important gene families suggests that the ''evolutionary shuffling'' of the genome has exploited the mechanics of excision and transposition that are now playing such an interesting role in molecular approaches to development.

This extended example yields two morals. First, recent discoveries concerning DNA mechanics reveal an immensely powerful, highly evolved, and extraordinarily complex system of controls for the programmed rearrangement of DNA. The mechanics of this programmed rearrangement do not depend on the peculiarities of particular cells (such as the precursors of antibody-secreting cells), but on ubiquitous features of DNA – e.g., that inverted repeats will, in the right circumstances make a stem and loop configuration that can be excised, transported, and reinserted into the same, or a different, DNA molecule. The second moral is that the very same mechanics enter into evolutionary histories. Prof. Wallace argues that there is no feedback from the somatic line into the germ line. I argue, in contrast, that molecular studies of the changes that take place in development have shed considerable light on the *general* mechanics of DNA rearrangement. The resultant rearrangements can affect the structure of DNA in the germ line and, on the available evidence, almost certainly do so. To this extent, the issue of feedback from somatic to germ cells is a red herring. The fundamental question is whether the mechanisms that are found in development enter importantly into the evolution of genetic programs.

Speculative scenarios of the sort that I have suggested may well prove wrong; the extreme evolutionary conservatism of the genome warns against exaggerating the importance of the mobility of various genetic units. In the end, the questions to be faced in this connection are factual. They deserve serious theoretical and experimental investigation. Professor Wallace's argument, though correct as far as it goes, does a disservice by needlessly discouraging such investigations insofar as it suggests that one cannot find a unifying theory comprehending the mechanisms of (DNA) evolution and those of development.

In the light of these findings, if we are to find a powerful theory on the basis of which to unify the treatment of development and evolution, it will have to work at a fairly high level of abstraction. This is characteristic of the approaches taken by Kauffman and Wimsatt. Yet there are rather considerable and quite interesting differences in their starting points. To bring out various similarities and differences between them, it will be useful to consider their two positions within a single section.

Kauffman and Wimsatt

Kauffman employs a very natural and promising abstraction as his starting point. Beginning from an overview of the character of the genome, he employs some radical simplifications, and comes up with a fundamental classification of the ways in which genes or, better, functional units of DNA, interact during the development of the organism. His threefold division – genes which code for an extra-nuclear product, genes which turn on, turn off, promote, or inhibit their neighbors, and genes which turn on, turn off, promote, or inhibit distant genes – allows one to construct, as it were, a generic mechanics. Primitive as it is, such a mechanics allows the derivation of some quite general claims – one can derive rules that tell a fair amount about how (virtually) any system of a reasonable size, constructed along these lines, would behave. And by following a series of natural steps which fill in more and more details about the construction of a particular genetic system, one can get increasingly refined predictions about the behavior that such a system is likely to exhibit in the absence of interfering phenomena. Since selection is one of a small class of interfering phenomena, this tool, properly deployed, should allow one to identify those systems which behave somewhat oddly, to focus on those, and to ask quite concretely what factors or forces make those systems depart from the generic or expectable properties that their general construction as genetic systems would lead one to expect. If this approach works, it should yield a first or second order analysis of the genomic system for controlling development on which natural selection acts – an analysis which, to my knowledge for the first time, promises to enable one to assert with reasonably justified confidence that certain systemic genetic features of the developmental system are due to selection while others are not.

At present, the principal power of Kauffman's approach is found in his theorems about the generic structure of genetic systems. The spirit of these theorems is rather parallel to that of R. A. Fisher's fundamental theorem of natural selection, which states that the rate of evolution is proportional to the additive variance in fitness. Fisher conceived that theorem as a general law of very much the same character as the second law of thermodynamics. Derived from abstract considerations, it applied to all systems of a Mendelian sort. It was derived using certain simplifying assumptions (e.g., that one was dealing with an infinite, panmictic population) which might not be entirely correct in real populations. The differences between real populations and the idealized ones for which the theorem was rigorously proved could then be taken care of by a series of first, second, and third order corrections. Meanwhile, the idealized case gave one a law that, subject to such corrections, pertained to all Mendelian systems.

Let me strike a note of caution. One consequence of Fisher's fundamental theorem is that the fitness of a population is always increasing on average. This may be true for infinite, panmictic populations, but when one takes into account the finitude of real populations, stochastic effects, environmental change, and a variety of other phenomena, fitness does not increase, on average, in all populations. Extinction is not always random, or due to catastrophes or loss of habitat. Thus, first order corrections alter at least one of the fairly fundamental consequences of Fisher's theorem. We must be prepared for a similar contingency in the elaboration of Kauffman's theory.

What is striking about the character of that theory is that it ascribes a fundamental underlying structure to virtually all genetic systems and derives its theorem from very elementary features of that structure. Furthermore (with the possible exception of some of the smaller viruses that violate the size constraints entering into Kauffman's calculations), so far as we know virtually all genetic systems conform to the elementary constructional features on which his models are based. The result is an abstract and general theory which, indeed, lends the same sort of abstract generality to Kauffman's as to Fisher's work.

Wimsatt proceeds from an entirely different set of abstractions rooted in a problem-solving analysis of development, characterized phenotypically. He treats developmental programs as solutions to the problem of building an organism suited to its environment, starting from a single cell. His developmental lock is independent of the mechanics of development and is intended to show that certain *strategies* for solving this problem are orders of magnitude more efficient than others. Other things being equal, an organism that uses the more efficient strategy is more likely to develop properly and to have adequate resources left at its disposal than one that uses an inefficient strategy. For this reason (assuming that Wimsatt's heuristics are sound), selection will favor those developmental programs that are well modelled by the developmental lock.

Issue	*Kauffman*	*Wimsatt*
Starting Point for Abstractions	Genetics. "Generic Mechanics" of genetic systems.	Heuristics. von Baer's Laws and the phenomena of generative entrenchment.
Descriptive Power	Weak. Genetic Architecture well described. Phenotypes and links to phenotypes undescribed.	Strong. Phenotypes, environmental features, and developmental stages classified as generatively important or unimportant.
Explanatory Power	Strong. Powerful "generic mechanics."	Weak. Generative entrenchment not grounded in underlying mechanics.
Evolutionary Implications	Strong. Internal dynamics of genetic systems have powerful evolutionary consequences.	Arguable. See text.

I am more inclined to find difficulties with Wimsatt's approach than with

Kauffman's. To motivate the (constructively intended) criticisms offered below, it will be useful to set forth a few comparisons of their approaches. These are listed in Table 1, on which the following comments are based.

Perhaps the best starting point is the claim that the explanatory power of Wimsatt's developmental lock is weak in comparison to that of Kauffman's genetic regulatory systems. Wimsatt himself, speaking of canalization, says that his "model does not explain canalization (one needs a model like that proposed by Kauffman in this symposium for that purpose), but it does allow us to model it" (p. 221). The reason for this is quite general, and is not restricted to the phenomenon of developmental canalization. While Kauffman's models pertain to the underlying mechanics of *genetic* change, Wimsatt's pertain to developmental programs of whatever type, whatever their source. (It is for this reason that he can treat features of the physical environment as generatively entrenched in developmental programs even though they are not "genetically innate" or otherwise inborn.) To drive home the importance of this point, it will be useful to emphasize one of Mayr's comments about the hierarchical nesting of developmental programs:

> A particular instinctive behavior act is, of course, never controlled directly by the genotype but rather is controlled by a behavior program in the nervous system that resulted from the translation of the original genetic program. It is particularly important to make this distinction for the open program. The new information acquired through experience is inserted into the translated program in the nervous system rather than into the genetic program because, as we know, there is no inheritance of acquired characters (Mayr 1976, p. 696).

As this quotation makes clear, a "behavior program" – and, for that matter, a developmental program in general – need not be a genetic program. Indeed, the stable features of the somatic or physical environment (e.g., a nest) that enter into one of Wimsatt's developmental programs may be at a great remove from the genes; in some cases, they may not in any sense be programmed into the genes (though they may in some sense be presupposed by the genetic program). Thus, it is a fact about greylag goslings that the first moving being larger than a gosling to make a sound in their presence is normally their mother; on Wimsatt's model this fact is part of the developmental program of greylag geese. But it is in no sense part of what is programmed into the genes of the greylag zygote.

This illustrates the independence of Wimsatt's developmental lock models from the underlying mechanics of development. It also explains the great descriptive and weak explanatory power of those models: generative entrenchment of phenomena (endogenous or environmental) can be gauged by an examination of their effects, but the grounds of that entrenchment and the consequences of distant changes on the effects of the entrenched feature – these are not built into, clarified by, or explained by the developmental lock.

This characterization brings with it, I believe, a problem of great importance to Wimsatt's overall project. The issue is difficult, and I am not entirely sure of my ground, but it appears that the developmental lock *in its general formulation* has virtually no evolutionary consequences. The reason for this is precisely the one that Mayr and Wallace focus on: there is no inheritance of acquired characteristics, at least to the extent that *genetic* programs of development are insulated from direct effects of the environment. If, thanks to Konrad Lorenz, a particular greylag goose imprints on something other than mother, the only effect on the *genetic* developmental programs of greylag geese is the failure of that goose (or, better, its genes and, thus, its developmental program) to be represented in the next generation. *Unlike DNA mechanics, including Kauffman's "generic mechanics," Wimsatt's developmental programs are not, in general,* genetically *heritable.* What this means for their role in evolution is terribly unclear, precisely because they include an extraordinary diversity of factors which are "inherited" from generation to generation in extraordinarily diverse ways, governed by extraordinarily diverse rules. At the very least, those rules are not the ones governed by the genetic theories of the current evolutionary synthesis.[8]

In terms of the historical sketch presented at the beginning of this paper, Wimsatt's developmental lock draws, effectively, on none of the resources of molecular biology. This need not, of course, be a drawback. However, insofar as an understanding of genetic evolution requires an intrinsic characterization of the genes (or the genetic material), one would anticipate inadequacies in generalized models like the developmental lock. The abstractions with which this class of models begin are very distant, indeed, from genetics.

This is not to deny the power of the abstractions built into models based on the developmental lock; rather, it is to claim that, at the very least, those models have to be specified further if they are to pertain to distictively genetic evolution. There is no reason not to attempt such a specification, and some reason to be optimistic about its prospects. There are, in turn, at least two directions in which such a program might be attempted. The first is to build up a hierarchy or heterarchy of interlocking developmental locks. There are at least two directions in which this is biologically plausible. As innumerable studies of development have shown, in metazoans the typical developmental pattern of a cell lineage fits developmental

[8]   This point of view was driven home to me by discussions at the Mountain Lake Conference on Development and Evolution. There are some echoes of it in the paper produced by that conference, Maynard Smith et al. (1985), particularly in the passages that motivate the tendency of the authors to assimilate all developmental constraints to genetic constraints.

224

lock models with considerable elegance. (One of the strengths of Kauffman's models is that they yield this phenomenon as a natural expectation.) Typically, a given cell type can switch fairly easily to at most three or four cell types – and, typically, once such a switch is thrown it is irreversible. But there are *many* cell lineages in a metazoan, and there is at lest some independence of development between the different lineages. Viewed in this way, the development of an organism is not modelled well by a single developmental lock, but might be better modelled by a moderately large number of interacting developmental locks with a certain amount of feedback among them. The modelling difficulties involved may well be severe, but the gain in biological realism is considerable.

Another natural way of utilizing interacting developmental locks captures considerably more of Wimsatt's insights about the non-genetic character of the generatively entrenched features driving a number of developmental programs. As the passage quoted from Mayr suggests, fleetingly, there are behavior programs built into the architecture of the nervous system (but also, I would add, morphogenetic programs determined or altered by the structure of the developing tissues and organs of a developing organism). Some of these, at least, once set up, exhibit considerable autonomy from the genetic programs that initially guided the formation of those systems, tissues, and organs (or their primordia). (One should add all the necessary caveats here about the epigenetic character of *all* development, but that does not alter the point.) Thus, there are also (partially?) *non-genetic* developmental programs interacting with genetic ones; many, perhaps most, of these programs will be well modelled by one developmental lock or another. Supposing that such a scheme (which will obviously be difficult to execute) were to work, it would remain a large and interesting question whether – or, perhaps, under what conditions – interacting developmental locks yield an overall pattern of development that is also well represented by a developmental lock.

The second natural extension of Wimsatt's models might be to apply them in interaction with, or as providing additional constraints on, such genetic models as those which Kauffman is developing. The difference between this suggestion and Wimsatt's own suggestion of combining his models with those of population genetics is very much like that between modelling life history strategies and performing molecular studies of the genetics of aging – both are legitimate enterprises, but they are likely to yield different products. In order to shed fundamental light on the evolution (and genetics) of development, I suspect that one must come to grips, somehow, with the *intrinsic* structure of the genetic system and the machinery employed in constructing the organism. For this purpose, little will be gained, I think, from the application of the developmental lock in population genetics (fruitful as that will surely be for other purposes).

I should add that I think that Kauffman's models would benefit from the attempt to constrain them by use of models built with the developmental lock. Wimsatt's comment (5), p. 221, is entirely apposite: there is at least some evidence that the genetic packages that control early development are, themselves, deeply generatively entrenched and extremely hard to alter. In terms of Kauffman's primitive characterization of genes as cis-acting, trans-acting, and making extra-nuclear products, these "packages" (my term) are highly evolved, extremely complex, and extremely well-protected structures. It is an interesting and important enterprise to attempt to understand how such cohesion is possible – and this is one of the things that Kauffman is doing with his models. Another useful enterprise that those models should allow one to undertake is to ask what the presence of such protected gene complexes or functionally cohesive genetic units as are apparently pertinent to early development does to the evolution of the architecture of a genetic system. And this question might well be aided by use of constraints derived from one version or another of Wimsatt's developmental lock.

Concluding Remarks

Does the argument presented in this paper allow one to draw any interesting conclusions regarding the best means of integrating scientific disciplines? Perhaps someone wiser than I could draw some clear and powerful morals, but the best I am able to do is to provide a few fairly weak rules of thumb. Others will have to evaluate their general usefulness.

(1)   Abstraction is a necessity. Whatever else may be involved in unifying or integrating distinct disciplines, the power of the integrating perspective, theory, or models is tied to the abstractions from which one begins.

(2)   In the formative phases of disciplinary integration, pluralism is of great importance. That is, the perspectives of each of the disciplines, the phenomena that they encompass, and a large number of alternative models and approaches must be kept in mind. Although preliminary evaluations of many of these approaches is usually possible and usually negative – most of them, after all, will fail to provide a lasting or satisfying integration – a number of them should be driven hard, if for no other reason, then in order to allow the formation of a comparative perspective allowing assessment of their relative strengths and weaknesses, accomplishments and failures. (Part of my concern about a stand like Wallace's is that it violates this rule of thumb.)

(3)   Where possible, models should attempt to be fully integrative. Thus, if the disciplines to be integrated include evolution, genetics, and development, even

though many or most of the specialized perspectives, results, or techniques of those disciplines will be set aside in the leading simplifications or abstractions on which the attempted integration is based, nonetheless, those leading abstractions should incorporate elements or perspectives from each discipline. (My criticisms of Wimsatt's developmental lock models turned, effectively, on an estimate that he had incorporated too little knowledge of contemporary molecular (as opposed to population) genetics into his developmental lock models.) Failing that,

(4) the next best choice, where possible, is to start from common or underlying mechanisms (as Kauffman's models do). In such cases and, indeed, generally,

(5) when constructing models one should start, wherever possible, from a general class of models which allows sequential specification and specialization, optimally on the basis of real data. Both Kauffman's and Wimsatt's families of models exhibit this virtue; they both have "natural" parameters that can be specified on the basis of empirical investigation of such matters as fitness effects of mutations that are expressed at different stages of development and the connectivity of particular genes. Indeed, one of the primary reasons for thinking that both of these families of models hold considerable promise is the likelihood that they can be tested and improved by a series of successive approximations in interaction with empirical determination of some of their key parameters.

Finally, (6) evaluation of the distinct approaches should not only be comparative, it should also be tightly connected to the particular objectives of the attempted integration. As a specific example, I will conclude this essay with a few words about Wimsatt's approach to the separation of the concepts of generative entrenchment and innateness.

Wimsatt argues persuasively that contemporary ethology has confused a number of distinctions in opposing "innate" and "acquired" traits. He usefully separates two main components in this mixture – a "genetic" one, connected with the idea that what is innate is inborn, and an "ethological" one, connected with generative entrenchment of traits affecting the phenotype. He is probably right that generative entrenchment is the more useful concept in ethological contexts, where the distinction between "open" and "closed" programs of behavior is of great importance, and where the developmental program (so long as it is stable) is inseparably epigenetic. One of the very many contexts in which the rich notion of generative entrenchment that Wimsatt is developing is likely to prove of use is in the integration of ethology and evolutionary theory. *With respect to a general integration of the disciplines of development and evolution*, however, I find his argument that generative entrenchment is the appropriate substitute for confused uses of "innate" and "acquired" unpersuasive. While it is surely true that development is thoroughly and completely epigenetic, it is also true that it is driven by *both* an inherited genetic system and the interaction of that system and its products with a nested series of wider environments. In light of the genetic theory of inheritance – which virtually no one involved in this work wishes to challenge – the genetic endowment of the zygote (however much it undergoes programmed change in development) constitutes one of the primary bases for the desired integration of development and evolution. Precisely because the concept of generative entrenchment is powerless to distinguish between genetic and non-genetic components or aspects of the (phenotypic) traits which are more or less generatively entrenched, Wimsatt's conceptual base is, on the available evidence, unlikely to serve as an adequate tool for the integration of genetics, evolution, and development. Or rather, more charitably, it is incumbent on him to show us otherwise.

Acknowledgments

This paper has been improved by discussions at the Conference on Integrating Scientific Disciplines and the Mountain Lake Conference on Development and Evolution. I am indebted to Bill Bechtel, Stuart Kauffman, Anne McNabb, Bruce Wallace, and Bill Wimsatt for specific criticisms, and to colleagues too numerous to mention for helpful discussion.

References

Bateson, William (1979). *Problems of genetics*. New Haven: Yale University Press. Reissue. Original 1913.

Campbell, John (1985). An organizational interpretation of evolution. In Depew and Weber (1985), pp. 133–167.

Depew, David and Weber, Bruce (1985). *Evolution at a crossroads: The new biology and the new philosophy of science*. Cambridge, MA: MIT Press.

Dover, Gabriel and Flavell, Richard (Eds.) (1982). *Genome evolution*. New York: Academic Press.

Hunkapiller, Tim; Huang, Henry; Hood, Leroy; and Campbell, John (1982). The impact of modern genetics on evolutionary theory. In Milkman (1982), pp. 164–189.

Hutchinson, G. Evelyn and Rachootin, Stan (1979). Historical introduction. In Bateson (1979).

Little, J. W. and Mount, D. W. (1982). The SOS regulatory system of *Escherichial coli*. *Molecular and General Genetics, 105*, 74–83.

Maynard Smith, John (1978). *The Evolution of sex*. Cambridge: Cambridge University Press.

Maynard Smith, John (Ed.) (1983). *Evolution now*. San Francisco: W. H. Freeman.

Maynard Smith, John; Burian, Richard; Kauffman, Stuart; Alberch, Pere; Campbell, John; Goodwin, Brian; Lande, Russell; Raup, David; and Wolpert, Lewis (1985). Developmental constraints and evolution. *Quarterly Review of Biology, 60*.

228

Mayr, Ernst (1976). Behavior programs and evolutionary strategies. In Mayr, *Evolution and the diversity of life*. Cambridge, MA: Harvard University Press, pp. 694–711. Reprinted from *American Scientist,* **62,** (1974), 650–659.

Milkman, Roger (Ed.) (1982). *Perspectives on evolution*. Sunderland, MA: Sinauer.

Nei, Masotoshi and Koehn, Richard (eds.)(1983). *Evolution of genes and proteins*. Sunderland, MA: Sinauer.

Sapp, Jan (1984). Cytoplasmic inheritance and the struggle for authority in the field of heredity, 1891–1981. Unpublished dissertation, University of Montreal.

Williams, George (1975). *Sex and evolution*. Princeton: Princeton University Press.

Editor's Commentary

While the evolutionary synthesis discussed in the previous section served for the most part to draw disciplines together, its effect with regard to developmental biology seemed to be one of segregation. By hardening into orthodoxy the Weismannian distinction between the germ line and the somatic line, it separated the two domains. Evolution considered what happened to the germ; developmental biology studied developments in the somatic line. In terms of Mayr's (1982) distinction between proximate explanations and ultimate explanations, developmental biology is concerned with proximate explanations but cannot address the ultimate question of origins. The notion that there is a connection between development and evolution, however, is an old one that was actively explored in the 19th century (by von Baer, 1828, and Haeckel, 1866) and in the early parts of the 20th century (by Garstang, 1922, and de Beer, 1930). Only recently has the interconnection between development and evolution again become the focus of much attention. Gould's *Ontogeny and Phylogeny* served to draw attention to some of the older issues at the boundary between evolution and development.

The split between evolutionary theorizing and developmental biology has been accompanied by another split, that between genetics and developmental biology. While questions of development and inheritance were closely linked at the turn of the century, the two diverged dramatically after the rediscovery of Mendel's laws and the development of the Morgan school, which produced a fruitful study directed at a "theory of the gene" independent of considerations of development. The older view, that what was inherited was an organized structure that differentiated in further development, was supplanted by the idea that particular units, genes, were the material of heredity which could be studied separately (Maienschein, 1985). Developmental biology, meanwhile, pursued its own concerns, for example, identifying the causal factors governing the development of major organ systems (Hamburger, 1980). While there were numerous calls to reunite studies of heredity and development through the 1930s, most had little positive

Bechtel, W (ed), Integrating Scientific Disciplines. ISBN 90-247-3242-5.
© *1986, Martinus Nijhoff Publishers, Dordrecht. Printed in The Netherlands.*

effect. Morgan's 1934 book *Embryology and Genetics* discussed both topics, but independently, revealing how separated the two disciplines had become. Greater ground for unity appeared during the 1930's (see Churchill, 1980, on de Beer) and after. The role of different genetic processes in determining the course of development became an active topic of interest in the theorizing of Goldschmidt (1940) and Waddington (1940 and 1957), among others (Gilbert, 1985). With the development of molecular genetics, the ability to study the process of gene expression and theorize about gene regulation made this a promising area of inquiry.

There is another aspect of the relationship between evolutionary biology (viewed as incorporating genetics) that provides the focus for these papers. This is the question of how developmental processes can be informative about evolutionary processes. As Burian notes, evolutionists have always been aware that selection works on the full life cycle of organisms, not just their adult forms, but it is only with the rise of molecular genetics that fuller investigations in this area have been possible. Recent knowledge about the complexity of the control processes governing gene expression has raised new questions about the potential relevance of developmental biology to the task of understanding the process of evolution. There remains a serious question, though, as to whether increased molecular understanding of the mechanism controlling gene expression will lead to a worhtwhile integration of developmental biology and evolutionary biology. The authors have disagreed on the issue. This, in fact, is one of the interesting features that makes this case ripe for analysis. As I argues in the Introduction, there is limited interest in integration for integration's sake. Before it is worth building bridges between disciplines, there must be something that can hoped to be learned from the activity. The sceptics deny that there is anything to be gained, at least for evolutionary theorizing, by focusing on developmental biology. Of course, one needs to ask what grounds the scepticism. It may be a matter of protecting one's turf – of evolutionary biologists not wanting the tools and theoretical frameworks in which they have invested their intellectual resources to be deflated by the introduction of new tools and new frameworks that would follow from integrating developmental interests into the synthesis. While some such motivation might underlie the negative appraisals of integrating developmental biology more into the evolutionary synthesis, the cases presented do not suggest that motivation. Rather, they are based on an argument that developmental processes do not figure centrally in the processes of concern to evolutionists.

The negative position on an integration of developmental biology into the evolutionary synthesis is developed in Wallace's paper. In it Wallace articulates and defends the view put forward by the evolutionary synthesis according to which development depends upon what is provided by evolution, but evolution is not influenced by the mechanisms governing development except indirectly insofar as they contributed to differential survival and replication. Germ cells determine somatic cells, but not vice versa. The conclusion that seems to follow from this is that the processes of concern to developmental biologists must utilize what is given by evolution, but have no impact on evolution (except as the developing organisms are subject to selection) and so are not of direct interest to evolutionary theory.

What Kauffman, Wimsatt, and Burian try to do is show that developmental processes end up having consequences for, and thus must be considered by, evolutionary studies. Before drawing such implications, however, they are offering new perspectives on developmental processes. In this respect their approach is different from what is seen in the other cases of interfield theorizing, for the new perspective is primarily being developed within one field and then shown to have impact on another.

Kauffman's developmental theorizing leads him to propose a revision to an important constraint on evolutionary models, namely the specification of the null hypothesis of what would happen without selection. Kauffman begins with a problem posed by developmental biology that would seem to require an evolutionary answer: the existence of a very complex regulatory system to control the expression of genes seems to require extensive selection to be maintained across, especially since evolution would, if anything, seem to disrupt the process. Kauffman resolves this puzzle by showing that a generic model of the developing system points to a surprising resolution of the question. A logical-mathematical characterization of such regulatory systems shows that such systems would tend to move into a few statistically typical patterns that would remain stable. In fact, selection forces would have a difficult time moving such a system out of one of these stable conditions. This answer, however, provides a new specification of the null hypothesis of what happens if there is no selection – a hypothesis that needs to be taken into account in evolutionary theorizing.

Wimsatt's approach is quite different. He begins with a generic model of what a developing system must be like: it is a system that goes through a series of stages, where it must realize the right structure at each stage in order to be able to proceed to the next. To account for this he introduces his schema of a developmental lock, in which development is conceptualized as imposing requirements as would a lock consisting of a series of dials, where what is indicated as the correct setting on each successive dial is altered by the setting of the previous dial. He then tries to use this model of development to make sense of a controversial distinction frequently used in evolutionary contexts, the distinction between innate and acquired traits. Wimsatt contends that his model of a developmental lock provides a notion closely

232

related to innateness, which he calls "generative entrenchment." This notion, he maintains, captures much of what ethologists have meant when they speak of a trait as innate, as reflected in their criteria of universality in a species, early manifestation in ontogeny, and widespread distortions if it malfunctions.

The connection between Wimsatt's conceptualization of development and evolutionary thinking is not limited to clarifying the use of the innate-acquired distinction in evolutionary biology. Wimsatt contends that the notion of generative entrenchment can account for many features of evolution, including the greater evolutionary conservatism of earlier developmental stages; the more pervasive consequences of earlier modifications; the persistence of 'frozen accidents'; the greater taxonomic, morphological, and functional generality of earlier developmental features; and the rarity of biological revolutions. Another important connection between development and evolution for Wimsatt stems from the difference between generative entrenchment and innateness. Wimsatt allows environmental features to be generatively entrenched insofar as they play regular and critical roles in development. As a result, Wimsatt's notion is not able to serve one traditional function of the concept of innateness, that of differentiating between traits "preformed" in the genome and those shaped by the environment. However, Wimsatt maintains that evolutionary thinkers were conceptually confused by joining the notion of genetically controlled to the features associated with being generatively entrenched. This has important implications for evolutionary theorizing, which has commonly focused on selection for genes. Selection causally operates on the phenotype, which is a product of the generative program. If environmental factors are generatively entrenched in this program, they too, and not just the genes are subject to selection.

It is just at this point, though, that Burian is dissatisfied with Wimsatt's analysis as a vehicle for bridging developmental and evolutionary biology. Burian argues that a true integration must focus on the genetic mechanisms of inheritance. Wimsatt's integration of evolution and development is more in the tradition of von Baer and Gould in trying to employ developmental patterns to understand evolutionary patterns. It does not specifically link developmental considerations with the genetical aspects of the synthetic theory. Burian, on the other hand, points to a phenomenon in terms of which one might integrate information about the genetic foundations of the synthetic theory and information about developmental mechanisms. This is the phenomenon of genetic reorganization, which affects development and evolution equally.

Kauffman, and Wimsatt have thus each developed theoretical models designed primarily for phenomena within the domain of developmental biology but which they claim have broad implications for evolutionary biology. Burian has pointed to an aspect of theorizing about genetics in developmental biology that also promises to have implications for evolutionary theorizing. Whether any of them are right in contending that evolutionary biology must attend to such developmental processes will ultimately be determined by whether these developmental facts sufficiently alter the processes of evolution that evolutionary theorizing cannot overlook them. If they do make such a difference, then the basis for an interfield theory between the evolutionary synthesis and developmental biology will be in store. There is, however, a significant difference between this interfield theory and others that have been studied previously or earlier in this volume. In those discussed by Darden and Maull (1977), one field addressed a question to another field. In the biochemistry case considered earlier in this volume, there was a clear common problem area where physiology and chemistry were brought together. In the case of the evolutionary synthesis, there had been competition between Mendelian geneticists and Darwinian evolutionists to explain the origin of species. In this case, neither the evolutionary synthesis nor developmental biology was addressing a question to the other. There was no common problem area or a problem both disciplines were competing to answer. Rather, developmental biology can be viewed as presenting itself to the evolutionary synthesis as offering perspectives which it feels the evolutionary synthesis requires. Whether there will be an interfield theory in this case will depend on whether evolutionary biologists see this offer of service as one that can advance their enterprise, or one that will only obfuscate their endeavor.

In concluding this section, we can turn briefly to the question of what kinds of social and institutional arrangements might accompany the kinds of interfield theorizing proposed by Wimsatt, Kauffman, and Burian if they are successful. It seems unlikely that renewed interest in this cross-disciplinary area will produce major new organizational or institutional entities. This is largely because the case can be understood as simply an attempt to expand the already active interdisciplinary research cluster that emerged from the synthesis so as to give a more prominent role to developmental biology. What one can expect, and has already begun to occur, is the holding of interdisciplinary conferences to explore the new status of developmental biology within the synthesis. The Dahlem workshop of 1981 (reported in Bonner, 1982), for example brought together molecular geneticists, invertebrate zoologists, population biologists, comparative anatomists, vertebrate and invertebrate paleontologists, theoretical and mathematical biologists, developmental geneticists, and developmental biologists who focused on neurobiology, invertebrates, insects, and slime molds. Bonner indicates four sets questions that were central to that conference: how do genes control development and how can these means of control contribute to changes in the course of evolution? what is the physical basis of the timing mechanism critical to development and how are these changed through evolution? what levels of control operate above that of the gene? what kinds of constraints are there that can limit the power of selection? This list indicates that there are a variety of questions at the interface of developmental biology and evolutionary biology that can provide a basis for further joint inquiry. At least two other conferences have occurred subsequently (see Goodwin, Holder, and Wylie, 1983, and Maynard Smith, et al., 1985). In addition to further interdisciplinary conferences other modes of cross-disciplinary activity are likely, including societies and journals devoted to the connections between development and evolution and perhaps the creation of some cross-specialty committees and joint degree programs at some major universities.

References

Baer, K. E. von (1828). *Entwicklungsgeschichte der Thiere: Beobachtung und Reflexion.* Konigsberg: Borntrager.

Bonner, J. T. (1982). *Evolution and Development.* Berlin: Springer-Verlag.

Churchill, Frederick B. (1980). The modern evolutionary synthesis and the biogenetic law. In E. Mayr and W. B. Provine, eds., *The evolutionary synthesis. Perspectives on the unification of biology.* Cambridge: Harvard University Press.

Darden, Lindley and Maull, Nancy (1977). Interfield theories. *Philosophy of Science, 44,* 43–64.

de Beer, G. R. (1930). *Embryology and evolution.* Oxford: Clarendon Press.

Garstang, W. (1922). The theory of recapitulation: A critical reassessment of the biogenetic law. *J. Linn. Soc. Zool.* **35,** 81–101.

Gilbert, Scott (1985). The rivalry of genetics and embryology: 1920–1940. Talk presented at Summer History, Philosophy and Social Studies of Biology Conference, St. Mary's College.

Goldschmidt, Richard (1938). *Physiological genetics.* New York: McGraw–Hill.

Goodwin, Brian; Holder, Nigel; and Wylie, C. C. (1983). *Development and evolution.* Cambridge: Cambridge University Press.

Gould, Steven J. (1977). *Ontogeny and phylogeny.* Cambridge: Harvard University Press.

Haeckel, E. (1866). *Generalle Morphologie der Organismen: Allgemeine Grundzuge der organischen Formen-Wissenschaft, mechanisch begrundet durch die von Charles Darwin reformirte Descendez-Theorie.* Berlin: Georg Reimer.

Hamburger, Victor (1980). Embryology and the modern synthesis in evolutionary theory. In E. Mayr and W. B. Provine, eds., *The evolutionary synthesis. Perspectives on the unification of biology.* Cambridge: Harvard University Press.

Maienschein, Jane (1985). Development and inheritance: Early 20th century divorce? Talk presented at Summer History, Philosophy and Social Studies of Biology Conference. St. Mary's College.

Maynard Smith, John; Burian, Richard; Kauffman, Stuart; Alberch, Pere; Campbell, John; Goodwin, Brian; Lande, Russell; Raup, David; and Wolpert, Lewis (1985). Developmental constraints and evolution. *Quarterly Review of Biology, 60.*

Mayr, Ernst (1982). *The growth of biological thought.* Cambridge: Harvard Univerity Press.

Morgan, T. H. (1934). *Embryology and genetics.* New York: Columbia.

Waddington, C. H. (1940). *Organizers and genes.* Cambridge: Cambridge University Press.

Waddington, C. H. (1957). *The strategy of the genes.* London: Allen and Unwin.

PART IV
Extending Cognitive Science

Introduction

ADELE ABRAHAMSEN
*Language Research Center, Georgia State University, Atlanta, Georgia
30303–3083, U.S.A.*

Scholars have always been curious about cognition, and have exercised that curiosity in so many different ways that there has never been a single discipline which had exclusive rights to its study. As disciplinary structure emerged and differentiated across the 19th and early 20th centuries, the study of cognition threaded its way through much of that structure. At times, a single individual brought multiple perspectives to bear in ways that were especially fruitful; at other times, disciplinary perspectives developed in relative isolation.

I pick up the story with the decade of the 1950s. This was one of the periods of relative isolation; the study of cognition had been parceled out among several different disciplines and fields which had little interchange or cross-influence, except for a common commitment to positivist frameworks such as behaviorism in psychology. Particularly in the United States, experimental psychologists had chosen to focus on the process of acquiring new behaviors. That is, learning was the one aspect of cognition that was prominently studied, and it was held that what was learned were responses to environmental contingencies. Some theorists did try to extend the framework by positing chains of covert stimuli and responses which could, in principle, be used to model what would soom be called "cognitive processes." However, the central research task was identified as uncovering the laws of learning itself, and for this work, the animal model was most frequently chosen to avoid the clutter of human behavior. Everything else was extensions and applications.

On the Continent, animal ethologists were just as concerned as the behaviorists with empirical investigation of observable behavior, but they focused on innate action patterns which were important in regulating the life of the various species, and so this field was not yet among those working on a slice of cognition. Displeasure with these two extremes of nature and nurture underlay the development of an interationist position within developmental psychology. Jean Piaget (1950) and Hans Werner (1948) proposed organismic theories which applied

Bechtel, W (ed), Integrating Scientific Disciplines. ISBN 90-247-3242-5.
© *1986, Martinus Nijhoff Publishers, Dordrecht. Printed in The Netherlands.*

the metaphor of biological growth to the developing individual. Unafraid of mentalism, and focusing on the human species, they found a burgeoning of cognitive phenomena beginning as early as the second year of life, and attributed this to an interaction between innate genetic programs and the individual's experiences in the world. Piaget in particular developed and formalized a general model of cognitive structure and growth, drawing on and intending to influence several different disciplines (including biology, mathematics, and epistemology as well as psychology). However, due to the encapsulating tendency of disciplines, his influence by the 1950s was still largely limited to "child psychology" and education.

There were other fields which in the past had dealt with questions of cognition, but were fairly quiet in the 1950s. Linguists had occasionally treated language as an activity of the human mind and considered its relation to other mental activities, but as the decade opened, the dominant approach to linguistics (Bloomfield's descriptivism) was as behaviorist as the psychology of the time. Students of language development were often willing to consider the role of cognition, but did not have the tools to make substantial progress. Philosophy's remaining contribution (characterizing the formal structure of idealized reasoning by means of symbolic logic) was encapsulated. Psychiatry was encapsulated; social psychology was encapsulated; neurophysiology was encapsulated; anthropology was encapsulated; and so forth through the disciplines.

Then cognition became fashionable. It can only be argued what the primary impetus might have been, as so many fields changed at once, and cross-disciplinary discussion blossomed. In linguistics, Chomsky (1957, 1968) asserted that his field was a branch of cognitive psychology and proposed powerful models of the structure of mental grammars, poised to fight those who would challenge him for his mentalism. In experimental psychology, some researchers rediscovered the mind, positing such mental entities as short-term memory and sensory information stores, as well as knowledge structures which could only awkwardly be modeled using stimulus-response units as the building blocks. Perhaps more reluctantly than Chomsky, these researchers too began to argue for a kind of mentalism, and began calling themselves "cognitive psychologists." Neisser's 1967 book, *Cognitive Psychology*, defined the new field and served as its primary text for some time. A journal of the same name was established. In a related development, psycholinguistics became more prominent as it became more Chomskian. Piaget's books began to be read more widely and became a second influence (in addition to Chomsky) within the fast-growing field of language development. In anthropology, Romney and D'Andrade (1964) published "Cognitive aspects of English kin terms," and the modern field of cognitive anthropology was born. In philosophy, some departed from ordinary language analysis (the counterpart to behaviorism within that field) and resumed the old philosophical concerns with mental phenomena such as intentionality (Dennett, 1978). Outside influences on neurophysiology led to the development of the interdisciplinary field of neuroscience, which took as one of its tasks determination of the neurological underpinnings of cognition and language. Finally, in research not easily assigned to one field, a few psychologists began their demonstrations that animals could acquire at least rudimentary use of symbols, opening the door to studies of animal cognition that would affect all fields dealing with either animal behavior or cognition.

Preceding all of these developments, an idea came on the scene that changed everything: the idea of information as an abstract entity, which could be instantiated and manipulated by machines (abstract or real) as well as animate beings. At the risk of oversimplifying, I would claim that this idea was foundational for all of the other changes; the other disciplines would not have developed and burgeoned in the way they did without it. The new field of cybernetics (Weiner, 1948) could have been (and in its historical antecedents in fact was) a mere curiosity of no real use, except that powerful tools accompanied its arrival: actual physical machines (digital computers) which are now already in their fifth generation of development, and the mathematical tools of information theory (Shannon & Weaver, 1949) and automata theory (Turing, 1936) which could be harnessed towards the modeling of thought. The availability of the machines (and of appropriate programming languages) engendered the new field of artificial intelligence, in which by combination of brute force and insight on the part of their programmers, machines were made to think. In other developments, automata theory had its most influential payoff in Chomsky's (1957) work on syntax, in which he sketched how one might construct an automaton that need only be "switched on" to begin generating all the sentences of a given language (but not finish, due to the device of recursion). Information theory had some direct applications within psychology early on (e.g., Miller, 1951), which set the stage for other ways of modelling cognition. All of these developments functioned to demonstrate that the study of cognition can be made tractable; one can model cognitive structures and processes, not just the behaviors which they control.

As an interdisciplinary cluster took form, some of the key figures took it upon themselves to name and institutionalize the cluster as "cognitive science." Though many disciplines contribute to the cluster, cognitive psychology, artificial intelligence, and to a lesser extent linguistics, have been at its core. I believe the reasons for this are primarily historical (the origins in cybernetics) and sociological (the size of the fields, the collaborations that worked, and so forth). Particularly salient, though not surprising if I am correct about the key role of information technology, is the fact that biologically oriented disciplines are to be found only on the periphery of the cluster. Notable instances are the fields focused on animal behavior, such as animal ethology and animal learning, and the field of neuroscience. Yet, there is potential for both of these fields to have an impact within the cluster.

In the case of animal studies, the demonstrations of primitive symbolic capacities in apes have renewed and revitalized interest in studying cognition comparatively, an enterprise that fell into relative obscurity after the excitement about Kohler's and Harlow's apes earlier in the century. To understand cognition itself, we must decompose and recompose it many times over by looking at babies and old folks and chimpanzees and the geniuses and the mentally handicapped of various species, as well as the traditional college sophomore and the more recent machine simulations of that perenniel subject of study. This variety is a tool that nature has supplied, a guide for locating its "joints" that is more powerful than we have cared to recognize. The comparative data also provide motivation to bring evolutionary and ethological perspectives to bear on questions of why cognition is as it is for particular species or social groups. In the first paper of this unit, Sue Savage-Rumbaugh and William Hopkins show how only this kind of broad, comparative approach can yield a deep enough account of such phenomena as communication and intentionality. Their methods and conceptual underpinnings are not those that have been typical of cognitive science, and they do not even use the word "cognition" to describe what they are studying. However, the outcomes of their research are important for cognitive science, and as an interdisciplinary research cluster, cognitive science is capable of absorbing their contributions. Whether or not it actually does so will depend on a number of factors, of the type outlined in the general introduction to this volume.

The second biological field on the periphery, neuroscience, has at least two roles to play in the cognitive science cluster. First, the tradition of studying the behavior of individuals with damaged structures is essentially another addition to the mix of cases needed to carry out a comparative study of cognition. Secondly, and more fundamentally, neuroscience promises to describe the mechanisms by which cognition occurs; it can bring explanation all the way down to the physical structures and the way they function. The best-known work of this type is concerned with the cerebral localization of speech and other higher functions. Though most cognitive scientists have a passing knowledge of this work, it has not had a strong impact. Two recent trends are potential exceptions. First, Zurif and his colleagues (e.g., Grodzinsky, Swinney, & Zurif, 1983) have used patterns of results from aphastic patients to argue for a different partitioning of language into component processes than has been assumed in the past. Second, models and findings from neuroscience have figured centrally in a new type of cognitive theory, connectionism. If connectionism succeeds in establishing itself, this should be a powerful influence towards bringing neuroscience further inside the cognitive science cluster.

Connectionism is too recent a trend to be considered in this volume, but the localizationist research, including Zurif's recent work, is considered in Richardson's paper along with the animal language research represented by Savage-Rumbaugh and Hopkins. Richardson points out that Descartes foreshadowed the peripheral status of both fields by rejecting their data as relevant to cognition. Cognitive science has, consciously or unconsciously, followed Descartes in this (though without his dualism). It will be of considerable interest to follow the next steps in the development of the cognitive science research cluster, and see whether it can overcome these Cartesian strictures and integrate animal studies and neurological studies more throughly with its other perspectives.

References

Chomsky, N. (1957). *Syntactic structures.* The Hague: Mouton.

Chomsky, N. (1968). *Language and mind.* New York: Harcourt, Brace & World.

Dennett, D. C. (1978). *Brainstorms.* Montgomery, Vermont: Bradford Books.

Grodzinsky, Y.; Swinney, D.; & Zurif, E. (1983). Agrammatism: structural deficits and antecedent processing disruptions. In M. L. Keane (Ed.), *Agrammatism.* New York: Academic Press.

Miller, G. A. (1951). *Language and communication.* New York: McGraw–Hill.

Neisser, U. (1967). *Cognitive psychology.* New York: Appleton-Century-Crofts.

Piaget, J. (1950). *The Psychology of Intelligence.* London: Routledge & Kegan Paul.

Romney, A. K., & D'Andrade, R. G. (1964). Cognitive aspects of English kin terms. *American Anthropologist, 66,* 146–170.

Turing, A. M. (1936). On computable numbers, with an application to the Entscheidungs problem. *Proceedings of the London Mathematical Society (Series 2), 24,* 230–265.

Werner, H. (1948). *Comparative psychology of mental development.* New York: International Universities Press.

Wiener, N. (1948). *Cybernetics.* New York: Wiley.

The Evolution of Communicative Capacities

E. SUE SAVAGE-RUMBAUGH[1] and WILLIAM D. HOPKINS[2]
[1]*Language Research Center, Georgia State University and Yerkes Regional Primate Center, Emory University, Atlanta, GA 30322, U.S.A.*
[2]*Language Research Center, Georgia State University, Atlanta, GA 30303, U.S.A.*

Introduction

Ape-language reports, regardless of the extent of language skills claimed for apes, present both a challenge and a problem for psychology as a field. The problem is that none of the ape-language studies fit neatly into the extant categories of research. They are not "animal learning" studies in that they do not look for basic principles or laws of learning nor do they not deal with a retricted type of learning such as discrimination, memory, etc. Instead, they involve all of these processes at once in a complex manner for the expressed purpose of communication (Savage-Rumbaugh, Rumbaugh, and Boysen, 1978).

Apes who acquire language skills do not behave as animals are traditionally expected to do in animal learning experiments. For example, reports of this work often do not describe how learning has occurred, only that it has occurred (Fouts, 1973; Gardner and Gardner, 1971; Terrace, 1979). Also, instead of the experimenter being required to draw inferences about the animal's world based on observations of the animal's observable behavior, experimenters draw inferences based on what their subjects have told them. (Gardner and Gardner, 1971; Savage-Rumbaugh, Pate, Lawson, Rosenbaum and Smith, 1980). This sort of report is viewed as alarming and discomforting by some (Umiker-Sebeok, 1980; Sugarman, 1983; Chomsky, 1972), as gratuitous to others (Epstein, Lanza, and Skinner, 1980; Bennett, 1978), and as exceedingly important by still others (Griffin 1975; Terrace, in press).

Ape language work does not fit appropriately within the traditional matrix of ethologically based studies of animal communication – wherein a limited repertoire, consisting of a maximum 100 "social signals" is the domain of study (Eibl–Eibesfeldt, 1970). In ethological studies of animal behavior, the methods of analysis of communicative behaviors center around the alterations produced in the "displays" of other species members. Animals engaged in such displays are not

Bechtel, W (ed), Integrating Scientific Disciplines. ISBN 90-247-3242-5.
© *1986, Martinus Nijhoff Publishers, Dordrecht. Printed in The Netherlands.*

viewed as literally capable of "telling each other things," but rather as unknowingly behaving in ways which function to "tell" other individuals things (Smith, 1977). Terms suggesting consciousness and intentionality are generally used only as a means of explanatory shorthand (Smith, 1977; Dennett, 1982).

In contrast to ethologists, psychologists have traditionally been interested in "purer forms of learning." This interest has surely been prompted by the continuing belief that it is such learning which accounts for the clear majority of human behavior, while genetic programs account for a good deal of animal behavior. From the view that there is a profound difference between the cognitive processes of animals and humans has come the field of cognitive psychology, devoted to the study of the structure and development of human intelligence. When psychologists move into the realm of animal behavior, the language rich procedures employed with human subjects become inadequate tools because of their emphasis upon verbal report. The central result of the inability to employ verbal paradigms with animals, was that psychology's view of the structure of animal mind became radically different from the structures attributed to the human case. The study of animals consequently has been split into the twin dichotomies of "animal learning," and "animal behavior," a distinction which in itself reflects the view that these are two separate processes. It is important to note that there is no such division in human psychology, that is, no distinction betwen "learning" and "behavior" is made. The issue of unlearned human behaviors, or behaviors which predispose the human being to attend to and learn certain phenomena more readily than others, and to organize this information in a particular way, is rarely studied experimentally within any sort of comparative framework.

Inherited Patterns have "Evolved" and Changed, but Learning Processes have Only "Expanded"

All vertebrate animals enter into their environment pre-equipped with a highly specific morphology, a somewhat less specific set of predisposed behavioral patterns (often referred to as fixed action patterns) and a rather loosely predetermined capacity to learn alternative behaviors when they encounter situations for which they have either an incomplete behavioral program or essentially no behavioral program. It is generally acknowledged that genes affect both morphology and fixed action patterns in a rather direct manner. However, the behaviors which are "to be acquired" by the organism are often viewed as being only indirectly within the province of the genes. That is, the genes can limit what an organism may learn, but they are not typically viewed as determining, in any specific sense *what* an organism will learn, or how it will learn, though they may be seen as generally delimiting its learning capacity (Lorenz, 1981). Because learned behaviors do vary to a great extent in vertebrates, even between organisms of the same species, they have not been viewed from the evolutionary perspective which has characterized the study of morphology and fixed actions patterns.

Learned behaviors are often viewed as falling beyond the realm of the genes. Consequently discussions of animal intelligence and human intelligence have lacked the evolutionary framework that has been the bedrock of scientific progress in fields such as comparative morphology, population biology, ethology, and more recently, sociobiology. Thus, while it is generally acknowledged that some animals are more intelligent than others, the *evolution* of intelligence is characterized only in very broad terms as proceeding from rigid to flexible, and the structure of animal intelligence is viewed, when it is studied at all, as a function of experience, not of the organism's evolutionary history.

Fixed action patterns, when viewed from an evolutionary perspective, illustrate the manner in which highly complex ritualized behaviors have evolved from simpler patterns observed in closely related species. For example, Lorenz (1971) discusses the tail beating behavior of many teleost fish and closely related species. Comparisons of these species reveal the remnants of simpler tail beating patterns, from which the more complex ritual of the teleost fish has evolved. By contrast with the ethological method of studying complex ritualistic displays by looking for their simpler antecedents, the psychological method of studying increasingly complex learned behaviors does not look to related species who may display similar but simpler learning skills. Instead, complex learned behaviors are typically viewed as a result of one common process which has become elaborted, but has not been altered in any fundamental way (Hull, 1943; Guthrie, 1935). To psychologists, basic learning processes are the equivalent to the "genome" of all higher order learned behavior and animals differ from one another only by their ability to demonstrate more elaborate associations.

By not looking for different types of learning processes in different species, and by not attempting to determine how more complex learning processes have evolved from simpler ones, psychologists eschew all opportunity to develop a framework of learning processes which could become unified with the approaches taken by the other life sciences. This mistake has altered the progress of the study of learning processes and sidetracked the study of learning for decades, causing it to focus upon questions that have very little relevance to the actual behavioral requirements that human beings or animals encounter on a daily basis.

Acquired versus Inherited Patterns of Behavior: Awareness of the Behavioral Goal

A significant distinction between acquired and inherited behaviors that is not often mentioned seems to be that the goal of the "display" pattern remains hidden from the individual organism who is emitting the fixed action pattern. Mayr (1971) refers to these inherited systems as "closed programs" which provide the organism with little capacity for learning, or modification of behavior when forced to alter under differing environmental conditions. By contrast, "open programs" provide the organism with a greater propensity for learning and responding to outside stimuli or changes in the environment which, in turn, improve the chances of survival. These acquired behaviors affect the success of that individual directly and are altered within his immediate life span, as opposed to the cross-generational effect of fixed action patterns. Because learned behaviors operate only within the life of a given individual, it becomes necessary that some sentience of the consequences of the behavior must occur. If the consequences of an organism's behavior are not monitored at the neurological level, it is difficult to understand how rapid behavioral change could occur.

For some constructive comparisons between fixed action patterns and learned behaviors, let us look at some larval care patterns in wasps as contrasted with the maternal care patterns of monkeys. Maternal behaviors are often viewed as "innate" in a wide variety of species, and certainly there would be a high premium placed on the adequate execution of such behaviors. If much were left to learning – it would seem that a lot of things could go awry since few opportunities for practice are available to most species.

The digger wasp (Baerends, 1941) displays the fixed action pattern of searching its larvae cavity before depositing food (caterpillars) in the cavity for the larvae. It drags the caterpillar to the edge of the cavity, drops it near the entrance, enters the cavity, inspects it, reappears head first, and pulls the caterpillar inside. "If one removes the caterpillar to a place some distance from the nest while the wasp is inspecting the cavity, it will search for it until it has found the caterpillar, bring it back to the entrance and repeat the entire sequence." (p. 218, Eibl-Eibesfeldt, 1970). This behavior can be elicited 30–40 times before the wasp wil finally pull the caterpillar into the cavity without inspecting it.

Contrast this difficulty in dealing with a disruption of a fixed action pattern with that of the squirrel monkey mother described by Rumbaugh (1965). Maternal behavior in these monkeys is far more complex than the larval care patterns of the digger wasp, yet it is the infant who is responsible for maintaining contact, by clinging, as the mother moves about. The mother, in normal circumstances, shows no tendency to emit behaviors desinged to keep the infant with her. She leaps rapidly about with little regard for the difficulties the infant may experience as she moves. Yet, when her infant's arms are bound so that it cannot cling, the mother, aware that the infant is not clinging, picks the infant up and carries it. To do so, she has to walk bipedally and her movement is impeded considerably. Prior to picking the infant up, she places herself near the infant so as to initiate clinging behavior from the infant, but when clinging does not occur, the infant is carried. In so doing the mother *displays a behavior that is not in the repertoire described for her species.* Moreover, bipedal infant carriage is not likely to be found in squirrel monkeys, or related species, even in vestigial form. Closely related species do not hold their infants, nor do more primitive primate forms such as tree shrews. Only more "advanced" forms of primates help in the support of their infants. Thus we see in the squirrel monkey mother an ability to compensate for lack of normal behavior in the infant. Not only is the mother able to do so, she seemingly comprehends the problem immediately and engages in a variety of experimental measures to deal with it. That is, she seems to have a comprehension of the goal – the baby is to stay with her, and she tries a number of ways to effect that goal, finally modifying her own locomotor patterns to an extreme degree to bring it about. In the case of the differ wasp, one is forced to wonder if it understands anything about the goal of "cavity inspection" – or if this is simply an act embedded in a chain of behaviors which are programmed to be executed without feedback as to their function or effectiveness.

Are Human Beings Aware of their Goals?

It is often pointed out by behavioral psychologists that human beings are not "aware of the consequences" of their behaviors and yet our behaviors are shaped quite effectively despite this lack of awareness. Often given is the example of the trick pinball machine, which gave points when users kicked it, swore, hit it, etc. These behaviors increased in frequency, although when asked to describe how they produced high scores, none of the users of the machine mentioned that any of these behaviors affected their score. (It should be noted that the machine did not give points just for being kicked, only for being kicked as a ball was progressing down its surface.) Although this example is an excellent one for demonstrating that human beings cannot always construct a lawful verbal statement which reflects the relationship between their behaviors and the consequences, the broader implication – that we do not need to deal with issues such as "awareness" – is unwarranted. The pinball machine example depicts a complex situation in which multiple cause-effect relationships are possible. Furthermore, the relevant ones are purposefully disguised from the pinball player to make it appear as though they are irrelevant.

248

Under these conditions it is not surprising that the subject should not express a verbal awareness.

The pinball machine example is very different from most of the conditioning studies in which the consequences of behavior are made extraordinarily evident and the human being so conditioned clearly displays behaviors which reveal expectancy of consequence. However, it is not only human beings who can identify their behavioral goals under conditioning schedules, other animals can do as well. For example, if a rat is trained to press a lever to operate a vending mechanism, the rat on a FR1 schedule does not wait till it hears the vendor open before approaching; it approaches as soon as it has finished the lever press, illustrating that it knows what will happen next and that it has pressed the lever with the expectancy of making the vendor operate. Lengthening the time between the lever depression and the vendor opening would, presumably, make the behavioral evidence of expectancy even clearer.

Let us contrast the rat's demonstration of an expectancy that a lever press will produce food with what a rat does after emitting a fixed action pattern. Eibl–Eibesfeldt (1949) observed that if rats were caught and killed immediately in traps, other rats who were present as one rat was caught did not avoid the traps themselves. They climbed right over the bodies of their dead compatriots and into traps themselves. Thus the consequences experienced by their companions had little relevance to them and there was no opportunity for them to acquire trap avoidance on an individual basis, not even on a trial one basis, since the first trial was the last. However, if but one rat was only partially trapped and died slowly in the trap while vocalizing, none of its companions could be similarly caught in traps. In this example we observe the value of the fixed action pattern – or prepotent learning mechanism, as it operates across generations. The rats do not need to learn the consequences of trap avoidance by trial and error, if instead they can learn from a signal, given off by a conspecific, to avoid the trap.

While rats do not seem to be able to learn from the experiences of their companions unless prewired mechanisms are controlling their responses, they do learn quite readily from their own experiences, suggesting that they are aware of relationships between their own actions and the results thereof – even in some cases when that relationship is between sets of events quite disparate in time, as Garcia and Koellign (1966) demonstrated with their classic studies of taste aversion. I suggest that our insistent avoidance of terms such as "awareness of behavioral goals" has caused us to unwittingly oversimplify what animals are able to learn and to overlook the significant distinction of knowledge of consequences which separates learned components of behavior from fixed action patterns.

Implications of the Concept of Awareness for Communicative Patterns

Although acquired patterns may replace or interface with fixed action patterns, the behavioral evidence seems to suggest that they often operate differently. When fixed action patterns of communication occur (normally between two members of the same species) the flow of communication and the ensuing hormonal synchrony which is brought about by the exchange of displays typically flows quite smoothly. When a receiver of a communicative pattern fails to respond, the initiator of a fixed action communicative pattern is unlikely to invent some other means of getting signals across. Fortunately, there are few misinterpretations of inter-animal fixed action communicative patterns. However, because errors so rarely occur there is also little opportunity to assess what effect, if any, the emission of given signal actually has upon the recipient apart from eliciting the next piece of the inter-individual behavioral chain. In the realm of fixed action inter-individual communicative patterns however, it appears that the participants have little sense of the "goal" of their communicative behaviors and act much as if each inter-linked unit in the chain is explicitly designed to evoke the next unit (Smith, 1977).

An example of this inter-individual inter-locking of communicative patterns may be seen in the flightless male cormorant (Eibl–Eibesfeldt, 1965), where we find the fixed action pattern of the male bringing seaweed to the mate on the nest. This has been interpreted as an appeasement communicative gesture, and indeed, if the seaweed is snatched away from the male as he approached the nest, he is attacked by the female. However, the fact that the male still approaches the female once he has performed the actions of retrieving the seaweed and starting toward her (whether the seaweed is snatched away by a curious ethologist or not) suggests that the male cormorant does not understand the significance of his own behavior of bringing seaweed. Bringing seaweed to the female is simply a communicative gesture that has become genetically programmed across time and the male need not concern himself with why he is doing what he is doing.

By contrast, when consequences are attached to acquired behaviors so as to produce an increased frequency of behavior, the indivudal emits behaviors which typically suggest that he has some appreciation of those consequences. For example, a chicken who has learned to put a small ball in a net (in imitation of the game of basketball) will, if it drops the ball, spontaneously search for it, instead of proceeding to the net without the ball. The chicken, in contrast to the cormorant, has learned that the ball is a necessary component for producing reinforcement and it monitors whether or not it has the ball (Breland & Breland, 1970). Perhaps if the seaweed were repeatedly taken from the flightless cormorant, it too would begin to monitor the presence or absence of seaweed, and in such a case this component of its behavior would clearly be acquired through the consequences emitted by the female (which would remain the result of fixed action patterns). It is, however, equally possible that, no matter how many times the seaweed is removed from the cormorant, it will not learn to retrace its steps and retrieve seaweed to replace the stolen bunch before approaching the female. If so, it would suggest that the male has no "awareness" of the function of brining seaweed – his genes would know why he carried the seaweed, but his brain would not.

With all our "laws of learning" it seems we have none that would clearly predict whether the cormorant would learn to retrieve the batch of seaweed or not, and thus we would have to conduct the experiment. If the cormorant does not fortuitously hit upon the idea of picking up the snatched piece of seaweed, he could surely be shaped to do so – through positive reinforcement of intermediate behaviors. However, it is difficult to see, how evolution could shape a behavior such as retrieving a piece of seaweed snatched away by a curious ethologist – even if the ethologist kept doing so for generations. Evolution does not shape behaviors in small increments as does learning, but rather builds new behaviors out of old behaviors. Moreover, it is also difficult to see how any non-positive technique (such as the punishment emitted by the female cormorant) could be used effectively to shape the behavior of retrieving the snatched seaweed. To shape the cormorant's behavior through positive reinforcement requires the hand of an individual conscious of the goal (i.e., teaching the cormorant to pick up seaweed) – not a likely case if the cormorant is left to his own devices without the aid of the psychologist.

Does Communication Become More Complex as the Six of the Cortex Increases?

Wilson (1975), after reviewing most of the published work on non-human communicative systems, concludes that all other nonhuman organisms, from bee to ape, have a relatively limited and fixed repertoire of communicative signals which are, for the most part, described as "closed systems." He remarks that "In the extent of their signal diversity, the vertebrates are closely approached by social insects, particularly honeybees and ants." (Wilson, 1975, p. 184). Is it really the case that all species other than humans have communicative patterns that can be reduced to a set of preprogrammed signs that are intimately linked to a broad bio-behavioral program shaped not by experience and knowledge of consequences, but by the successful reproduction of past generations?

We suggest that a simple counting of the "number of different signals," as presented by Wilson, underestimates the complexity of many mammalian systems, as contrasted with that of the social insects. This is because the mammalian communicative systems are more variable, more intergraded, and more amenable to alteration through learning than are the insect communication systems. The attempts to list a specific number of discreet signals reflects the inadequacy of the human observer when faced with complex mammalian species, rather than the inadequacy of the communicative systems themselves. For example, vervet monkeys are predisposed to emit alarm calls for certain predators, such as eagles. At first, the infant will emit alarm calls toward all flying birds; however, eventually the proper context of the call is learned so that alarm calls are emitted only for eagles. It is not only the proper stimulus of the call that appears to be learned, but also its function, i.e., that of causing other animals to flee. Thus, it is apparent that a mixing of learned and unlearned patterns occurs so as to permit an innate behavior (the sound itself) to become intermingled with learned behaviors and to occur in response to learned stimuli. In such cases, the animal does become aware of some of the consequences of its actions and in fact, it appears that the genes program certain aspects of behavioral chains with particular slots for learning. Thus, the vervet comes equipped with an alarm call toward flying objects, but it needs to learn which flying objects. Since fleeing of other group members occurs only when calls are given in the correct context (i.e. the dangerous eagle flying object), the consequences of the call, as well as the context, can be learned by the young monkey. (However, knowing when to call, and knowing that others will flee is not equivalent to knowing that one is "telling" others that one has seen an eagle, as will be discussed later.)

Another example, at the human level, involves the nonverbal facial and postural exchanges during social interactions between individuals in a group. During such exchanges, communicative turn-taking is often conveyed by certain facial expressions or postures. Thus, if an individual wishes to assume the role of speaker in a group, he or she will often begin to move in synchrony with the speaker and to exhibit behaviors and postures that mirror those of the speaker. To do so seems to both signal the listener's desire to assume the speaker's role and to facilitate the exchange of roles from speaker to listener. It is only through experience in the appropriate contextual situations that individuals learn the proper facial expressions and/or cues to initiate conversation, alter the exchange, and create a coordinated communicative exchange (Kendon, 1973). It is through a learning process similar to that experienced by the young vervet monkey that human speakers learn the proper social rules for initiating, continuing, and ceasing and interactive exchange and the consequences for not following established social rules of communication. However, even though an individual has learned the rituals of participating in coordinated conversation, this is not to say that he or she is aware of using these behaviors for communicative purposes. To the contrary, such nonverbal mannerisms, while they clearly function to regulate communicative interchange, can rarely be adequately or even partially verbalized by the individuals who are utilizing them.

Of course, once the behavior patterns are pointed out to the human beings who are using them, often through the use of video tape, these individuals become aware of the significance of their previously "unconscious" behaviors and use them in new and altered ways. This is why courses in "nonverbal communication" are so popular among business executives. Vervet monkeys would not similarly benefit from viewing tapes of their own behaviors. They are, by contrast to our own species, trapped in a field of semi-awareness, not fully cognizant of communicative intent.

Higher Order Communications or Communicating About Communicating

Having distinguished innate communicative patterns from acquired patterns and illustrated how these two modes of communication can interact, it is important to point out that abstract information about events removed in time and space can be communicated through innate patterns. The classic example is, of course, the "language of the bees." At least one species of bees (Apis mellifera), possesses an elegant system which enables it to inform its fellow hive members of the precise location of distal food sources on a trial one basis (von Frisch, 1965). However, the bees do not appear able to learn to inform each other of classes of information beyond that dictated by the genome. Moreover, they do not appear to know the outcome of their signals (i.e. whether the other bees go to the food source being depicted by their dance or to the food source being depicted by the dance of another bee, or even anywhere at all).

By contrast, primates clearly demonstrate complex learning of social signals and anticipation of the outcome of such signals (they threaten subordinates, but not dominants for example – suggesting that they also differentiate outcomes based on context and recipient). Still, it has not been demonstrated that even primates are spontaneously able to form abstract symbol systems with reference, or that they are able to communicate about communicating. Knowing how to communicate is different from knowing that one is engaged in the act of communicating.

An instructive example is to be found in the "whining" behavior of young children, which passes from the realm of unintentional communication to that of intentional communication (or knowing that one is communicating) with some very characteristic behavioral changes. For example, the first author was watching a 6 month old infant attempting to reach out for a bottle on a loaded cart as the stewardess passed in the aisle of a plane. The infant extended his hand, but was just inches away from grabbing a bottle. He then began ot fuss, more and more loudly, never taking his eyes off the bottle to look at the stewardess or his mother. However, his cries increased in frequency and intensity until they both noticed him and bent over to help him obtain the bottle. Even as he was being given the bottle he continued to fuss until it was in his hand. He never took his eyes off the bottle to note that others were preparing to help him obtain his goal. The infant's fussing and reaching behaviors served to communicate his goal to his mother. However, the infant did not seem to be aware of the the communicative function of whining and reaching. Rather these behaviors were elicited by his desire to have the small bottles that he saw and his inability to reach them. It is instructive to contrast the behavior of this 6 month old infant with that of a 2 year old child running ahead of his mother (who was loaded with packages) into the drugstore. The youngster bumped into the turnstyle and hurt his head. He stopped, rubbed his head, his face turned red, and he walked away from the turnstyle, clearly angry at it. He looked at a few toys, then looked around for his mother who was just then catching up to him. As soon as she was inside the door and within hearing distance, the child burst into loud sobs and pointed to the turnstyle.

It seems as though the 6 month old had learned that fussing helped to get things, and that louder fussing was needed if one's wishes were not quickly filled. However, he did not look toward his mother to see if she heard him, or to see if she was observing his gesture. He seemed unable to switch his attention from the bottle to the mother and back, while maintaining his goal. Moreover, he seemed not to understand that before his mother could help, she needed to perceive his problem herself. The older child not only understood these things, he also knew that it was pointless to fuss, or to indicate the turnstyle, unless he first had his mother's attention. He clearly knew the difference between whining because one felt like whining, and whining for social purposes, whereas the 6 month old child was still learning this distinction.

Communicative behavior of the sort exhibited by the older child is often viewed as the communicative achievement which sets humans apart most dramatically from other animals. The fact that we cannot communicate with people from a different culture with a different language, seems to emphasize to us the enormity of learning in our species. It appears to us that animals similarly displaced do not encounter our problems – at least not in a way that we can discern. However, communicative behavior is, at best, simply a subset of all possible behaviors. It has the unique property of being directed toward other animates – most often these others are members of the same species and consequently are themselves predisposed to decode and respond to these communicative patterns in particular ways. Most communicative patterns found in animals are viewed as only partially learned. These patterns are seen as having "slots" into which specific stimuli are predisposed to fit and while the given stimulus may vary, the overall plan does not. For example, the young vervet monkey discussed earlier may learn to give an alarm call first for all birds, then only for eagles. However, the call itself changes only minimally and if the monkey were asked to learn a completely new call for an invented danger — such as an electric fence — there is no evidence that it could do so. Moreover, it is not clear that even the genetic boundaries of a partially learned call, such as the eagle alarm call, would permit it to be flexible enough to be used for something such as an electric fence.

The human communicative system is widely viewed as having no "slots," as completely open, and it is typically presented as the standard against which the communicative systems of all other species must be judged (Wilson, 1975). It is commonplace for humans to view language as a different order of communicative behavior from that found in all other species. Language has even been viewed by some philosophers as synonymous with thought (Fodor, Bever, and Garrett, 1974). Nearly every serious scholar, with the notable exceptions of Griffin (1975) and Skinner (1957), who has addressed the topic of human communication has felt compelled to set it apart from all forms of communication seen elsewhere in the animal kingdom.

Traditionally, these two types of communicative behaviors (intentional and nonintentional) have been divided into the groupings of verbal and nonverbal — with the attendant presumption that the nonverbal system seen in most nonhuman species is not flexible or open in the sense that virtually anything can be communicated. Where learning does play a role, it is typically viewed as being of a highly constrained nature. By contrast, the verbal communication of the human species is characterized as "open" and presumably able to express as many things as the human organism can conceive. An upspoken distinction, which also goes along with the oft repeated dichotomy sketched out above, is that the human system is also characterized by awareness and intentionality. In other words, the human being can communicate "on purpose," so to speak, while the animal systems are characterized by the lack of these features and thus animals are not viewed as communicating "on purpose," but rather as the result of experienced needs, emotions, etc. (Lancaster, 1969; Lieberman, 1975).

This strong dichotomy between human and animal communication is supported by the frequent observation that human beings themselves engage in both intentional and unintentional communications and that these two sorts of communications are controlled by different neuronal processes. For example, in the case of a human being who has experienced a unilateral thalamic lesion, the ability to smile spontaneously in appropriate social situations is lost on one side of the face.

However, the ability to smile upon command – a non-social smile if you will, is retained by both sides, suggesting that the conscious cortex has learned how to produce and expressive behvaior which is typically mediated by a more primitive system. However, whether the smile is a social smile – or an intentional cortically mediated smile – we would not want to say that the person was unaware of smiling. Indeed, in a social context, the person is clearly aware of their emotions and of the feeling of smiling – but not of having "decided" to smile.

Awareness that One is Communicating

Confusion between human and animal communication often arises over just this point, and much of the problem lies in the use of the word "awareness," (with reference to human communication) in two different senses. To be aware is, in one sense of the word, to be cognizant of sensation. In that sense we are always aware of bodily emotions and perceptions except when injury has interrupted neuronal processes. In this sense, animals also are surely aware of their ongoing behaviors and sensations. However, awareness in the human individual is also used to imply a knowledge of behavioral alternatives, a concomitant knowledge of having chosen one of these alternatives as opposed to other, and an ongoing conscious monitoring of the progress toward a chosen alternative. The last feature is illustrated when one has chosen a particular route to point B on a map and one is noting the towns which are being passed in order to determine if one is actually on the correct highway. With this conscious monitoring goes a readiness to engage in other routes should that one be perceived as ineffective. This sort of awareness is also surely within the capacity of higher mammals and is to be contrasted with that strange experience we have all had of having driving to a familiar location that we had not intended to go to while our thoughts were elsewhere. On arrival, it is not possible to remember what happened along the way and thus we say we were "unaware" of where we were going, even though we were negotiating all the decision points quite well.

More important, however, is the fact that "awareness" as applied to the phenomenon of human communication also implies something we would not attribute to animals – and this is the awareness that communicative acts are *behaviors about behaviors* (Griffin, 1975; Crook, 1983). An easy way to visualize this distinction is to think of the prespeech conversational babbling which human youngsters engage in. During such "conversational exchanges between mother and infant" there is a reciprocal exchange of roles and the intonation patterns of the infant typically follow that of the mother in a responsive manner. However, at this stage, it appears that the infant is not, in fact, saying anything – but rather is producing the vocal exchange as a behavior in its own right. Later though, the mother will comment that the infant now knows what he or she is saying. At this stage, the vocalizations may still be unintelligible to a novice observer and they may even be unintelligible to the mother herself – yet as a result of the surrounding behavioral context, they will be recognized as intentional communications, albeit rather poor in quality. That is, the act of uttering alone will no longer fulfill its own function – the infant will insist that specific sorts of action be taken in response to utterances and thus they will have become behaviors about behaviors.

Moreover, not only does the infant now distinguish behaviors which are their own raison d'être and behaviors which are about other behaviors, let us call these "intentional informative communicative acts," he or she recognizes that these informative communicative acts are primarily about the behavior of other individuals. That is, they are produced for the purpose of altering the behaviors of others in specific ways – ways which are best characterized as suiting the goals of the infant. Typically the goal of the adults about the infant, and the goals of the infant, are rather different. The adults enjoy observing the infant, exchanging highly organized linguistic information, completing complex tasks, planning for future occurrences, etc. while the infants wants to be played with, wants to eat the cookie that is out of reach, wants to be carried about, etc. Intentional informative communicative acts are used by the infant to control and orient the adults' behavior away from their own goals and toward his or her goals.

Along with the appearance of such intentional communicative acts, comes the concomitant awareness that other individuals also produce behaviors about behaviors and that they make choices regarding which behaviors to engage in. Thus they may or may not respond to the infants communicative acts. If they do not respond the infant may perceive the lack of response as resulting from one of two events; either the other party did not understand or perceive the intentional informative communication – or the other party received the communication, but chose not to respond. When the adult chooses not to respond, such will typically be apparent to the youngster because the adult will first acknowledge receipt of the communication and then indicate his or her unwillingness to comply. Premack and Woodruff (1978) term the onset of this level of communicative capacity the achievement of a "theory of the mind," that is, the attribution of mind to other individuals. However, nothing quite so esoteric as the mind needs to be evoked.

The awareness that one is communicating can clearly be seen when child, or ape, begins to use symbols to announce what it will do. Prior to this time, symbol use occurs on demand or in response to environmental cues – as when a child is prompted to say "Bye-bye" as visitors depart out the door uttering this comment to him or her. The child learns what to say in this context long before he or she learns to use "Bye-bye" to announce his or her own departure. Similarly, the chimpanzees Sherman and Austin learned how to name items shown to them by their teacher long before they were able to use the name of an item to announce that they were about to act upon that object in some way. Similarly, they learned to make requests of their teacher to tickle, bite, grab, etc., long before they were able to use these words to announce that they were getting ready to tickle someone else. It is only with the emergence of the ability to announce one's goals or actions, or one's past experiences, that there is evidence of a true awareness that one is communicating. The emergence of the capacity to use symbols to announce one's intended actions appears to occur quite naturally in the child – with little help from the parents. However, the emergence of this skill in Sherman and Austin was preceded by 2 years of training during which they learned to request objects appropriately, to name objects, and to give objects that others asked for. Once these skills were in place, the ability to announce an intended action appeared spontaneously (Savage–Rumbaugh, Pate, Lawson, Smith, & Rosenbaum, 1983).

Another subject, Kanzi (*Pan paniscus*), a pygmy chimpanzee, has differed from Sherman and Austin with regard to the complexity of both his pre-verbal communicative skills and his symbol acquisition skills (Savage-Rumbaugh, 1984b). Kanzi has developed symbol using skills in a spontaneous manner without specific training, much in the sense human children develop language (Savage-Rumbaugh, Sevick, Rumbaugh, and Rubert 1985). With his spontaneous acquisition of language skills, Kanzi has developed many of the linguistic skills that took years to teach Sherman and Austin. In this same manner, Kanzi also began to use symbols to announce what he was about to do and where he wanted others to go. The rapid emergence of such announcement skills appeared to be promoted by Kanzi's environment. All of the foods Kanzi eats are placed outdoors each day in various locations within a 55 acre woods (See Savage–Rumbaugh, in press, for detailed explanation). Each day, Kanzi forages with his teachers to various locations within the woods. The foods which Kanzi obtains depend upon what he announces that he wants to obtain, or where he says he would like to go. For example, Kanzi may request a food such as "apple" by selecting the symbol for apple on his portable symbol board and then lead his teachers to the appropriate location in the woods where apples can be obtained. Shortly after Kanzi demonstrated that he knew both how to announce where he intended to go and how to get there, he began to evidence some additional interesting behaviors. He would often tug on the symbol board which the teacher was carrying *in order to* say that he wished to "talk." That is to say that Kanzi not only announced what he wanted to do, he also announced that he wanted to tell us what it was he wanted to do.

By using a combination of gestures and symbols, Kanzi can adequately communicate to his teacher about locations or foods he would like to visit, or communicate about locations that the traveling group is about to approach. The fact that Kanzi differs from both Sherman and Austin in his pre-verbal and initial symbol acquisition exemplifies the enormous differences which can occur in the learning of complex patterns even among closely related species. It also reveals that the nature of the environment in which communication occurs can be manipulated to foster the development of complex communicative skills.

Do the types of communications achieved by apes such as Sherman, Austin, and Kanzi differ from the sort of "symbolic communications" reported for other nonhuman species? Epstein, Lanza & Skinner (1980) trained two pigeons, Jack and Jill, to communicate with each other by pecking lighted keys during a communicative exchange. Epstein's attempt was to equate the intra-species key pecking exchange to the communicative exchanges regarding hidden foods reported for Sherman and Austin (Savage–Rumbaugh, Rumbaugh, & Boysen, 1978a, b). In this study, only one chimp was allowed to see what type of food was hidden in a container, but neither could obtain the food unless both chimpanzees could identify the container's contents. Therefore the chimpanzee who saw the food being placed in the container had to tell the other chimpanzee what type of food was used on any given occasion.

Chimpanzee versus Pigeon Communication

A number of people have challenged the assertion that apes are capable of human-like intentional communication (Umiker–Sebeok and Sebeok, 1980). One of the most serious challenges has come from Skinner himself who has, with his colleagues Epstein and Lanza, attempted to mimic the communicative behaviors of apes by teaching pigeons to engage in behaviors that are made to appear similar to those of apes (Epstein, Lanza, and Skinner, 1980). In one of the studies by Epstein, et al., a pigeon named Jack was taught to peck a key when he observed a colored light in his cage. Jill (a male) was taught to peck a series of keys when he saw Jack peck a key. The tongue in cheek "interpretation" given to these behaviors was that Jack was telling Jill the color of the light in his cage. The real interpretation that was promoted by the study was that the chimpanzees Sherman and Austin were doing little more than Jack and Jill in a study which demonstrated that Sherman could tell Austin (or vice versa) the food that he had seen hidden in a container.

Can we equate the abilities demonstrated by the two pigeons, Jack and Jill, with those demonstrated by Sherman, Austin, or Kanzi? The real question revolves around whether or not apes indeed know that they are communicating and whether they know what information they are conveying. In order to determine if the chimpanzees Sherman and Austin indeed knew that they were telling each other what food was hidden in the container, their keyboard was removed and replaced with an alternate symbol system. They were given *no training at all* with this alternative symbol system – which consisted of manufacturers' packaging labels as symbols. If Sherman and Austin were merely engaging in a rote chain of behaviors, their ability to tell each other what they had seen hidden in the container would surely break down in the absence of the keyboard. However, if the chimpanzees knew that they needed to give one another information, they would try to find an alternative means of communicating the same information. Manufacturers' labels function perfectly well to identify the contents of packages, so they could likewise be used to identify the contents of food hidden in containers, if Sherman and Austin indeed understood that they were communicating and what they needed to tell each other. When this test was run, both Sherman and Austin immediately switched from the lexigram symbols on their keyboard to the manufacturers' packaging labels pasted on pieces of lexan (Savage-Rumbaugh, 1984a; in press). They unequivocally demonstrated that they knew they were communicating and they knew what they were communicating about. It is self evident that pigeons would not be able to compensate for the removal of keys from their chambers in Skinner's study.

Conclusion

The issue is not whether or not we can shape an organism's behavior to mimic those of another species but, rather, what are the underlying processes guiding the organism's learning or behavior, and how does this relate to the specific species and its experience or environment. Recent work, such as that begun by Olton, Collinson, and Werz (1977), D'Amato and Salmon (1984), and Rumbaugh and Pate (1984) has begun to address more complex questions of animal inteligence. Along with this new effort has come an interest in defining the nature of an animal's representation – for example, the interest in whether or not a rat has a Euclidean concept of space. Given that communication is but one subset of all behaviors, what we hope to offer is a preliminary framework from which the study of increasingly complex communicative and learned behaviors can be studied from an evolutionary perspective.

What we do need to suppose is the evolutionary advance of an ever increasing ability to monitor the results of one's actions. First only immediate effects would be monitored, then more indirect longterm effects, and eventually the effects of one's behavior on the behaviors of others. Finally we would see an ability not only to monitor the effects of one's own actions, but also the effects of the actions of others. From this would grow quite naturally a desite to control the actions of others for one's own ends – and from this desire, a communicative system capable of representing to others the actions you would have them take. Generally, one can determine the effectiveness of one's own behavior by ensuing events which affect the individual directly – however, intention communicative acts must be monitored by judging a change in the behavior of another and determining whether or not that change corresponds with the change which the intentional communicative act was intended to bring about.

Acknowledgment

This research was supported by grants from the National Institute of Child Health and Human Development (HD–06016) and from the Division of Research Resources, National Institutes of Health (RR–00165).

References

Baerends, G. P. (1941). Fortpflanzungsverhalten und Orientierung der Grabwespe. In I. Eibl–Eibesfeldt (Ed.), *Ethology, The Biology of Behavior*. New York: Holt, Rinehart and Winston.

Bennett, J. (1976). *Linguistic behaviour*. Cambridge, MA: Cambridge University Press.

Breland, K. & Breland, M. (1966). *Animal Behavior*. New York: MacMillan.

Brookshire, K. H. (1976). Vertebrate learning: evolutionary divergencies. In R. B. Masterson, C. G. B. Campbell, M. E. Bitterman, & N. Hotton (Eds.), *Evolution of brain and behavior in vertebrates*. Hillsdale, New Jersey: Erlbaum.

Chomsky, N. (1972). *Language and Mind*. Enlarged Edition. New York: Harcourt Brace Jovanovich.

Crook, J. H. (1983). On attributing consciousness to animals. *Nature, 303,* (5912), 1114.

D'Amato, M. R. & Salmon, D. P. (1984). Cognitive processes in cebus monkeys. In H. L. Roitblat, T. G. Bever & H. S. Terrace (Eds.), *Animal cognition*. Hillsdale, New Jersey: Lawrence Erlbaum Associates.

Dennett, D. C. (1982). Intentional systems. In J. Haugeland (Ed.), *Mind design*. Cambridge, MA: The MIT Press.

Eibl–Eibesfeldt, I. (1949). Uber das Vorkommen von Schreckstoffen bei Erdkrotenquappen. *Experimentia, 5,* 236.

Eibl–Eibesfeldt, I. (1965). *Nannopterum harrisi (Phalacrocoracidae)*. Brutablosing. Encycl. cinem., E596. Publ. zu wiss. Filmen, 1 A, 303–306 Gottigen (Inst. wiss. Film).

Eibl–Eibesfeldt, I. (1970). *Ethology, the biology of behavior*. New York: Holt, Rinehart and Winston.

Estes, W. K. (1970). *Learning theory and mental development*. New York: Academic Press.

Fodor, J. A.; Bever, T. G. & Garrett, F. (1974). *The psychology of language: An introduction to psycholinguistics and generative grammar*. New York: McGraw–Hill.

Fouts, R. S. (1973). Acquisition and testing of gestural signs in four young chimpanzees, *Science, 180*, 970–980.

Frisch, K. v. (1965). Die Tanzsprache und Orientierung der Bienen. In I. Eibl–Eibesfeldt (Ed.), *Ethology, the biology of behavior*. New York: Holt, Rinehart, and Winston.

Garcia, J. & Koellign, R. A. (1966). Relation of cue to consequence in avoidance learning. *Psychonomic Science, 4*, 123–124.

Gardner, B. T. & Gardner, R. A. (1971). Two-way communication with an infant chimpanzee. In A. M. Schrier & F. Stollnitz (Eds.), *Behavior of nonhuman primates*. (Vol. 1). New York: Academic Press.

Griffin, D. R. (1976). *The question of animal awareness: Evolutionary continuity of mental experience.* New York: The Rockefeller Press.

Guthrie, E. R. (1935). *The psychology of learning.* New York: Harper.

Hull, C. (1943). *The principles of behavior.* New York: Appleton.

Kendon, A. (1973). The role of visible behaviour in the organization of social interaction. In M. von Cranach & I. Vine (Eds.), *Social communication and movement: Studies of interaction and expression in man and chimpanzees.* London: Academic Press.

Lancaster, J. B. (1969). Primate communication systems and the emergence of human language. In P. C. Jay (Ed.), *Primates: Studies in adaptation and variability.* New York: Holt, Rinehart and Winston.

Lorenz, K. Z. (1971). *Studies in animal and human behaviour.* Cambridge, Massachusetts: Harvard University Press.

Lorenz, K. Z. (1981). *The foundations of ethology.* New York: Springer-Verlag.

Marler, P. & Vandenberg, J. G. (1979). *Handbook of behavioral neurobiology.* New York and London: Plenum Press.

Mayr, E. (1971). *Populations, species, and evolution.* Cambridge, MA: Harvard University Press.

Olton, D. S.; Collinson, C.; and Werz, M. A. (1977). Spatial memory and radical arms maze performance of rats. *Learning and memory,* Vol. 8, 289–314.

Premack, D. and Woodruff, G. (1978). Does the chimpanzee have a theory of mind? *The behavioral and brain sciences, 4*, 515–526.

Rumbaugh, D. M. (1965). Maternal care in relation to infant behavior in the squirrel monkey. *Psychological reports, 16*, 171–176.

Rumbaugh, D. M. and Pate, J. L. (1984). The evolution of cognition in primates: A comparative perspective. In H. L. Roitbalt, T. G. Bever, and H. S. Terrace (Eds.), *Animal cognition.* Hillsdale, New Jersey: Lawrence Erlbaum Associates.

Savage-Rumbaugh, E. S.; Rumbaugh, D. M.; and Boysen, S. (1978a). Symbolic communication between two chimpanzees (Pan troglodytes), *Science, 201*, 641–644.

Savage-Rumbaugh, E. S.; Rumbaugh, D. M.; and Boysen, S. (1978b). Linguistically mediated tool use and exchange by chimpanzees (Pan troglodytes), *The behavioral and brain sciences, 4*, 539–554.

Savage-Rumbaugh, E. S.; Pate, J. L.; Lawson, J.; Smith, S. T.; and Rosenbaum, S. (1983). Can a chimpanzee make a statement? *Journal of Experimental Psychology: General, 112*, 457–492.

Savage-Rumbaugh, E. S. (1984a). Verbal behavior at the procedural level in the chimpanzee. *Journal of the Experimental Analysis of Behavior, 41*, 223–250.

Savage-Rumbaugh, E. S. (1984b). *Pan paniscus and Pan troglodytes:* Contrast in preverbal communicative competence. In R. L. Susman (Ed.), *The pygmy chimpanzee.* New York: Plenum Press.

Savage-Rumbaugh, E. S.; Sevick, R. A.; Rumbaugh, D. M.; and Rubert, E. (1985). The capacity for animals to acquire language: Do species differences have anything to say to us? *Philosophical Transactions of the Royal Society of London, 308*, 177–185.

Savage-Rumbaugh, E. S.; Rumbaugh, D. M.; Rubert, E.; McDonald, K.; and Murphy, J. (in press). Spontaneous language acquisition by a pygmy chimpanzee. (*Pan paniscus*), *Behavioral Neuroscience.*

Savage-Rumbaugh, E. S. (in press). *Ape languages: From conditioned responses to symbols.* Columbia University Press.

Seyfarth, R. M.; Cheney, D. L.; and Marler, P. (1980) Monkey responses to three different alarm calls: Evidence of predator classification and semantic communication. *Science,* **210,** 801–803.

Skinner, B. F. (1957). *Verbal behavior.* New York: Appleton–Century–Crofts.

Skinner, B. F. (1984). Selection by consequences. *Science,* **213,** 501–504.

Smith, W. J. (1977). *The behavior of communicating.* Cambridge, MA: Harvard University Press.

Sugarman, S. (1983). Why talk? Comment on Savage-Rumbaugh et al. *Journal of Experimental Psychology: General,* **112,** 493–497.

Terrace, H. L. (in press). In the beginning was the "name." *American Scientist.*

Terrace, H. S. (1979). *Nim.* London: Eyre Methuen.

Umiker–Sebeok, J. and Sebeok, T. A. (1980). Questioning apes. In T. A. Sebeok and J. Umiker–Sebeok (Eds.), *Speaking of apes.* New York and London: Plenum Press.

Wilson, E. O. (1975). *Sociobiology, the new synthesis.* Cambridge, MA: Harvard University Press.

Language, Thought, and Communication

ROBERT C. RICHARDSON
*Department of Philosophy, University of Cincinnati, Cincinnati, Ohio
45221, U.S.A.*

The Cartesian Gambit

In a well-known, nearly infamous, passage in the *Discourse on Method* (1637),
Descartes issues this argument:

> ... if there were machines which bore a resemblance to our body and imitated our actions
> as far as it was morally possible to do so, we should always have two very certain tests
> by which to recognise that, for all that, they were not real men. The first is, that they could
> never use speech or other signs as we do when placing our thoughts on record for the
> benefit of others. For we can easily understand a machine's being constituted so that it
> can utter words ... for instance, if it is touched in a particular part it may ask what we
> wish to say to it; if in another part it may exclaim that it is being hurt, and so on. But
> it never happens that it arranges its speech in various ways, in order to reply appropriately
> to everything that may be said in its presence, as even the lowest type of man can do. And
> the second difference is, that although machines can perform certain things as well as or
> perhaps better than any of us can do, they infallibly fall short in others, by the which
> means we may discover that they did not act from knowledge, but only from the
> disposition of their organs. For while reason is a universal instrument which can serve for
> all contingencies, these organs have need of some special adaptation for every particular
> action (Haldane and Ross 1973, p. 116).

The same argument was mobilized a decade later in his correspondence with the
Marquess of Newcastle in arguing that we ought not "attribute understanding or
thought to animals" (Kenny 1970, p. 206). Though he again granted that the use
of language was the surest sign of the presence of thought in any creature, and
granted that animals did "express" their passions, he remained intransigent in his
insistence that such expression of the passions was no indication of the presence of
thought. Similarly, though he granted that "animals do many things better than we
do", he thought this explicable on the grounds that they act "naturally and
mechanically"; indeed, he clearly contends that the inflexibility and specialization
of their actions betokens their mechanical origin.

Bechtel, W (ed), Integrating Scientific Disciplines. ISBN 90-247-3242-5.
© 1986, Martinus Nijhoff Publishers, Dordrecht. Printed in The Netherlands.

Thus, the 'Cartesian gambit' presents a dual-faceted challenge to those who would attribute significant cognitive abilities to nonhuman animals. The charges reflect a determinate, if speculative, conception of the structure of thought and its relation to language. Moreover, they reflect an ambitious and well-defined conception of the nature of mechanistic explanation, its role in the explanation of behavior, and its limits.

The first leg of the Cartesian gambit – the use of articulate language as a sign of thought – is in fact the less central of the two. It is a simple diagnostic, reflecting a Cartesian insistence on a fundamental split between the mentalistic and the mechanistic. It is a mild historical irony that it is this leg of the gambit that has assumed the greatest prominence in more recent discussions. Indeed, Chomsky's insistence that the character of language use in humans – in particular, its innovative applications, relative freedom from stimulus control, and flexibility in applying to novel situations – is little more than a reflection of the Cartesian emphasis on language use as diagnostic of thought, weaned from its Cartesian underpinnings.

The second leg of the Cartesian gambit is grounded less in Cartesian mentalism than in a commitment to mechanism. Descartes wrote in corresponding with Henry More (1649):

> there are two different principles causing our motions: one is purely mechanical and corporeal and depends solely on the force of the spirits and the construction of our organs, and can be called the corporeal soul; the other is the incorporeal mind, the soul which I have defined as a thinking substance (Kenny 1970, p. 243).

It is reasonably clear that the "mechanical and corporeal" principle affirms that the mediation of action by the nerves is essentially a type of reflex action. The point is even clearer in *The Passions of the Soul*:

> ... it is easy to conceive how sounds, scents, tastes, heat, pain, hunger, thirst and generally speaking all objects of our other external senses as well as of our internal appetites, also excite some movement in our nerves which by their means pass to the brain; and in addition to the fact that these diverse movements of the brain cause diverse perceptions to become evident to our soul, they can also without it cause the spirits to take their course towards certain muscles rather than towards others, and thus to move our limbs (Haldane and Ross 1973, p. 338).

In its application to the Cartesian strictures on the attribution of thought, the role of appeals to reflex action is of a piece with appeals to principles of parsimony. Starting with the fact that, in humans, the mechanisms of the brain are adequate to initiate action without the intervention of thought or will, Descartes presses that the appeal to such mechanisms are *sufficient* for the explanation of animal behavior.[1] In short, Descartes charges that no behavior of animals *requires* explanation in terms which attributes thought to them. Mechanical causes alone are sufficient to explain their behavior. The mechanical nature of the underlying causes is in turn revealed in the fact that animal behavior is highly specialized; that is, the abilities manifested in one area do not carry over to other apparently related areas, and the behavior manifested is strongly stereotyped. He thought human behavior, by contrast, was guided by generalized cognitive capacities (the Understanding); and, accordingly, abilities or deficiencies in one area are correlated with abilities or deficiencies in others. Furthermore, the appeal to generalized capacities means human behavior is correspondingly modifiable, whereas the specialization of animal behavior is revealed in its lack of responsiveness to novel situations.

Animal Language

The first leg of the Cartesian gambit has been vigorously defended in recent years by Noam Chomsky, who likewise has pressed that some of the salient properties of human language – in particular, its relative freedom from stimulus control – is symptomatic of deep and profound differences between animal communication and human language (see Chomsky 1966 and 1968). Sue Savage-Rumbaugh, as well as other researchers in animal communication, attempt to provide us with something of an antidote to this leg of the Cartesian challenge by displaying the variety and complexity of communicative abilities in nonhuman primates. It would be an error of considerable magnitude to deny that human languages have properties lacking serious analogues in animal communicative systems. This shows itself most clearly, perhaps, in the freedom from stimulus control emphasized by Chomsky. Animal displays in natural settings are typically, though certainly not invariably, "stereotypes"; that is, they are reasonably uniform within the population or species, they are evoked by a limited number of stimuli, and are invariably evoked by stimuli within the appropriate class. Linguistic communication in humans, by contrast to some expressions of emotion, is typically not stereotyped. It would be an error of equal magnitude to underestimate either the complexity of the natural communicative systems displayed by animals or the capability of nonhuman primates to master yet more complex communicative systems.

---

[1]　If language is a sign of thought, that will be, in turn, only because the use of language does possess a responsiveness to variables other than the immediate situation and our passions – both of which Descartes explicitly excludes. It is no accident that the revolution in our understanding of language initiated by Chomsky has at its core an attack on the behaviorist portrayal of linguistic skills.

266

Though a sober assessment of research into 'animal languages' must avoid each of these errors, it is doubtful that fixing on more superficial characteristics of human languages is likely to provide a deep rationale in support of the Cartesian gambit. For example, though variability in human language use is striking, the most significant traits of human language center on the syntactic properties which show that the resources of classical behaviorism are incapable of capturing human linguistic abilities (see Chomsky, 1957, or Leiber, 1975). For like reasons we should be reluctant to accept that research into animal languages is likely to provide a clear resolution to the Cartesian challenge without an examination of the deeper properties of such communication systems. Each of these doubts has a common basis. The former, which will be the focus of subsequent discussion, derives from a conviction that the second leg of the Cartesian gambit, rather than the first, is foundational. The latter, which will be our immediate concern, derives from the fact that behavioral studies alone undermine theories of mental processing.

These latter doubts are best approached by turning to the methodological problems which arise in attempting to argue for the continuity between the forms of communicative behavior, and in thus trumping the Cartesian gambit. To see this, we need to recall that the first leg of the Cartesian gambit does *not* treat language as "criterial" for thought in a Wittgensteinian sense: there is no single behavioral effect which, on Cartesian grounds, would be logically or conceptually *sufficient* for the attribution of thought. The relation is rather that of a symptom to an underlying cause. Any appeal to a use of language, or an apparent use of language, is defeasible: if there is an alternative mechanistic explantion of the behavior, then, as with any behavior at all, Descartes will opt for the mechanistic explanation to the exclusion of a mentalistic one. In a more familiar and more modern guise, this is an appeal to a variety of the principle of parsimony, commonly referred to as Lloyd Morgan's canon (1894):

> "In no case may we interpret an action as the outcome of the exercise of a higher psychical faculty, if it can be interpreted as the outcome of the exercise of one which stands lower in the psychological scale."

The force of Morgan's canon can be briefly illustrated by turning to work by Jonathan Bennett (1976). Bennett proposes to provide an analysis of language *use* which emphasizes its communicative functions, grounding, he contends, all attributions of linguistic intent on behavioral criteria (see Bennett 1976, §2). His aim is to provide an analysis of "systematic communicative behavior" in the richest forms it takes, where this manifests communicative intent. The criterion he defends is this:

... when an [utterer] *U* utters [a sentence] *S* he means something by it if he utters *S* intending (i) to produce some change in [a hearer] *A* by (ii) inducing a in *A* a belief for which (iii) the uttering of *S* constitutes intention-dependent evidence (Bennett 1976, §53).

Setting the details aside, the crucial element in the criteria adduced by Bennett is the emphasis on communicative intent: the speaker must intend to get the hearer to believe or do something by inducing a belief which the hearer has reason to believe in light of the speaker saying what he or she says.

Savage-Rumbaugh is similarly prepared to accept an emphasis on communicative intent. Her insistence on the anticipation of the effect signals have on others (generally, though not always, conspecifics) as indicative of sophisticated communication is wholly compatible with Bennett's emphasis. Communicative intent, within this scheme, requires an awareness of the target's reponse and an ability to monitor the influence of one's own actions on others. The latter ability, which shows itself in Bennett's commitment to the 'teleological' dimension of communication, is the justification for Savage-Rumbaugh's appeal to behavioral modifiability, rather than fixed action patterns.[2]

Bennett's overall program in fact falls far short of its avowed goal, since the behavioral criteria he appeals to are either too weak to serve as indicators of mentality or too strong to be purely behavioral. Specifically, to gain a purely behavioral criterion, Bennett needs a behavioral criterion for the assignment of intention and belief. His own analysis in terms of teleological notions turns out to fail because it either presupposes mentalistic notions or is too weak to be criterial. Exploring this rather confusing web would take us too far afield. Let me settle for pointing to the moral: Bennett's attempt at a behavioral reduction fails for the reason all such attempts fail. There is no litmus test for the attribution of linguistic intentions any more than there is for any other mental ascriptions. As a result, it becomes implausible in the extreme to expect a uniform and infallible set of behavioral criteria sufficient for the attribution of mentality. We have instead an inference to the best explanation.

This consequence serves to highlight the force of a number of restrictions on the attribution of communicative intent, restrictions which are strongly reminiscent of Morgan's Canon, and which should be acknowledged if one is to make an interesting and compelling case for the attributions of communictive intent to

---

[2] Such modifiability seems commonplace. In many species of birds, there is a considerable variation in local song; and in others, the development of the song characteristic of the species or local population is strongly dependent upon exposure to the song. It would be interesting to inquire how far the emphasis by principals in the debate on flexibility and learning derives from a misbegotten reliance on a distinction between 'nature' and 'nurture'.

268

nonhuman animals. "One should", as Bennett says, "always prefer the lower-level or more economical of two unrefuted hypotheses about the mental causes or background of behaviour" (Bennett 1976, §50). In explaining the force of the principle, Bennett introduces a number of useful and quite specific constraints. Among them, four are worth noting.

(1) Stay with singular beliefs as far as possible, excluding as far as possible general beliefs and beliefs about what is not present (Bennett 1976, §32).

It is enough to explain my dog's welcome by the fact that she believes I will take her for a walk now, rather than that she believes I always walk her when I get home.

(2) Minimize *specific* content where *generic* content will suffice to explain the behavior (Bennett 1976, §36).

After the walk my dog expects to be fed, but it would be unwarranted to claim she expects to be fed Purina Dog Chow.

(3) Minimize heavily theory-laden attributions, especially when theory is lacking as in non-linguistic subjects (Bennett 1976, §36).

My dog doesn't have *any* beliefs about the mechanics of motion, even though she can catch a ball. The fourth principle defended by Bennett is designed to isolate "fraudulent teleological explanations." In its least problematic form, it runs something like this:

(4) Eliminate teleological or mentalistic explanations when there is a mechanistic explanation for the same phenomena.

This last principle is particularly appealing when the mechanistic explanation can be adopted with no loss of ease in applying the explanation. Bennett gives the following example (Bennett 1976, §22). There is evidently a lake in the Yukon named, aptly, "Stable Lake". It is a volcanic basin, fed only by rainfall and losing water only by evaporation. Along its shore, just above water line, alpine lilies grow, and just under the shoreline a population of shellfish. If the water level came to be too high, the lilies would die. If it became too low, the shellfish would go. Now, the stability is really remarkable: even in very dry years, the lake protects the shellfish by keeping the water up; and even in very wet years, the lake protects the lilies by keeping the water down. Of course, the lake doesn't actually *protect* either of them. The crucial maneuver in undermining the "intentionalist" explanation, though, rests on *finding* an appropriate mechanistic one. In this case it's simple. The lake has a shoreline which looks like this (Figure 1) in cross-section.

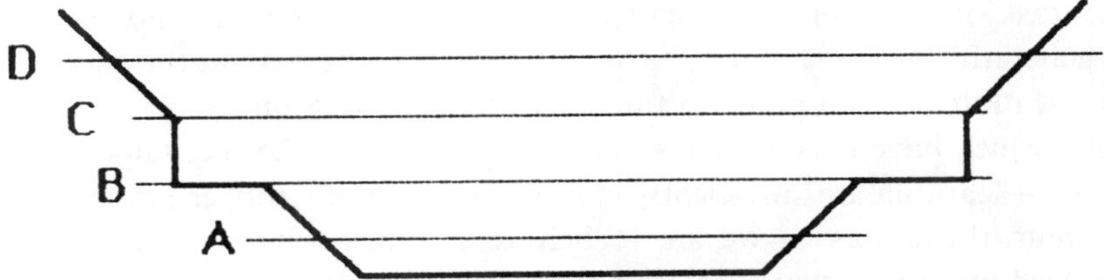

Figure 1. Schematic representation of the shoreline of "Stable Lake". The lilies live above D, and the shellfish below A. When the lake goes below level B, surface area is dramatically reduced, and the rate at which the water falls is correspondingly reduced. When the lake goes above C, surface area increases dramatically, which correspondingly decreases the rate at which the water rises. The net effect is to maintain a relatively stable level, leaving the biotic forms unthreatened.

The shellfish all lie below A and the lilies above D. Even though the water line is normally between B and C, in an unusually wet year it will rise above C. This has the effect of substantially increasing the surface area, and consequently the rate of evaporation, hence "stemming the tide" and saving the lilies. If the level falls below B, the surface area is significantly reduced, with the result that the loss from evaporation is curtailed. The shellfish are safe. Each of the populations is secure, but from wholly mechanical causes. The possibility of such a mechanistic explanation shows, as Bennett urges, that the "intentionalist" explanation is at best a manner of speaking and at worst just plain wrong-headed. The point is generalizable. When there is a tractable mechanistic explanation, an explanation in terms of "purpose" or "intention" is a mistake.

Such principles as Bennett's do double duty in requiring us to minimize the strength of mentalistic attributions and in reminding us that the criteria we do in fact apply are weak and fallible. Thus, to press the first and most central moral, no appeals to specific behaviors will *require* the Cartesian to opt for the attribution of mentality in the absence of a demonstration that an underlying mechanistic explanation is not to be had. This is a difficult task. Savage-Rumbaugh has argued that we are not in a position to answer the question of the linguistic competencies of our primate cousins at Yerkes. She writes:

> We do not have evidence of symbolic representation unless it is demonstrated that cognitive operations proceed even when an individual is presented only with symbolic information. ... In fact, no theoretical framework has been developed within the field of animal psychology to distinguish between conditioned discriminative responses and symbolic representational responses (Savage-Rumbaugh 1980, p. 51).

270

In this case the task of disarming Morgan's canon has an analogue in Chomsky's argument for generative grammars. To see how the issue might arise in the case of animal communication, we can point to the disputes over the syntactic properties of the languages which have been taught to apes. If it turns out that the languages they are *capable* of learning are sufficiently complex to require stronger grammars in order to explain them – as, if we are to believe Chomsky, human languages cannot be captured by phrase structure grammars – then, at least, we would have grounds for thinking their linguistic competence comparable to our own. If it turns out that weaker grammars are sufficient, then we ought to explain their linguistic abilities, at least in the interim, in those simpler terms. Fortunately, I don't have to enter into this dispute, save to point out that the available work has not yet mandated the need for more complex grammatical systems.

Let me turn instead to a second moral. Though Bennett, in unison with Savage-Rumbaugh, is skeptical about the attribution of complex intentions to nonhumans (see Bennett, 1976, §62; and Savage-Rumbaugh, 1980), it is in fact important and interesting to see that some cases of animal communication, despite being strictly non-linguistic, come reasonably close to meeting even Bennett's standard for communicative behavior, and certainly support the attribution of complex mental processing. Thus, Savage-Rumbaugh describes the use of packaging labels by Sherman and Austin, and the remarkable work with Kanzi, whose communicative skills promise to excel even those of his compatriots at Yerkes. The cases certainly show a sensitivity to the responses of others. Indeed, by my lights, they show an ability to monitor the effects of one's behavior on others and manipulate them. In other words, it shows a "communicative intent". Here we have the sort of case which will make it reasonable, if not quite compelling, to shift to explanations in terms of higher level, cognitive capacities.

We should not overstate the conclusion. The evidence does *not* defeat the first leg of the Cartesian gambit. While it is possible in principle to gain such evidence, behavioral studies of the sort utilized by Savage-Rumbaugh are not *sufficient* to show that the underlying mechanisms are homologous. The cognitive abilities of Kanzi and others *are* striking, and their communicative abilities are remarkable. But whether there is communication is not at issue, any more than is their intelligence. What is at issue is whether the cognitive mechanisms responsible for their 'linguistic' behavior are qualitatively similar to those responsible for human verbal behavior or whether they merely mimic human abilities over a limited range.

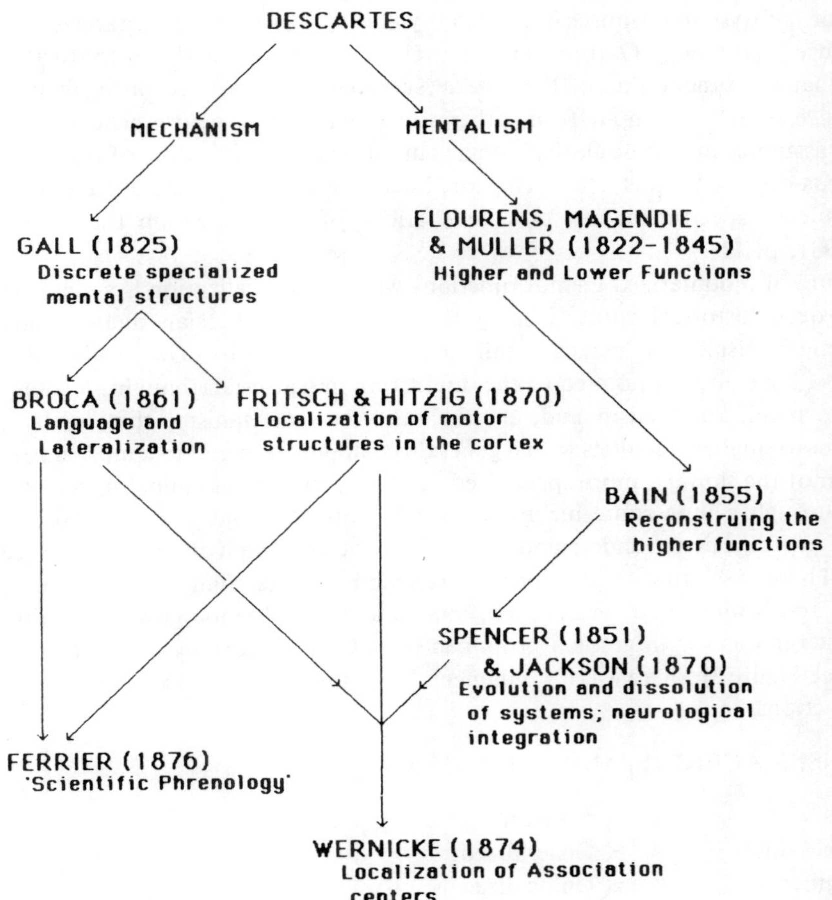

Figure 2. A diagrammatic representation of intellectual influences within the neurosciences.

Mental Modules and Brain Mechanisms

I doubt that behavioral studies alone are liable to resolve this issue. That does not make it irresolvable, though it does mean we must look elsewhere for a resolution. The sort of evidence we might seek can be illustrated by its relevance to the second leg of the Cartesian gambit, and indirectly to the first. Recent work in cognitive psychology, as well as much within the neurosciences, has been committed to a "modular" conception of cognitive organization. Applied to the case of language use, this is one which supposes "the language faculty is decomposable into a complex of subsystems whose interaction yields the observable character of language use" (Bradley, Garrett, and Zurif 1980, p.   ) and which further supposes that the functioning of the several subsystems are carried on in relative independence of specific input from other subsystems.[3] The neurosciences often couple the assumption of modularity with a commitment to localization of function. In the words of Eric Kandel, "all behavior, including higher ... mental functions, is localizable to specific regions or constellations of regions within the brain" (Kandel 1981, p. 11).

The picture of modularized mental functions with strict localization is a classical one, with deep historical roots, tracing finally to both Cartesian dualism and Cartesian mechanism. This makes it enlightening and useful to take an historical excursion.[4] I have already pointed to the dual strain in Cartesian thought, leading, on the one hand, to dualism and, on the other, to mechanism; that is, to a commitment to 'higher' faculties with a generalized application and to a mechanistic explanation of the 'lower', more specialized, abilities. By the beginning of the 19th Century (through a long route including Port Royal), this had given rise to two distinctive approaches to understanding the role of the brain in behavior. (See Figure 2) There was, first, a tradition represented by such figures as Flourens, Magendie, and Muller; and, in contrast, there was the "phrenological" tradition whose most famous (or infamous) proponent was Gall. According to the former theorists (see Figure 3), there was a fundamental difference between the 'higher' and 'lower' functions.

FLOURENS, MAGENDIE, MULLER (1822-1845)

Higher Functions (Will, Judgment)	⎰ Cortical ⎨ Generalized ⎱ Unlocalized
Lower Functions	⎰ Subcortical ⎨ Specialized ⎱ Localized

Figure 3. Localization of Mental Functions: Motor functions were ascribed to anterior structures, and sensory functions to posterior structures.

---

[3] For recent discussions, see Noam Chomsky, *Rules and Representations* (New York: Columbia University Press, 1980); also J. Fodor, *The Modularity of Mind* (Cambridge: MIT Press and Bradford Books, 1983).

[4] Some of my portrayal is admittedly eccentric, and is in need of greater detail to gain accuracy. I will here sacrifice some accuracy for simplicity. For those who desire a full treatment (one to which I am indebted), the best source is R. Young (1970); for a briefer treatment, see Young (1968).

The 'lower' functions – in particular, sensory and motor functions – were taken to be subcortical in origin, specialized in their application, and strictly localized. The 'higher' functions – paradigmatically, will and judgment – were, by contrast, taken to involve cortical mediation (this is usually a dualist view in more senses than one), to be applicable in a variety of domains, and to lack any specific location. The emphasis among these theorists was on the "unity" of the nervous system, and Flourens is particularly noteworthy for his attempts to demonstrate by systematic ablation that the cortex did not house any specific higher functions. One result was particularly important in aiding their opponents, however. This is the so-called "Bell–Magendie law". What was shown was that there was a "specific action" within the spinal nerve roots, so that anterior nerves were specialized for motor functions and posterior nerves for sensory functions. This was, to be sure, restricted to subcortial structures, but it was not long before the general distinction between anterior and posterior structures as being, respectively, motor and sensory was embraced by John Hughlings Jackson in England and by Meynart on the continent (with the latter directly responsible for the adoption of the view by Wernicke).

But we're getting ahead of the story. For there was an alternative tradition committed to localization and to the existence of discrete mental structures. This was, of course, promoted by Franz Joseph Gall. Gall's views may be usefully summarized as a function of four correlated variables (Figure 4).

The basic postulate was that the size of cortical structures was proportional to some basic ability; that is, to the degree to which some "fundamental power" was realized in the individual or the species. The basic abilities were supposed to determine, in a fairly direct manner, actual performance. Gall wrote, "The moral and intellectual dispositions are innate; their manifestation depends on organization; the brain is exclusively the organ of the mind; the brain is composed of as many particular and independent organs as there are fundamental powers of the mind" (Gall 1835). Development was correspondingly taken to insure that the cranium would respond to brain development. Though Gall made a fundamental error in thinking that the structure of the cranium would reflect the size of underlying organs, the basic assumption was carried over essentially intact within the localizationist tradition.

It is but a small step to the work of Broca, with his emphasis on neuro-anatomy rather than cranial measures. What Broca claimed to show us was that there is localization of language functions in the cortex (and, more specifically, that they are localized in one hemisphere of the cortex). This was a crucial step; for with this demonstration, it became clear that at least some 'higher' functions promised to have a fully mechanistic explanation in terms of specific neural action, just as sensory and motor functions admittedly had. Coupled with this, there was a

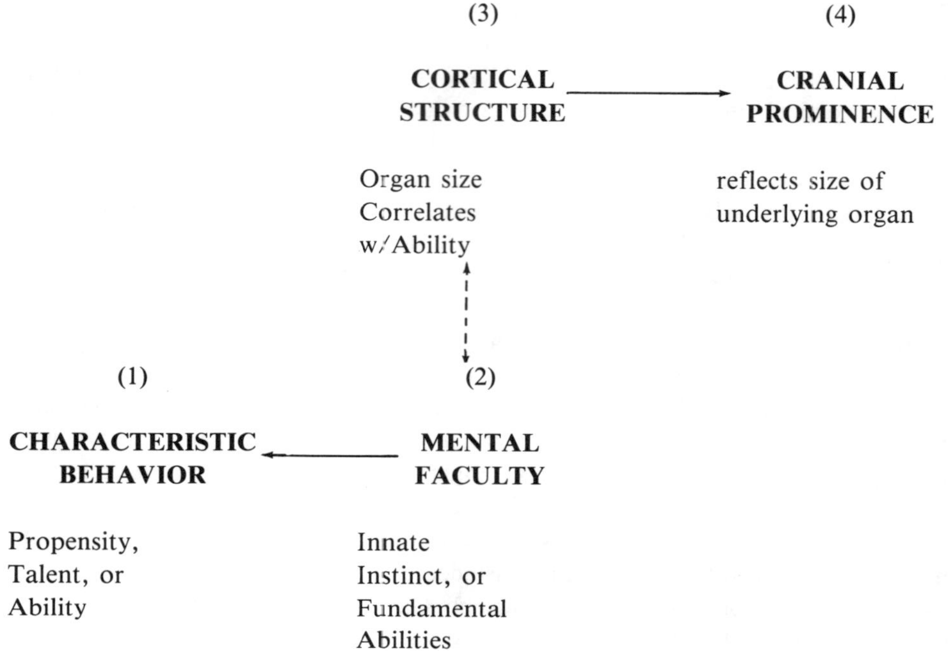

Figure 4. The Phrenological Scheme. There are four correlated variables. The crucial variable is (3) the size of the underlying cortical structure (organ). The correlation between (1) and (4), which formed the empirical basis for phrenology, is supposedly explained by, and evidence for, the connection between (2) and (3).

reinterpretation of the nature of 'higher' mental functions at the hands of Bain, a British follower of John Locke. What became something of a commonplace was that the more complex, 'higher', mental phenomena were but constructs built of sensations by the laws of association. The resulting collapse of the distinction between 'higher' and 'lower' functions was the birth of the 'new phrenology'', as Ferrier was later to dub it.

In the hands of Herbert Spencer and John Hughlings Jackson, the localizationist school, now bred from the mechanism of Gall and the associationism of Locke, was revitalized again by its union with evolutionary thought (see Figures 2 and 5).

In their hands, five doctrines became dominant. *First,* all nervous centers were taken to be essentially sensory-motor connections (See Jackson 1931, Vol. II, p. 63). Crucially, this included the higher functions: Jackson in fact thought his work showed the motor constitution of the will. *Second,* the centers of sensory-motor

SPENCER (1855) AND JACKSON (1870):
(1) Sensory-Motor Constitution of Nervous Centers
(2) Centers of Coordination Localized
(3) Continuity of Systems: No Physiological Discrimination of Higher and Lower
    Centers
(4) Disintegration of Coordination with Ablation of Motor Centers
(5) Epiphenomenalism and Parallelism

Figure 5. Summary representation of Spencer and Jackson.

coordination were taken to be strongly localizable (see Spencer 1881). *Third*, any discrimination between motor functions and 'higher' functions was banished. *Fourth*, the effect of damage or ablation of the neural centers was a loss of the sensory-motor coordination thereby mediated. And, *fifth*, conscious phenomena were relegated to a secondary position as mere 'epiphenomena'.

In coordination with these events, the work of Fritsch and Hitzig showed, by techniques of electrical stimulation of the cortex, that there was a degree of specialized localization for motor structures in the cortex. This served, on the one hand, to reinforce the rising localizationist trends directly, and, on the other to reinforce the allied views that the cortex subserved sensory-motor coordination and that the basic dichotomy reflected in the Bell–Magendie law could be projected to the division between posterior and anterior cortical structures.

The culmination of this tradition came with the work of Ferrier in England and Wernicke in Austria. Ferrier carried the results of Fritsch and Hitzig further, locating no less than fifteen motor and sensory areas, together with centers for voluntary movement. This was supposed to exhaust our cognitive abilities. As Ferrier said, "the cerebral hemispheres consist only of centers related respectively to the sensory and motor tracts, which connect them with the periphery and with each other" (1876, pp. 256–257). Wernicke, similarly, carried the "connectionism" and "associationism" into the analysis of 'higher' functions. Specifically, he promoted a systematic explanation of linguistics faculties and the aphasic syndromes arising from their disruption as a result of the disruption of association pathways. The most striking results lay in the analysis of language comprehension and production. It is enlightening to see how this work developed, how it embodies associationism and localizationism simultaneously, and how these two views combine to undermine the cognitivist strain engendered by Descartes. By 1874, he had uncovered a form of aphasia distinct from that which had been observed by Broca – one which particularly affected the comprehension of speech. He embraced a trichotomous division among the aphasias (Figure 6):

276

PURE APHASIAS affect only speech, with no paralysis or intellectual impairment.

(1) SENSORY APHASIA (Wernicke's Aphasia)	(4) MOTOR APHASIA (Broca's Aphasia)	(7) CONDUCTION APHASIA
A. Loss of "acoustic memory representations" for speech.	A. Loss of "motor memory representations" for speech.	A. Deficiency in the ability to mimic: loss of "word concept"
B. Loss of word-sound associations, and therefore of the meaning of the spoken word.	B. Maintains speech comprehension, though with some deficit for complex constructions.	B. Maintains speech comprehension
C. Articulate speech is intact: copious and grammatical but also inappropriate; naming is also impaired.	C. Loss of articulate speech, with (occasionally) the exception of a few substantives; a lack of logical particles.	C. Articulate speech, though with some interchange of words.

Figure 6. Forms of aphasia according to Wernicke (1874).

It later became necessary to acknowledge several additional forms of pure aphasia:

ADDITIONAL FORMS RECOGNIZED BY WERNICKE (1885, 1906):

(5) PURE WORD-MUTENESS:

A. Disruption of afferent fibers.
B. Impaired use of speech but not of writing.
C. Ability to vocalize intact.

(2) PURE WORD-DEAFNESS:

A. Disruption of efferent fibers.
B. impaired understanding of speech, but not of written texts.
C. hearing still intact.

(3) TRANSCORTICAL SENSORY APHASIA:

> A. Loss of comprehension of written word, but with the ability of mimic maintained.
>
> B. Hence, without a general or specific loss in hearing.

(6) TRANSCORTICAL MOTOR APHASIA:

> A. Reduction (though not a total loss) of capacity for spontaneous speech, but with the ability to mimic maintained.
>
> B. Hence, without a significant motor deficit.

Figure 7. Additional forms recognized by Wernicke (1885, 1906).

The specifics will not substantially affect our discussion. What is important is that, as Wernicke then saw it, there were a variety of "pure aphasias" – that is, aphasias which affected limited abilities – each with a characteristic syndrome. The total matrix of syndromes was in turn interpretable in terms of a single scheme:

1. *Wernicke's schema of the psychic reflex:*

    (S) Acoustic Imagery (sensory input)
    (M) Motor Imagery (motor output)
    (B) Concept Center
    (A) Discharge -- site of conceptualization
    (Z)Ideational Processes (goal planning & integration)

2. *The Lichtheim Aphasia Schema:* the Varieties of Aphasia.
    (1) Nuclear Speech Deafness
    (2) Peripheral Conduction Speech Deafness
    (3) Central Conduction Speech Deafness
    (4) Nuclear Aphasia
    (5) Peripheral Conduction Aphasia
    (6) Central Conduction Aphasia
    (7) Conduction Aphasia

Figure 8.

The scheme provides an associationistic model for speech comprehension as a development from a simpler scheme due to Lichtheim. Learning a language was a matter of establishing functional associations between perceptual elements resulting from the lowered resistance – we would now say facilitation – following repeated contiguous presentations. Each of the pure aphasias was due to the disruption of one point in the association chain. Thus, a disruption of the pathway from S to A would impair comprehension while leaving auditory faculties intact. In short, it would be a case of Wernicke's Aphasia. Similarly, a disruption of the pathway from Z to M would impair expression of thought, while leaving comprehension intact. This would be a case of Broca's Aphasia. The other pure aphasias were given a similar treatment.

The significance of the existence of pure aphasias is particularly clear in Wernicke's writings. The very fact that there are cases in which the symptomology is exhibited is taken to show that there are centers responsible for the specific functions and for those functions alone. That is, there is a commitment to a *modularization* of function at two levels. On the one hand, motor and sensory centers for language are held to be distinct. On the other, the language processing centers are held distinct from other cognitive functions. They are specialized in their application, and also independent of one another. In the terms of Charles Spearman, this is an "oligarchic" model of mental functioning. The distinctness of these centers, in turn, relies on the assignment of *unique and specialized* functions which can be carried out in relative *insulation* from other centers.

The discipline did not, of course, stop here. There is now considerable discussion concerning the precise characterization of the deficits involved in the aphasias, whether, for example, Broca's aphasia is a syntactic or phonological deficit. This is one very interesting part of the story I'll have to sidestep, since I want to get on to a look at the place of Edgar Zurif's work in this scheme of things. If the suggestions which his research indicates are right, it is a significant departure from the tradition, and will reshape our understanding of language and cognition. It should be emphasized that the conclusions are, at this stage, suggestive only. Zurif acknowledges the modularist assumption of the tradition, pointing out that the integration of neurology and theories of linguistic processing rests on "the assumption that the faculty of language can be broken down into a complex set of interacting constituent systems", including at least phonological, syntactic, and semantic modules. He then isolates the central issue, "whether the processing of sentence structure is performed independently of the semantic consequences of such structure" (Zurif 1980, p. 307).

The modularity assumption has clear merits, both methodological and empirical. To take a simple and well-known case, there is evidence that sentence-parsing is a procedure carried out in relative isolation from other, 'higher', cognitive functions. If an experimental subject is set the task of discriminating words from non-words (a behaviorist has her push a button, a cognitivist allows her to talk), the latency for the decision can be shortened if the target word is preceded by a related word. This "priming" in some way facilitates access to the target word. Swinney presented potentially ambiguous priming words in contexts that strongly biased their interpretation. Thus, using "bug" as the priming word, Swinney presented these sentences:

> They found several spiders, roaches and other *bugs* in the room. They uncovered several micrphones, transmitters, and other *bugs* in the room.

What he found was that in each case, the presentation of "bug" primed associates of *each* homograph (see Swinney 1979, pp. 645–659). Thus, priming with either sentence would facilitate the recognition of both "spy" and "fly", despite the lack of semantic associations. The conclusion would appear to be that "much of linguistic analysis [is] ... an 'ignorant' process" carried on in relative isolation from other cognitive structures (see Pylyshyn 1983, pp. 147–151).

280

Yet as with many such models, it is no simple matter to discern whether the attraction of the model is an artifact of the methodology or due to the actual organization of language. It is this issue which is so elegantly addressed by Zurif's work. In the classic studies of aphasic patients – in Broca's work, even more clearly in Wernicke's, and in much recent work as well – the conclusion is drawn that the organization of language in the brain allowed for the isolation of those mechanisms responsible for comprehension and expression in such a way that the existence of "pure" aphasics showed the functional independence of those mechanisms from one another (and from yet other mechanisms revealed in alexia and agraphia). The principal support for this view was the observation that specific structural deficits in language use and comprehension – most notably the "little words" indicative of grammatical structure – accompany expressive (Broca's) aphasia, while such defects in grammatical structure are not evident in receptive (Wenicke's) aphasia. The following exchange from Wernicke's clinical work may be taken as reasonably characteristic of the latter form of aphasia:

> DR. WERNICKE – Good Morning. How are things going?
> MRS. ADAM – Thank you, things are fair.
> DR. W. – How old are you?
> MRS. A. – Things are going all right, thank you.
> DR. W. – How old are you?
> MRS. A. – Do you mean how I Hei [neologism] ... How I hear [hore]?
> DR. W. – I would like to know how old you are.
> MRS. A. – Yes, that I don't know at all. What my name [Wie Ich so heissen schwiere] ... What I'm called – hear (hore).
> DR. W. – Would you perhaps give me your hand?
> MRS. A. – I don't know how I [Presents not trace of comprehension].

Zurif and his colleagues have argued that, even in the case of patients suffering from Broca's aphasia, there are collateral and parallel deficits affecting comprehension and, consequently, that we are obliged to acknowledge that comprehension and production are not easily separable as cognitive functions. The solution he once favored was not an abandonment of modules, or even of an oligarchic model, but a redefinition of them. Thus, the existence of collateral and parallel deficits shows at most that the module for syntactic manipulation is implicated in a variety of functions, and that there is a module for syntactic manipulation.[5] His most recent work carries this yet one stage further, arguing for collateral and parallel deficits in non-linguistic cognitive functions (more generally in retrieval), and indicates that

---

[5] It would be good to examine a case of Wernicke's aphasia with the same issues in mind. If there is a comparably sophisticated treatment here, I do not know about it.

in the so-called "Broca's aphasia" there is a loss of some general information-processing capacity (see Grodzinsky, Swinney, and Zurif 1983).

It is an interesting question how far such work threatens componential models. Several options can be explored, including: (1) reformulating our conceptions of the functions of the modules, thus retaining the basic componential structure; (2) allowing for diminished functional integrity within the subsystems, and thus accepting that the system is not subject to the degree of decomposability, or modularity, initially supposed; and (3) shifting from a modular approach to a more "connectionist" or "holist" scheme, emphasizing the hierarchical dependencies of cognitive functioning.

Animal Language and the Structure of Thought

I want, in concluding, to shift to look briefly at the implications of research in the neurosciences, and of Zurif's work in particular, to that of Savage-Rumbaugh. A componential model of cognitive abilities minimally incorporates substantial independence of functioning for the several modules underlying cognitive processing; in practice, it is commonly taken to implicate an oligarchic structure as well. The more traditional Cartesian approach, underlying the second leg of the Cartesian gambit, is committed to a monarchic structure: there is a hierarchically organized system, with more specialized 'modules' dependent on 'modules' whose application is domain independent.

Insofar as a componential approach to the understanding of language functioning is vindicated, it appears necessary to accept that the underlying cognitive organization responsible for language production and comprehension involves a number of highly specialized linguistic mechanisms. While it is surely to be expected that we would find related mechanisms in other organisms (as is evidenced by the lateralization discovered in macaque monkeys, with differential capacities for handling the various calls), a modular approach would make arguments for the continuity of linguistic communication between humans and nonhumans suspect exactly insofar as the comparison fails to respect phylogenetic distance. Specialized adaptations, in this case to social exigencies, would not be expected to correspond at any level of depth among unrelated or distantly related species. It would be expected that we would find analogues to human communication, thus conceived, but homologies should be few and not especially deep or robust.

Insofar as a componential approach is softened in the sorts of directions which seem indicated by Zurif's research, arguments from the continuity of various forms of communication on the basis of behavioral similarity may be more defensible, and useful as a cure for *a priori* skepticism, but still lie open to doubts over whether the similarities fail to run especially deep. The question needing to be asked is still whether the underlying organization responsible for the communicative behavior is similar, or whether, as Descartes was willing to bet, the communicative behavior of the "brutes" was due to mechanisms of a different sort from those underlying human language use.

In conclusion, we may note one last point of historical irony. Though the two legs of the Cartesian gambit were intimately related, it now looks as if the plausibility of attacks on the first 'mentalistic' leg of the gambit – attacks of the sort advanced by Savage-Rumbaugh – are diminished insofar as we are given reason to think the second leg fails. That is, if the mental is modular in organization, then language is likely to be a "specifically" human trait, since the mechanisms for it will be "hard-wired" in the human species. But this hardly would prove much comfort to a latter-day Cartesian.

References

Bennett, J. (1976). *Linguistic behavior.* Cambridge: Cambridge University Press.

Bradley, D.; Garrett, M.; and Zurif, E. (1980). Syntactic deficits in Broca's aphasia. In D. Caplan (ed.), *Biological studies of mental processes.* Cambridge: MIT Press.

Chomsky, N. (1957). *Syntactic structures.* The Hague: Mouton.

Chomsky, N. (1966). *Cartesian linguistics.* New York: Harper & Row.

Chomsky, N. (1968). *Language and mind.* New York: Harcourt, Brace, Jovanovich.

Chomsky, N. (1980). *Rules and representations.* New York: Columbia University Press.

Ferrier, D. (1876). *The functions of the brain.* London: Smith & Elder.

Fodor, J. (1983). *The modularity of mind.* Cambridge: MIT Press.

Gall, F. (1835). Critical review of anatomico-physiological works with an explanation of a new philosophy of moral qualities and intellectual faculties. In Winslow Lewis (trans.), *On the functions of the brain and each of its parts,* Vol. VI. Boston.

Grodzinsky, Y.; Swinney, D.; and Zurif, E. (1983). Agrammatism: structural deficits and antecedent processing disruptions. In M. L. Keane (ed.), *Agrammatism.* New York: Academic Press.

Haldane, E. S.; and Ross, G. R. T. (1973). *The philosophical works of Descartes.* Cambridge: Cambridge University Press.

Jackson, J. H. (1931). *Selected writings.* Ed. James Taylor. London: Hodder and Stroughton.

Kandel, E. (1981). Brain and behavior. In E. Kandel and J. H. Schwartz (eds.), *Principles of neural science.* Elesevier/North Holland.

Kenny, A. (1970). *Descartes: Philosophical letters.* Oxford: Clarendon Press.

Leiber, J. (1975). *Noam Chomsky: A philosophical overview.* New York: St. Martin's.

Pylyshyn, Z. (1983). Syntax as an autonomous component of language. In M. Studdert-Kennedy (ed.), *Psychobiology of language.* Cambridge: MIT Press.

Savage-Rumbaugh, S. (1980). Do apes use language? *American Scientist, 68,* 49–61.

Spencer, H. (1885). *The principles of psychology.* London: Longman's.

Swinney, D. A. (1979). Lexical access during sentence comprehension: (re)consideration of context effects, *Journal of Verbal Learning and Verbal Behavior, 18,* 645–659.
Young, R. (1968). The functions of the brain: Gall to Ferrier (1808–1886), *Isis, 59,* 251–268.
Young, R. (1970). *Mind, brain, and adaptation in the 19th century.* Oxford: Oxford University Press.
Zurif, E. (1980). Language mechanisms: A neuropsychological perspective, *American Scientist, 68,* 305–312.

Editor's Commentary

There is an important parallel between the topic of this unit and that of the previous one, since both consider possible extensions to already existing interdisciplinary research clusters. However, there is also a noteworthy difference. While the evolutionary synthesis brought together disciplines that fell within the domain of the more general discipline of biology, cognitive science is an interdisciplinary research cluster that spans several different disciplines. The two disciplines that have been most central to this interdisciplinary research cluster have been artificial intelligence, a sub-discipline of computer science, and cognitive psychology, a sub-discipline of psychology. Other disciplines have also figured in cognitive science, including linguistics, neuroscience, anthropology, and philosophy. Hence, cognitive science is an interdisciplinary research cluster that integrates disciplines identified at even a very broad level.

It is clear that, while cognitive science arose somewhat later than the evolutionary synthesis, and has attempted to reach across higher level disciplinary boundaries, it has by now established itself as a powerful interdisciplinary research cluster. While there are few departments of cognitive science, there are well-organized multi-disciplinary programs at a significant number of universities. There is a major interdisciplinary society, the Cognitive Science Society, and several interdisciplinary journals (e.g., *Cognitive Science* and *The Behavioral and Brain Sciences*). In both the academic units and the interdisciplinary journals and societies, artificial intelligence and cognitive psychology have played the central roles. While there is some conceptual basis for this, since these are the two disciplines most readily able to contribute to an empirical investigation of cognitive phenomena, probably the more telling reasons are historical and social; it was primarily those in psychology and artificial intelligence (and to a lesser extent linguistics) who founded the society, named the field, started most of the journals, and so forth. (There has been an additional society founded by philosophers and philosophically minded psychologists, the Society for Philosophy and Psychology, which has provided an

Bechtel, W (ed), Integrating Scientific Disciplines. ISBN 90-247-3242-5.
© *1986, Martinus Nijhoff Publishers, Dordrecht. Printed in The Netherlands.*

additional interdisciplinary forum, but it has played a far less central role in the shaping of the interdisciplinary cluster.)

One other feature of cognitive science differentiates it from the evolutionary synthesis. While it is possible to point to a particular theory as the product of the synthesis (even if there is no one universally accepted statement of this theory), there is no central theory to cognitive science. There has been, however, some commonality at the meta-level to the theories developed in cognitive science. For the most part, they accept the view of the mind as an information processing system and try to model various cognitive phenomena in terms of operations being performed on representations of information. But this meta-level assumption is not on the same level as the synthetic theory of evolution which is often presented as the product of the synthesis and the vehicle for integration of disciplines within the synthesis.

Rather than examining the origins of this already functioning research cluster (Abrahamsen's Introduction to this unit provides a brief overview of this history), the focus of this unit has been on ways the scope of cognitive science might be further extended to incorporte contributions from yet other disciplines which, conceptually, would appear to play an important role in a fully-articulated cognitive science, but so far have not figured centrally. Currently, most work that goes by the name "cognitive science" has focused on typical adult humans, using either the experimental techniques of cognitive psychology, the simulation techniques of artifical intelligence, or the analytic techniques of linguistics to determine the processes of cognition. The theoretical models that have been developed reflect the rationalist perspective in philosophy (Chomsky has been the most explicit in recognizing this rationalistic approach).

One of the consequences of the rationalistic orientation adopted in current cognitive science has been the focusing of interests on certain kinds of problems and techniques and a lack of interest in others. Given the nativist emphasis in classical rationalism, it is not surprising that learning has not played a central role in cognitive science (the most extreme rejections of learning are found in Fodor, 1981). But Richardson's discussion of the "Cartesian Gambit" draws attention to two other areas that have not been focal in cognitive science. Descartes, employing the gambit of first using language as an indicator of thought and then contrasting cognitive functions with mechanistic ones, denied both that machines, including brains, and animals of other species could think. While contemporary cognitive scientists are not proponents of the Cartesian Gambit, since they generally are not dualists and reject at least the second, antimechanistic leg, (the use of computer models to simulate cognitive function itself commits one to a mechanistic analysis), they have continued to accept some of the consequences of the gambit. Thus, many cognitive scientists continue to shun the use of neuroscience or comparative psychology in developing their theoretical frameworks. Of course there have been prominent exceptions to this generalization. In particular, there have been cognitive scientists who have thought that much could be learned from neuroscience (see Posner, 1978) and neuroscience is nearly always included in lists of the contributing disciplines to cognitive science. Yet, it has generally not been regarded as a promising source of information for guiding cognitive theorizing. In fact, principled arguments have been made (e.g. by Fodor, 1974 and Pylyshyn, 1984) that since the interesting regularities of cognition will only be found at the cognitive level, not at the neuroscience level, neuroscience will not be of much relevance for studying cognition. (I have argued against these claims in Bechtel, 1984.)

Recently, though, in a number of areas cognitive scientists have begun to consider other ways in which neuroscience might be applied to developing models of cognitive activity. The recent proposals of connectionist or parallel distributed processing models, for example, were inspired by reflections on the character of neurological processing (see Hinton and Anderson, 1981, and Rumelhart and McClelland, in press). Partly as a result of taking their inspiration from the idea of neural networks, these connectionist models reject the rules and representations type of architecture used in more traditional cognitive science. They treat the generalizations that have been described in terms of inferences being made on propositions as emergent (see Rumelhart, 1984). In addition, they also reintroduce a concern with learning. Thus, in many respects, these models represent a departure from the rationalistic approach of traditional cognitive science and capture some aspects of the associationistic tradition (see Bechtel, 1985). Yet another area in which neurological information has been brought into studying cognitive processes is in the area of memory where distinctions between different memory processes are being proposed on the basis of studies of neurological deficits (see Graf, Squire, and Mandler, 1984).

The contributions in this unit explore how both comparative psychology and a particular approach to neuropsychology that focuses on the deficits that result from lesions might make greater contributions to the cognitive science research area. It is important to recognize why these areas are currently outside the central focus of the cognitive science research domain. As noted above, the focus of cognitive science has been on normal adult human cognitive functioning. The introduction of comparative studies with other organisms and with humans with brain lesions involves an extension of the domain of the research area. As Shapere (1984) argues in his discussion of domains, classifying phenomena together as part of one domain requires *demonstrating* important relationships between the phenomena. Extension of cognitive science into these areas also involves a change in the research techniques

288

and, as is brought out in the papers of Savage-Rumbaugh and Hopkins and of Richardson, requires challenging some of the theoretical postulates of the cognitivist approach and the development of new theoretical frameworks.

It is important to bear in mind that the problems which generate the theoretical frameworks discussed in this unit did not begin with problems recognized within cognitive science. Rather, they resulted from a recognition of tensions or conflicts in studies within the other domains or between approaches in these domains and those used in cognitive science. The frameworks that are developed are, in part, to overcome the conflict and provide a useful point of intersection for developing further work. In this respect the proposals of this unit provide a different foundation for interfield theorizing than we have seen in the previous units. There is also, however, a significant difference between the two cases. The work of Savage-Rumbaugh and Hopkins takes its starting point from outside of cognitive science and produces a framework that might be of use to cognitive science in extending its inquiry. In the work of Zurif as described by Richardson, on the other hand, the motivation came from cognitive science. It was a desire to bring work from neuropsychology to bear on issues of cognitive science that lead to building a interfield theoretical framework. Given these differences in starting points, the products of the two endeavors are also quite different. The framework of Savage-Rumbaugh and Hopkins has the status of a proposals that may in the future prove useful within cognitive science whereas, with Zurif's work, the results of psychobiology are already proving informative for cognitive science. I will consider each of these briefly in turn.

Savage-Rumbaugh and Hopkins note early in their paper a radical division separating human psychology from animal psychology, due in large measure to the use of verbal paradigms in the studies in human psychology (this is step one in Richardson's Cartesian Gambit). However, overcoming this division is not their primary motivation. Rather, their desire for a new theoretical framework stems from the fact that ape language projects, such as the one in which they are engaged, do not fall naturally either within the domain of animal learning studied by experimental psychologists or within the domain of animal behavior studied by ethologists. Some psychologists engaged in ape language projects have dealt with this by simply borrowing the framework from studies of human language. Interestingly, Savage-Rumbaugh and Hopkins do not follow this lead, but rather try to shape a theoretical framework for ape language studies that first overcomes the dichotomy of behavior and learning that has divided animal studies. The connections with human studies only emerge in the context of the framework they craft for this other challenge.

To develop this framework, Savage-Rumbaugh and Hopkins appeal to yet another discipline, evolutionary biology, but in a different way than that connection has been used in animal ethology. Rather than applying evolutionary considerations only to fixed behavior patterns, they propose to consider the evolution of learning processes. Thus, they invoke Mayr's (1974) distinction between open and closed problems (for another use of this distinction in building interdisciplinary connections, see Wimsatt's paper in the previous unit), and consider what kind of operations are needed to employ an open program. (Savage-Rumbaugh and Hopkins' interest in the closed/open program distinction is different than Mayr's. Mayr was concerned primarily with distinguishing when each strategy would be evolutionarily advantageous.) For acquiring communicative behavior, Savage-Rumbaugh and Hopkins propose a progression of requirements from being aware of the goals of one's communication (which is necessary in order to be able to revise strategies so as to satisfy these goals), to being aware of how one communication can alter the behavior of others, to finally being aware of how others use communication to alter one's own behavior. As an organism is able to develop these different levels of awareness, it is able to learn and employ more complex communicative intentions.

By developing this framework, Savage-Rumbaugh and Hopkins are able not only to provide a coherent framework for understanding ape language projects, but also to offer a valuable framework for using these studies to better understand human linguistic function. As they note, human language activities have often been seen as simply the product of our great learning capacity. These activities have also been viewed, especially within cognitive science, as simply highly formal activities. While the notion of communicative intention does, in fact, have roots in studies of human language use, notably those of Austin (1962), Searle (1969) and Bates (1976), it is hardly the dominant notion shaping the study of human language. Rather the focus has been on the formal manipulation of linguistic units. While it is still unclear how much can be learned about the formal manipulations underlying human speech from ape language projects, by developing the notion of intentional communication, Savage-Rumbaugh and Hopkins have provided a framework that might be valuable for understanding human language use. Through understanding what communicative intentions can be developed in other species, and possibly what communictive intentions cannot be developed, we can gain a perspective on and understanding of the phenomenon of language that the first leg of the Cartesian gambit took to be a uniquely human phenomenon. Whether it will result in major contributions to cognitive science will partly depend on whether ways can be found to interface this framework with those now used to study language in cognitive science.

Richardson describes a case where the second motivation for developing a new cross-disciplinary framework, that of reconciling differences between theoretical frameworks from different disciplines, was a major factor. He traces a long history of neuroscience research that has tried to identify the mechanisms underlying cognition, and language use in particular (which constitutes an attempt to trump the second leg of his Cartesian Gambit). The early history of this research, growing out of both the localizationist traditions of Gall and the associationist tradition of Locke and Bain, culminated in Wernicke's program of localization of association centers coupled with a connectionist model linking these centers. At this stage in its history this program had denied the anti-mechanism of the second leg of the Cartesian Gambit but it was not in a position to interface with the approach of modern cognitivism. Like modern cognitivism, it approached the mind as a modular system (where distinctive operations are assigned to each module), but the two disciplines divided the modules quite differently. Whereas the neuropsychological lesion tradition followed the associationist tradition in defining the modules in terms of positions in sensory-motor chains, cognitive science, following linguistics, has differentiated its modules by the kinds of processing performed on what might be the same content or representations. Thus, cognitive scientists proposed modules for phonetic processing, syntactic processing, and semantic processing. While there was no obvious problem within either the cognitive science itself or in the neuropsychological research area (the research tradition of Wernicke continues to be developed in work like that of Geschwind, 1979), there was an obvious inconsistency between these two approaches to modelling language processing.

For cognitive scientists interested in developing models of language comprehension and production, the lesion-produced aphasias were obviously a phenomenon of interest, but before they could employ the results of lesion research to develop their theorizing, a connection between the conceptual frameworks of the two traditions had to be developed. The research of Edgar Zurif was of central importance in developing such a framework, for he provided a way to reinterpret the lesion data in terms of the kinds of processing units cognitivists were employing. A major accomplishment of this approach was to show that, contrary to the interpretation offered within the associationist tradition, Broca's aphasics do suffer comprehension deficits. These deficits, however, are only manifest when syntactic indicators are critical to the interpretation of the sentence (Bradley, Garrett, and Zurif, 1980). This suggested that the function of Broca's region might be in syntactic processing. With this reinterpretation, a framework was established in which neurological lesion data could be applied to developing and elaborating the kind of models being developed in cognitive psychology. Note that there is a parallel between the situation confronted by Zurif and that which Darden (this volume) describes Dobzhansky as confronting, in that both found competing explanations of the same phenomenon in different fields. Zurif's resolution of the situation, however, is radically different from Dobzhansky's. Dobzhansky developed a framework that employed the mechanisms proposed by both fields and added yet a further mechanism to explain the origin of species. Zurif, on the other hand, adopted the solution provided by one field and reinterpreted the data of the other in accord with it. However, as Richardson indicates, Zurif's work has not stopped at this resolution. His more recent work may lead to another reconceptualization of the modules of cognition, as Zurif is suggesting that the module disrupted in Broca's aphasia may not be restricted to language processing, but may be involved in other automated processes as well.

We see in these two cases two different foundations for building interfield connections. Savage-Rumbaugh and Hopkins' work stems from problems that were confronted outside of cognitive science but the resolution they offer is one that is of potential interest for redirecting thinking within cognitive science. In the work of Zurif discussed by Richardson, the problem stemmed from an inconsistency between the frameworks of two disciplines, requiring as a solution a theoretical framework integrating the two fields. How much further integration of animal studies and neuropsychological studies with cognitive science will result from these frameworks is not something that can be answered now. We will have to monitor the future to see whether they become influential. In any event, the work of Zurif and Savage-Rumbaugh is not likely to radically alter the institutional character of this interdisciplinary research cluster. One should expect cognitive science to continue in much the same direction – not trying to establish itself as a distinct discipline, but providing a nexus for separate disciplines to interact. However, there is a question of how broad this interdisciplinary research cluster can become before it will itself fission into two or more groups that tend to become isolated in the way disciplines traditionally have. This would be a disappointment to many of the creators of cognitive science, but it may not be avoidable.

References

Austin, J. L. (1962). *How to do things with words*. Oxford: Oxford University Press.
Bates, Elizabeth (1976). *Language and context: The acquisition of pragmatics*. New York: Academic Press.
Bechtel, William (1984). Autonomous psychology: What it should and should not entail. In P. D. Asquith and P. Kitcher (eds.), *PSA 1984*. Volume 1. East Lansing, MI: Philosophy of Science Association.
Bechtel, William (1985). Contemporary connectionism: are the new parallel distributed processing models cognitive or associationistic? *Behaviorism*.

Bradley, D.; Barrett, M.; and Zurif, E. (1980). Syntactic deficits in Broca's aphasia. In D. Caplan (ed.), *Biological studies of mental processes*. Cambridge: MIT Press.

Fodor, Jerry A. (1974). Special sciences (Or: The disunity of science as a working hypothesis). *Synthese, 28*, 97–115.

Fodor, Jerry A. (1981). The present status of the innateness controversy. In J. A. Fodor, *Representations*. Cambridge: Bradford Books, MIT Press.

Geschwind, Norman (1979). Specializations in the human brain. *Scientific American, 241,* 180–199.

Garf, Squire, and Mandler, (1984). The information that amnesic patients do not forget. *Journal of Experimental Psyhcology: Memory and Cognition, 10,* 164–178.

Hinton, Geoffrey E. and Anderson, James A. (1981). *Parallel models of associative memory*. Hillsdale, NJ: Erlbaum.

Mayr, Ernst (1974). Behavioral programs and evolutionary strategies. *American Scientist, 62,* 650–659.

Posner, Michael I. (1978). *Chronometric explorations of mind*. Hillsdale, NJ: Erlbaum.

Pylyshyn, Zenon (1984). *Computation and cognition: Towards a foundation for cognitive science*. Cambridge: Bradford Books, MIT Press.

Rumelhart, David E. and McClelland, James L. (in press). *Explorations in the microstructure of cognition. Volume 1. Foudations*. Cambridge: MIT Press.

Searle, John R. (1969). *Speech acts: An essay in the philosophy of language*. Cambridge: Cambridge University Press.

Shapere, Dudley (1984). Remarks on the concepts of domain and field. In D. Shapere, *Reason and the search for knowledge*. Dordrecht: Reidel.

PART V
Infusing Cognitive Approaches into Animal Ethology

Introduction

JAMES L. PATE

*Department of Psychology, Georgia State University, Atlanta, Georgia
30303–3083, U.S.A.*

The three papers on cognitive animal ethology are by psychologists with varying backgrounds, ranging from rather traditional primatology and comparative psychology to modern human cognitive science. Thus, the perspectives are radically different and reflect very different aspects of the current approach to animal cognition. Comparative psychology, despite the title of one of John Watson's books, has never been at the center of psychology, but this condition may be less damning than it appears since, in fact, one may claim, to use Donald Campbell's terminology, that psychology has no center. However, the emergence of animal cognition can be viewed as an attempt to bring the study of animal behavior into the mainstream of psychology if not its center.

It is clear to the most casual observer that there has been tremendous growth in animal research in which cognition plays a substantial part. Consider the state of psychology, or more particularly animal psychology, twenty-five years ago, i.e., 1960. Hull had died in 1952. Skinner had not reached the peak of his popularity but was certainly a major factor in animal research. Tolman died in 1959, and even though his system included cognitive concepts, they were rather different from the concepts of today. In 1949, Harlow published research on learning sets, and transpositional phenomena had been described. While few quarrled with the observations, there was considerable doubt about the appropriateness of the explanations of these phenomena. In fact, much of the work in animal research from 1930 to 1960 involved attempts to explain these phenomena in the traditional behavioristic terms of that day. Today, however, these phenomena or similar ones and the cognitive explanations of them would be accepted readily. Thus, in 1960, animal research was very different from that of today.

Although the change in animal psychology has been gradual, a comparison of the animal psychology of today with that of only twenty-five years ago impresses one with the magnitude of the change. While some may view the emergence of cognitive animal research as a return to mentalism or even the anecdotal introspection of

Bechtel, W (ed), Integrating Scientific Disciplines. ISBN 90-247-3242-5.
© *1986, Martinus Nijhoff Publishers, Dordrecht. Printed in The Netherlands.*

George John Romanes, I believe that it is neither, and the papers in this section constitute potent evidence against that view. What has changed is not the rigor of theorizing or doing research but the evaluation of the plausibility of the theories. While Tolman's view was not widely accepted, it would be judged to be much more plausible today than in the 1950's, and the present papers are in the Tolmanian tradition, i.e., inferences about underlying processes are made on the basis of objective observations. Notice that I have not mentioned correctness of the explanations, and there is good reason for omitting correctness as a factor influencing our acceptance of an explanation. Consider the number of theories and even observations that had little or no influence on science at the time the theory was stated or the observation was made. A theory that was not accepted in one period may be accepted readily at another period. Perhaps there is a scientific analog of behavioral preparedness which might be called theoretical or systematic preparedness. The theories that are proposed must be related to our knowledge – a concept discussed by Mason. However, the relation between new theories and our existing knowledge is, to use the technical term, fuzzy. Theories must advance our understanding, but they must not be too far ahead of the times.

Mason claims that "we begin the study of behavior by relying implicitly and profoundly on anthropomorphism" (Mason, 1985, p. 8). Rumbaugh and Sterritt claim that studies of intelligence are anthropocentric. In some respects we are certainly anthropomorphic and anthropocentric, but can these tendencies be avoided or should they be avoided? Both Mason and Rumbaugh and Sterritt believe that there is no problem. Mason applies the term "knowledge" to relatively low level organisms, leading to the question of the pragmatic advantage of extending the concepts of knowledge or intelligence to such creatures. Do the concepts help explain their behavior?

Psychology has historically emphasized methods at the expense of theories and processes, and that theme is reflected in Bechtel's commentary on this section, where he notes that the methods of human cognitive psychology have not been applied to the study of animal cognition. It can be argued that the proper study of cognitive animal ethology or of psychology concerns the processes exhibited by animals rather than the methods of investigation. In fact, one must adapt one's methods to the particular species being studied rather than attempting to utilize the same methods or even the same tasks across species. All of the papers in this section contain some aspects of this approach and consequently constitute important advances in our understanding of cognitive processes.

Behavior Implies Cognition

WILLIAM A. MASON
*Department of Psychology, and The California Primate Research Center,
University of California, Davis, California 95616, U.S.A.*

Introduction

My background and orientation is that of a psychologist with particular interests in animal behavior, and in what may be described as the evolution and natural history of minding. Because we are concerned here with the problems of communication between different scientific disciplines, it may serve a useful purpose if I begin by reviewing some general assumptions which I, as a working scientist, accept rather uncritically and which I believe are also accepted, at least implicitly, by a majority of empirical scientists, whether their principal interests are in behavior, or in some other field. I will treat these as areas of consensus, although I have never sought any direct evidence that this is the case. In my experience, scientists do not spend time discussing what they regard as self-evident truths. Instead, they proceed as though those with whom they form a community of interests hold the same basic attitudes and beliefs, until events prove them wrong. When that happens, of course, the problems of communication become acute, for it is seldom clear precisely where or why consensus breaks down. In the second section, I will consider some properties that are peculiar to behavior. They raise some difficult methodological and conceptual issues that are the subject of heated discussions within the behavioral sciences and are a continuing impediment to communication with other scientific disciplines. Next, I will present a perspective on behavior which is not new, but is different, I believe, from that held by scientists in other disciplines (and probably most non-biologically oriented behavioral scientists, as well). An essential element in this proposal is that behavior is most usefully regarded as an expression of the "organism-as-a-whole", rather than as a discrete, self-contained "system", analogous to the traditional systems of physiology. Finally, as an extension of this perspective I hope to show that cognition is an integral aspect of behavior, that it has been a major functional dimension in the evolution of behavior, and that comparative investigation of cognitive processes is for this reason one of the enduring interdisciplinary tasks that the scientific study of behavior must confront.

Bechtel, W (ed), Integrating Scientific Disciplines. ISBN 90-247-3242-5.

Areas of Consensus

As to commonly held assumptions, it seems clear to me that any scientific enterprise accepts the existence of a world beyond our senses and desires – an ultimate reality – that is orderly, structured, and in some degree intelligible to human beings. Although we can only come to understand and make sense of that reality in human terms, the human apparatus for knowing has evolved within and has been constrained by that ultimate reality, thus leading to a kind of historical assurance that a limited isomorphism exists between the world as we experience it and believe it to be, and the world as it really is. The principal concern of science is to improve that isomorphism, to create better approximations between our ideas about the nature of ultimate reality and that reality itself. The principal evidence that such improvement is possible, and has in fact occurred, is pragmatic, in the sense that we have become better able to interpret external events, to predict or control them, and to use our knowledge to make things happen in ways that are consistent with our plans and desires. Although ultimate truth is unattainable, we seem to be capable of improving our approximations of it.

A second point of presumed agreement is that scientific ways of knowing the world are elaborations, refinements, and extensions of everyday ways of knowing. Scientific understanding begins with common sense and the scientific enterprise is inevitably dependent on common sense as it proceeds. By common sense, I mean simple observation, everyday language, ordinary logic, and the unexamined beliefs, values and expectations that are assimilated from the culture, shaped by individual experience, and to some extent rooted in the biological makeup of the human species. Common sense is the basis for naive realism, the idea that the world is actually as it seems to be, and it is as indispensibly and inevitably a part of the scientist's ways of dealing with the world, as it is for anyone else. Any claim that a person has, by reason of his or her scientific vocation, managed to transcend the limitations of common sense, that such an individual has become something a bit more than merely human, naturally deserves to be treated with deep skepticism.

There are a set of procedures, however, which the scientific enterprise uses to overcome some of the shortcomings of common sense, and this is the third area of agreement. I am referring to what is usually called the scientific method, although scientific attitude would be a happier term. Its main tenets are that phenomena are to be given first priority, and that information about them (observations, data, etc.) must be gathered and presented in a way that is amenable to scrutiny by the scientific community, thus providing the basis for replication, confirmation, revision, and the like. In essence, independent confirmation of facts is the basis of scientific "objectivity". The emphasis in the behavioral sciences on such matters as operational definitions, inter-observer reliability, quantification of observations, and statistical analysis reflects, in part (but only in part) a concern with establishing or increasing objectivity.

A fourth assumption that I believe is widely accepted is that complete objectivity is never possible. "Facts" are theory- and value-laden. They are not immutable "elements" in the natural scheme of things, but the products of an elaborate set of interactions between a community of scientists (with all that implies in terms of theoretical expectations, accepted modes of observation, conventions about rules of evidence and about what constitutes "good" data, etc.), and some manifestation of that ultimate reality that they are attempting to understand. In psychology some thirty years ago it was fashionable to distinguish between empirical researchers, whose task was supposedly to collect "facts", and "theorists", whose job it was to evaluate their significance and to incorporate them into some larger integrative structure. The implication was that this was a natural and appropriate division of scientific labor. There is no denying that scientists differ in their interests and proclivities, and, surely there can be no objection to individual scientists' decisions to pursue empirical research or theory with whatever emphasis or mix they choose; the idea that fact and theory can be categorically distinguished, however, is seriously misleading.

Some Problematic Properties of Behavior

The long-range aim of psychology and related disciplines is usually described as providing a scientific account of what individuals do, and how and why they do what they do. The general orientation and methods of approach seem to me to be much the same as those of nonbehavioral sciences. At the same time, however, the subject matter of psychology and allied fields is distinct in some respects, and it poses some unusual descriptive and explanatory problems.

Some of these can be illustrated if we adopt the stance of pretending to observe for a brief period some reasonably complex and lively animal as – unaware of us – it goes about its normal activities in a natural setting. We find it lying in the shade, eyes half-closed. It raises its head, stretches, scratches, yawns, looks about, stares briefly at an insect crawling by, then gets up and, moves slowly to a nearby stream where it starts to drink. Another animal of the same species now appears, moving hesitantly toward the animal we have in view. It immediately ceases to drink, fixes its gaze on this second individual, moves stiffly a step or two in that direction, gives a harsh vocalization, then dashes toward the newcomer, who rapidly withdraws from the scene. In a few minutes another conspecific enters our purview. It moves smoothly and directly to within a few inches of our subject; they sniff and lick each other around the muzzle, and lie down side by side.

If we examine our own behavior as observers of this episode we might note that the focus of our attention has been an individual. This is our phenomenological bedrock, so to speak, the "agent" to which various actions are referred. We might also note that those actions convey a compelling impression of unity of organization and of purpose. We notice what the animal does is dependent on and articulated with the situation in obvious and varied ways. We find it resting in a shady spot; it goes to another place for a drink, and in so doing moves around a sizeable rock directly in its path, instead of climbing over it; its behavior is altered slightly and momentarily by an insect that chances by, and it seems to respond in quite different ways toward two other members of its own species, which, to us, appear virtually undistinguishable.

Nothing very remarkable has transpired, and if someone were to ask us what we had seen we might offer something like the following as a straightforward, spontaneous, "objective" account: "I came upon this animal while it was dozing. After yawning and stretching and scratching, its attention was drawn to an insect crawling by and it stared at it for awhile. A short time later it became thirsty, detoured around a rock to reach the stream and started to drink. While it was drinking another animal intruded, was threatened, and chased off. A minute or so later a second animal appeared and they greeted each other affectionately." No one could regard this recounting as fanciful or embroidered. Even so, a purist might object to it as going well beyond an "objective" description. After all, it could be claimed that such terms as "attention", "threatened", "greeted affectionately", and even "thirsty" are as much interpretive as they are descriptive. If we took this criticism seriously and tried to purge our account of all such objectionable language, we would ultimately find the task impossible (Purton, 1978). The reason is that at an immediate personal level – which is the one we have adopted here – our perceptions of behavior confound form, function and causation. This is inherent in the inevitable tendency to refer what we observe to the organism as agent or actor.

To observe in a way that makes sense to us, and to provide an intelligible description to someone else, we find it necessary to use concepts and terms that imply something about the animal's needs, perceptions, memories and the like. We find it necessary to attribute to the organism internal states or processes (i.e. mental events) that we obviously cannot observe. This fact sharply distinguishes the behavioral sciences from other scientific disciplines. Furthermore, it is highly significant that we label these processes with terms that are drawn from our ordinary language and everyday experience. The whole process is reminiscent of what Polanyi (1959, 1964) calls *indwelling*, a mainly unconscious projection of our own attitudes, feelings, and beliefs into objects and actors and events of the external world, particularly the world of behavior. To understand the behavior of another to the point where we can begin to study it scientifically requires that we understand it first in a simple, intuitive, noncritical way (Margolis, 1976; Rapoport, 1976). In short, we begin the study of behavior by relying implicitly and profoundly on anthropomorphism.

The argument that such terms can be (and usually are) seriously misleading is irrefutable. For this reason a great deal of attention in psychology has been devoted to cleaning up our language, to reducing surplus meaning and making our descriptive vocabulary more "operational" and precise (e.g. Bachrach, 1972; Mandler & Kessen, 1975; Pratt, 1939; Underwood, 1957). Although such efforts are clearly important, it would be unwarranted to assume that they will eventually make it possible to describe behavior in the kind of language that is appropriate in physics, chemistry or molecular biology.

The descriptive language of the behavioral sciences is incorrigibly mentalistic and anthropomorphic at a fundamental level because compelling analogies exist between the observer and the observed. This creates both a privilege and a temptation, which are present whether the observed is another human being or a member of a different species. In either case, we perceive intentions, emotions, recognition, conflict, and the like. To be sure, we can be taught that we do not, in fact, directly "perceive" these characteristics, but "infer" them. By the same token, we can be taught that our perception that an object does not change its form, in spite of changes in the retinal image of that object, is an "inference", even though the phenomenon of shape constancy is not an inference at all in the ordinary sense, but a "fact" of our immediate experience. Although the language in which such phenomena are described is a matter of legitimate concern, the more important issue is how they are to be explained.

This issue constitutes a critical difference between the scientific and nonscientific approaches to behavior, and a perennial impediment to communication with other disciplines. To the lay person (and I include most nonbehavioral scientists here), it is obvious to the point of banality that an individual eats "because" it is hungry, solves difficult problems "because" it is intelligent, flees "because" it is frightened, fights "because" it is angry, and so on. No one could deny the presence of such processes or states *in some form*. Behavior is unintelligible without them. Furthermore, the weight of the empirical evidence – both descriptive and analytic – is overwhelmingly in support of the inference that something functionally similar to "motives", "feelings" and similar mental processes or events are very widely distributed in the animal kingdom and play an important causal role in behavior. So what is scientifically objectionable in this commonsensical approach to behavior? What is objectionable is to assume (as common sense would have us do) that such terms refer to processes that are fully understood, either in ourselves or in the other organisms with which we analogize, and that they have substantial explanatory value. To accept them as adequate general explanations of behavior, rather than to treat them as first-order descriptions and crude explanatory stop-gaps, puts the scientific study of behavior exactly in the place that Skinner has been so concerned to avoid, in which the machinery of minding becomes a facile invention, arbitrarily modified in the name of plausibility, as the specific occasion seems to demand (see Dennett, 1978, for a more extended discussion of this issue).

Although this is the favored recourse of common sense, I do not see it as a serious risk for contemporary behavioral science. Mental processes and states are widely and explicitly recognized to be essentially hypothetical and to lack the unitary qualities that such mentalistic terms as hunger, fear, memory, perception, or learning imply. The complexities of the causal nexus are widely recognized. Elegant techniques and sophisticated methodologies have been developed to explore these processes systematically, and such innovations continue apace. A great deal has been discovered about the machinery of minding, and with each new development, it becomes less likely that the homunculi which Skinner abhors (Dennett, 1978) will play a part in the explanatory process.

A Perspective on Behavior

Is it not reasonable, therefore, to anticipate that these refinements and extensions of our knowledge of minding will eventually proceed to the point where dependence on intuition and mentalistic terms are no longer necessary? Clearly, that is the classical reductionist hope. Insofar as the performance of the organism-as-a-whole in a "real life" setting is what we seek to explain, however, it seems unlikely that this will ever come to pass. Reductionism is, of course, a methodological necessity. To understand a process, some dissection and identification of its constituents is essential. Partly as the result of these analytic efforts, however, it has become clear that what we call behavior is not a system "within" the organism; instead, it is an aspect of (or a particular way of looking at) the functioning organism-as-a-whole. Any behavior of moderate complexity is the product of more-or-less integrated interactions among many such subsystems. More-or-less should be underscored. Integration is a matter of degree precisely because it is the outcome of interactions among many constituent subsystems, and not a response to some supreme "control center" (another of Skinner's homunculi). In this respect, the order and integration of the organism may be likened to that in an ecosystem, in which functional stability is maintained through complex, reciprocal, organizational constraints operating within the system as a whole. In the organism, as in the ecosystem, these interrelationships are inherently probabilistic. This is not to say that all subsystems within the organism are equally relevant to behavior of course, particularly, in complex animals, in which division of labor and hierarchic organization appear to be essential to efficient functioning. The idea, rather, is that the assumption of independence of any subsystem is no more than a convenient abstraction. The unity which is apparent is also real – in the sense that species could not evolve and individuals could not survive unless they were able to maintain some degree of balance, coherence and internal harmony among the workings of their constituent parts – but it is probably far from complete at any level.

One of the tasks for the science of behavior is presumably to provide adequate models of the behavior of whole organisms. This is essentially a synthetic endeavor in which the results of analytic investigations from many disciplines are creatively reassembled. As we turn to this task, I see no way of proceeding except by confronting once more our profound reliance on anthropomorphic biases and assumptions. It is our human experience and our human perspective that allow us to distinguish between mere movement and directed action, to perceive the connection between the event that occurred last week and the performance we observe today, to infer that another is hungry, or angry, or afraid. These are the kinds of observations, the phenomena, that our scientific models must account for, and it is clear that our descriptions of the structure of behavior are couched in terms that are distinctly mentalistic and anthropocentric.

If our empirical efforts are successful we may know with certainty that "hunger" or "sexual arousal" or "pain" cannot be aligned with unitary physiological states; furthermore, we may even be able to go some distance in describing the diverse material bases contributing to such states. As new data accumulate they will certainly alter our views of the organism, which is to say, our views of how behavior is organized. This information will not obviate our need for mentalistic terms, however; it will only reinforce the growing understanding that mentalistic terms are *fundamentally* abstractions referring to partly covert organismic functions. We will continue to use such terms because they provide best-fit approximations of the causal/functional "elements" that appear to be most relevant to the organization of behavior. They give focus and direction to behavioral research. Goodness of fit will no doubt be improved by refinement or modification of these concepts as additional information becomes available, as, for example, Beach (1976) has done with respect to sexual behavior or Moyer (1968) with regard to aggression, but they will remain anchored to observed performances at the phenomenological level. This must be so, I suggest, because behavior is an emergent aspect of the (more-or-less)

304

integrated functioning of the organism-as-a-whole and cannot be definitively
"reduced" to a lower ontological level (e.g. brain, endocrine system).

Behavior and Cognition

By belaboring some of the peculiar and perplexing aspects of behavior as a field of
study, I hope to have paved the way for this final section. What we call behavior,
I have argued, is one aspect of the functioning of the organism as a whole. Although
some organismic constituents will obviously be more influential participants in this
emergent process than others, there is no distinct system within the organism with
which behavior can be uniquely and completely aligned. What distinguishes
behavior as an object of scientific study is a focus on how mobile ("freely moving")
organisms carry out the basic tasks involved in staying alive and perpetuating their
kind, and the elaborations and refinements that have evolved that contribute to
these ends. Living systems, according to traditional textbook definitions, can be
distinguished from nonliving systems by certain extremely general functional
attributes. They display irritability (sensitivity to environmental events), the ability
to maintain their own integrity (homeostasis, defense, repair of wear and tear on
the system), motility (minimally expressed in the ability to change shape), and the
capacity for reproduction in some form. Behavior is obviously an important aspect
of the processes through which these vital functions are carried out. As the principal
(not exclusive) interface in the organism's transactions with its surroundings, it is
a final common path subserving many functions. How this is accomplished provides
the distinctive focus of behavioral research.

From this standpoint, it is clear that behavior is about something, and that what
it is about chiefly and in its most basic form, is action. As distinguished from mere
motility, action is based on information the organism has about its real or potential
requirements and about the state of its environment, and it is carried out in such
a way as to increase the likelihood that these requirements will be met. It is for this
reason, that we relate behavior to an agent and perceive it as adaptive and goal-
directed. To the extent that behavior *qua* action is based on the actor's knowledge
of its world and its own needs, it implies a cognitive process.

There is no reason to be squeamish about applying the term cognition to
nonhumans. I have argued that the description of behavior is necessarily couched
in "as if" functional terms. To assert that behaving organisms are capable of
cognition is to claim nothing more than at a molar level their behavior is organized
such that it presents itself to us as if it were informed by their knowledge of the
world. In a sense that is more than merely metaphorical, the paramecium that

approaches a mild source of mechanical stimulation and reverses its direction as the stimulation becomes more intense is demonstrating a form of knowing. Furthermore, the reaction to a given stimulus is not an invariant reflex, but often depends on the protozoan's physiological state (Jennings, 1906). These observations convey the fundamental properties of behavior as a biological phenomenon: action based on information and need. Behavior is an interesting and distinct scientific domain only when it is viewed as the molar outcome of a group of interrelated activities which revolve around the generic issues of knowledge and value.

It follows from this perspective that the structure or organization of the environment is represented in the organization of behavior. This is an inevitable consequence of natural selection; but to say that behavior implies representational processes says nothing about what aspects of the environmental structure are represented or how. A visual image rapidly increasing in size represents an impeding collision, and a sweet taste represents sugar, but the image could be caused by a passing shadow, or the sweetness by saccarhin. It also follows, therefore, that actions based on such representations are necessarily predictive or probabilistic. Behavior always points ahead, for it is the consequences of behavior that count. Actions are taken "as if" the information received is veridical, but whether it is or not will only be determined after the fact. Of course, mistakes are often made, even in complex and sophisticated organisms, and ways of making better predictions by reducing uncertainty or changing actions based on erroneous information, have played an important role in the evolution of cognitive functions (Brunswik, 1955; Mason, 1980; Maturna, 1970; von Foerster, 1968).

Within this perspective, the study of cognitive processes in animals does not appear as a special domain within comparative psychology and ethology, but as the central unifying theme in all their endeavors. What behavior is about is knowledge and value. What organisms know, how they come by this knowledge, what they value and how their values are arrived at or changed – these are the kinds of generic issues that are the foundation for all empirical efforts in these fields. To be sure, the generalized "as if" functions become more and more concretely embodied and specific as the data accumulate, and as this happens, the conceptual links between different endeavors may be lost from view. Investigations of feeding behavior in rats, of "releasing stimuli" in fish, of the ontogeny of song in sparrows, and the acquisition of social skills in rhesus monkeys, may be far apart in terms of vocabulary and methods, but they share a common concern about what sort of knowledge organisms possess, and how they obtain and use it.

The perspective I am advocating here may at least serve as a reminder of these commonalities to the practitioners, and make clear to those from other disciplines that beneath the diversity of techniques and specific details lies a common theme.

In addition, a more explicit acknowledgment of the central role of cognition in the organization of behavior may help to dispel the prejudice that minding refers to functions that are restricted to a handful of privileged evolutionary newcomers, when, in fact, they are close to the threshold of life itself. Given the capacities for sentience and directed movement and the exigencies confronting all living systems, the evolution of minding in some form is virtually assured. The questions of whether minding is a legitimate field of biological study or why it has appeared in the living world are scarcely worth debating. On the other hand, the origins of the refinements and elaborations that have taken place in the evolution of minding, the material bases of these innovations, the nature of their complex workings, and the sources of their causal efficacy and power, are issues that will occupy us for many years to come.

There is nothing radical in this perspective. It was clearly presented in a book written nearly 70 years ago by Margaret Floy Washburn, entitled *The Animal Mind* (Washburn, 1917). The major differences between her position and the one advocated here are not in perspective, but in the degree of sophistication that developments since her time have made possible. In the intervening period we have given a thorough hearing to radical behaviorism and have concluded that, in spite of its powerful merits as an analytic approach, it cannot provide the necessary theoretical underpinning for a complete science of behavior. We have gone beyond such simple absolute dichotomies as innate versus acquired, organism versus environment, sensation versus perception, cognition versus conation, and human versus animals. We no longer insist that the "laws" of learning or perception or memory, worked out on pigeons or rats or monkeys or college sophomores, be applicable universally and without qualification. We have made great strides in developing effective methods for the systematic investigation of such cognitive processes and achievements as memory, concept formation, perceptual illusions, problem-solving strategies and cognitive maps, in a broad range of species. And we have had the benefit of the critical insights of thinkers who were unwilling to dismiss the possibility that mental processes are susceptible to empirical research (e.g. Dennett, 1978; Manicas & Secord, 1983; Natsoulas, 1984). In spite of these developments, one thing has not changed since Washburn wrote: "Knowledge regarding the animal mind, like knowledge of human minds other than our own, must come by way of inference from behavior" (p. 4). That gap will never disappear, but the ground is a good deal firmer than it was in her time for the inferential leap from the knower to the known.

Acknowledgments

I am grateful for comments from M. W. Andrews, A. Anzenberger, A. G. Butler, S. P. Mendoza, C. R. Menzel, and the editor, William Bechtel. Support was provided by grants RR00165 and RR00169 from the National Institutes of Health, to the Yerkes and the California Primate Research Centers, respectively.

References

Bachrach, A. J. (1972). *Psychological research* (3rd ed.). New York: Random House.

Beach, F. A. (1976). Sexual attractivity, proceptivity, and receptivity in female mammals. *Hormones and Behavior, 7*, 105–138.

Brunswik, E. (1955). Representative design and probabilistic theory in a functional psychology. *Psychological Review, 62*, 193–215.

Dennett, D. C. (1978). *Brainstorms.* Montgomery, Vermont: Bradford Books.

Jennings, H. S. (1976). *Behavior of the lower organisms.* Bloomington: Indiana University Press.

Mandler, G. & Kessen, W. (1975). *The language of psychology.* Huntington, New York: Robert E. Krieger.

Manicas, P. T. & Secord, P. F. (1983). Implications for psychology of the new philosophy of science. *American Psychologist, 38*, 399–413.

Margolis, J. (1976). Countering physicalistic reduction. *Journal of the Theory of Social Behaviour, 6*, 5–19.

Mason, W. A. (1980). Minding our business. *American Psychologist, 35*, 964–967.

Maturana, H. (1970). Neurophysiology of cognition. In P. Garvin (Ed.), *Cognition: A multiple view* (pp. 3–23). New York: Spartan Books.

Moyer, K. E. (1968). Kinds of aggression and their physiological basis. *Communications in Behavioral Biology, 2*, 65–87.

Natsoulas, T. (1984). Gustav Bergmann's psychophysiological parallelism. *Behaviorism, 12*, 41–69.

Polanyi, M. (1959). *The study of man.* Chicago: University of Chicago Press.

Polanyi, M. (1964). *Personal knowledge: Towards a post-critical philosophy.* New York: Harper & Row.

Pratt, C. C. (1939). *The logic of modern psychology.* New York: MacMillan.

Purton, A. C. (1978). Ethological categories of behaviour and some consequences of their conflation. *Animal Behaviour, 26*, 653–670.

Rapoport, A. (1976). General systems theory: A bridge between two cultures. Third annual Ludwig von Bertalanffy memorial lecture. *Behavioral Science, 21*, 228–239.

Underwood, B. J. (1957). *Psychological research.* New York: Appleton-Century-Crofts.

von Foerster, H. (1968). From stimulus to symbol: The economy of biological computation. In W. Buckley (Ed.), *Modern systems research for the behavioral scientist* (pp. 170–181). Chicago: Aldine.

Washburn, M. F. (1917). *The animal mind* (pp. 1–37). New York: MacMillan.

Intelligence: From Genes to Genius in the Quest for Control

DUANE M. RUMBAUGH[1] and GRAHAM M. STERRITT[2]

[1]*Georgia State University and Yerkes Primate Center, Emory University, Atlanta, Georgia 30303–3083, U.S.A.*
and
[2]*University of Colorado at Denver, Denver, Colorado 80202, U.S.A.*

Introduction: Intelligence

Intelligence is an elusive concept that will be with us for the indefinite future. It is a highly problematic term, born of the observation that individuals differ in their accomplishments, skills, and ability to profit from experience with sufficient stability that certain predictions can be made to pragmatic advantage. Understandably, the concept is anthropocentric. It was formulated in response to observations by humans of humans. That origin has complicated the formulation of a perspective concerning the sources of human intelligence. Our perspective is the comparative framework, which holds that human attributes had their origins in evolution that might be traced in part even today in extant animal forms.

Intelligence is surely linked to evolution of the brain's cortex. It has several dimensions or factors; it is not unitary in concept or in function. Intelligence is viewed here as having many dimensions which probably cluster differentially. Different aspects of intelligence almost certainly are interdependent in ways yet to be defined.

Human intelligence is of greatest interest, especially the kind that is typified by the precocious child, the student who skips grades in school, the member of Phi Beta Kappa, the successful inventor, and even the skillful criminal who plans his crimes so well as to avoid apprehension! This is the kind of intelligence that allows for many degrees of freedom of choice in how one's energies might be directed and spent.

We live in the era of neo-Darwinism and sociobiological theory. Science attempts to achieve a rational and precise perspective on life in its various adaptations. Although not all attributes of living things are necessarily "optimal" for the purposes of adaptation, there seems to be little question that intelligence is deeply rooted in adaptive processes which underlie reproductive success. If one survives, one might reproduce; if one does not survive, one does not reproduce.

Bechtel, W (ed), Integrating Scientific Disciplines. ISBN 90-247-3242-5.
© *1986, Martinus Nijhoff Publishers, Dordrecht. Printed in The Netherlands.*

Reproduction, however, is but one of the consequences of successful adaptation.

Success in meeting natural challenges such as social competition might yield more than just survival and reproductive success. It might yield, for lack of a better term (and perhaps there is none better), ''quality of life.'' The good life, from the anthropocentric perspective, includes not just food and drink, not just space, not just protection from the elements, not just an existence. Rather, the good life includes the best possible food and drink, the best possible space, and as much of it as one wishes – in short, ready access to resources for the full and varied satisfaction of many nuances of our various biological and psychological appetites. Because one of the appetites is sex, and because sex is fundamental to reproduction, it is obvious that if a species competes successfully for necessary and hedonistically valued resources, it will have at least the option of enjoying high reproductive success. Consequently, to the degree that intelligence is important to successful competition for resources, it becomes more likely that survival, reproduction, and even the enhancement of intelligence itself are likely to prevail in combination. (In the case of some intelligent humans, to be sure, the option of not reproducing has been exercised – though not necessarily through celibacy.)

There are, of course, other kinds of adaptive processes and capacities which, unfortunately, also are termed intelligent. They are found in the behaviors of insects and animals, and they facilitate adaptation and hence survival. The majority of these behaviors, however, are not manifestations of intelligence by anthropocentric standards. Instead, they are what might be called *biologically smart* behaviors. Examples of such behaviors in animals are those which allow for migration, nesting and brooding, care of young, or for copulatory competence as dictated by opportunity and not by existensive prior practice. They allow for the efficient learning of what is food for a given species and where it can be found, what predators are, and the like. They provide for facile learning of how to be a successful bird, or a successful antelope, lion, or alligator. The genetic preparedness for biologically smart behaviors is becoming clearer.

But not all animal behavior is reducible to these biological dictates and facilitations. There are other behaviors that are spawned from the roots of intelligence as defined even anthropocentrically. The evidence would imply that the roots of human intelligence are likely to be found, at least fragmentarily, in animals, particularly those closely related to us. What are they?

The most valid and efficient measure of human intelligence is that based in vocabulary, in the knowledge of word meanings and their use. The meanings of words which are used referentially, rather than just perfunctorily or performatively, are likely to be based upon relationships experienced by humans as words are used and heard. For obvious reasons, word usage is of limited value in assessing animal "intelligence." Only the great apes and some sea mammals, it would appear, have capacity for semantics with arbitrary symbols. As revolutionary and exciting as it has been to define that capacity, and we have been involved in that definition, the ability for semantics by these animals is profoundly restricted, by human standards, but it is there.

What might animals have, analogous to vocabulary, that would allow meaningful comparative measurements? One prime possibility is the capacity to infer *cause-effect* relationships and then to act appropriately to the end of enhancing *control* over the environment.

Our emphasis on the special significance of abilities to control the environment is similar to that of White (1959) in "Motivation Reconsidered: The Concept of Competence." Among precedents for his views, White cites Goldstein (1939) as having assumed "one master tendency, that toward self-actualization, of which the so-called visceral drives are but partial and not really isolated expressions, and which can find expression also in an urge toward perfection – toward completing what is incomplete, whether it be an outside task or the mastery of some function such as walking" (White, 1959, p. 313). White also notes (p. 515–6) that,

> Being interested in the environment implies having some kind of satisfactory interaction with it. Several workers call attention to the possibility that satisfaction might lie in having an effect upon the environment, in dealing with it, and changing it in various ways. Groos (1901), in his classical analysis of play, attached great importance to the child's 'joy in being a cause,' as shown in making a clatter, 'hustling things about,' and playing in puddles where large and dramatic effects can be produced. 'We demand knowledge of effects,' he wrote, 'and to be ourselves the producers of effects.' Piaget (1952) remarks upon the child's special interest in objects that are affected by his own movements. This aspect of behavior occupies a special place in the work of Skinner (1953), who describes it as 'operant' and who thus 'emphasizes the fact that behavior operates upon the environment to generate consequences.'

White concludes that a wide variety of motivations and interests are subsumed in the "effectance" motivation of the individual (p. 323). He notes (p. 324) that,

> Of all living creatures, it is man who takes the longest strides towards autonomy. This is not because of any unusual tendency toward bodily expansion at the expense of the environment. It is rather that man, with his mobile hands and highly developed brain, attains an extremely high level of competence in his transactions with his surroundings. The building of houses, roads and bridges, the making of tools and instruments, the domestication of plants and animals, all qualify as planful changes made in the environment so that it comes more or less under control and serves our purposes rather than intruding upon them. We meet the fluctuations of outdoor temperature, not only with our bodily homeostatic mechanisms, which alone would be painfully unequal to the task, but also with clothing, buildings, controlled fires, and with such complicated devices

312

as self-regulating central heating and air-conditioning. Man as a species has developed a tremendous power of bringing the environment into his service, and each individual member of the species must attain what is really quite an impressive level of competence if he is to take part in the life around him.

Campbell (1974) developed an evolution-based, action-oriented view of the perception of causation. He noted that,

> overlapping the activity theory of causation is a biological evolutionary perspective. From an evolutionary perspective, humanity's strong and stubborn psychological predisposition to infer causal relations can be seen as the product of a biological evolution of brain-mind processes which has resulted in a psychic unity concerning causation. This unity is in general adaptive, because it will often validly reflect causal processes that can and do occur in the real world; but it will often not fit the world perfectly. . . . The perspective is evolutionary because it assumes a special survival value to knowing about causes and in particular about manipulable causes.

Later in the same paper Campbell comments:

> This evolutionary perspective brings us to an important distinction between mere correlation and truly causal relations, between [some species capable only of] forecasting and [other species capable of] diagnosing manipulable causes which will enable us to change the world. In our evolutionary scenario it is [the latter] species that would develop brain-mind systems giving causal perceptions and causal inferences diagnostic of *manipulable* relationships. We human beings are pre-eminently such a species, and our intuitive concept of causality is a product of an evolutionary adaptation. . . . (p. 29)

Rumbaugh and Pate (1984) present evidence that indicates qualitative changes in various measures of learning and transfer of training, paralleling the phylogenetic gradient of brain complexity from the more primitive primates to humans. But they do not by any means deny the continuities across species in "knowing." These continuities, especially the ubiquitous role of knowing as an instrument for effecting control over the environment, were pointed to by Popper (1963).

> ... my remarks are applicable without much change, I believe, to the growth of pre-scientific knowledge also – that is to say, the general way in which men and even aminals acquire new factual knowledge about the world. The method of learning by trial and error – of learning from our mistakes – seems to be fundamentally the same whether it is practiced by lower or by higher animals, by chimpanzees or by men of science (p. 216).

Animals do not publish, at least not in the journals we read, but in other respects they must behave in ways that bear certain formal similarities to the behavior of human scientists, if the individual of whatever species is to adapt its behavior to the environment efficiently enough to survive. More specifically, the individual must "remember" in some sense. Anderson (next paper) has pointed to the primacy of memory in cognition and in the achievement of control. We would say that it does not matter whether the "memory" is conscious or not, as long as it is retained in a form that enables comparisons among memories, and differential adjustments of behavior corresponding to different memories. In effect, the animal must conduct "experiments," albeit usually adventitiously, but then much scientific progress results from adventitious observations! In these experiments the organism must experience the differential consequences of different behaviors in one context, or of the same behavior in different contexts. It must remember the differences, if any, and must modify its subsequent behavior – i.e., "exercise control" – if it is to profit adaptively from the "knowledge" gained from these experiences. This ability to exercise control over one's behavior in accord with the remembered differential consequences of different actions and contexts is perhaps the essence, or at least an important aspect, of the continuity in knowledge, across lower organisms to scientists, recognized by Popper in the foregoing quotation.

In order to develop our theory of intelligence, we need to consider motivation, which is a key concept in psychology. It is a unifying concept in that it can account for a wide variety of behaviors of many species. Species are driven to action by needs; rewards appropriate to specific motivational states shape corresponding behaviors. In the Control Theory that we develop in the next section, we will indicate how a motivation to control one's environment may be a driving force in the development of intelligence.

Control Theory

The model for Control Theory that we advance in this paper rests on the following assumptions:

Assumption I

The most elemental physiological and behavioral processes are designed to maintain homeostatic requisites to life. The balance of electrolytes, the processes of ingestion and the elimination of wastes, preferences for environmental conditions that do not tax the thermo-regulating mechanisms are examples of states and needs based upon homeostasis. To the degree that the homeostatic states are disturbed, strong drives can be induced. Behaviors which are associated with efficient reduction of drives are selectively reinforced. The probability that these behaviors will recur, given reinstatement of the drives with which they coped, is increased because the reestablishment of homeostasis is inherently reinforcing.

Assumption II

The evolution towards more complex life forms has been coupled with the preference for excercising personal control over outcomes rather than experiencing outcomes that under random control or under the control of other individuals or agencies. There is a continuum from the simple origins of adaptive behaviors, referenced in Assumption I, to more complex behaviors which are based on more elaborate cognitions. These more complex behaviors have in common the fact that *they achieve perceived control* in terms of cause-effect relationships between *what is done* and *what is the consequence*. It is only to the degree that these more complex processes come into play that the term *intelligence* is used in this model.

Species have been selected to varying degrees for their proclivity to infer, and to draw conclusions regarding the relative validities of, cause-effect relationships. The operations of such processes have, at present, unclear origins in animal behavior; however, it is suggested that whenever an animal prefers a context in which it instrumentally produces contingencies, over another context in which the identical outcomes occur but not as consequences of responses, there is evidence of *control* motivations operating.

Evidence for the operation of control motivations comes from various sources. (i) Various species prefer to control delivery of outcomes rather than to have the same outcomes occur in ways that are not under their personal control. Osborne (1977) reviews the literature on "contrafreeloading," the finding that under some circumstances animals of various species act as if they prefer to "earn" reinforcements rather than to get them "free." We suspect that this effect will be especially strong in the more advanced mammals, and in particular primates (see the discussion below concerning Koffer and Coulson). (ii) We suspect that the ability to predict is necessary but not sufficient for satisfaction of motives to achieve control, and in this connection we find it interesting that even rats prefer contexts in which, if unavoidable shocks are to be experienced, shock occurrences will be predictable as opposed to random (Abbott, 1985). Daly (1985) notes that organisms generally choose situations in which rewards are predictable over those in which they are unpredictable. She showed that the Daly and Daly (1984) mathematical model, DMOD, successfully forecasts the times when that will occur, as well as correctly predicting those (less frequent) occasions when organisms will choose unpredicatble rewards. Important to our thesis is the fact that the Dalys' model treats unpredictability as a consistently aversive factor, which is always weighted negatively in their formula. Unpredictability combines with other factors to forecast correctly the unusual circumstances when unpredictable rewards will in fact be chosen, as well as predicting the more usual instances in which predictability is chosen, but the factor of unpredictability *per se* always remains a negative vector in their successful model, just as our theory would lead us to expect. (iii) The ability to predict is *adaptive* for "lower" as well as for "higher" species. Predictive capabilities may be possessed even by honeybees, as suggested by the finding that when the distance to a food site is increased systematically, honey bees will fly as though they predict how much further they will be required to fly each day to arrive on target (Gould, 1983). (iv) Also consistent with the possibility that many species show "expectancies" or predictions to the future based on past experience is the fact that members of a wide variety of species will be disturbed if there is a sudden diminution in the quality of incentives delivered by contrast to those reliably obtained in the past contingent upon instrumental acts. (v) Research by Held and Hein (1963) illustrates the importance of the dimension of *active control* of experience, in this case with respect to the "perceptual learning" which may occur naturally early in the life histories of members of most mammalian species. As Hebb (1949) suggested, such types of "perceptual learning" may be fundamental to many other kinds of learning abilities.

It will be fruitful to elaborate on this last piece of evidence. Held an Heind raised kittens in total darkness from birth, bringing them into the light only for a brief, carefully contrived periods of special visual experience. When brought out for these special visual stimulation sessions, one member of each pair of these kittens rode in a miniature gondola car which was hung on one end of a horizontal beam which could rotate around a central pylon inside a vertically striped cylinder. The other kitten wore a collar attached to the other end of the beam. Movement of the pair of kittens around the pylon, past the vivid stripes, constituted the primary visual stimulation experienced by these kittens for the first few weeks of postnatal life.

In each pair of kittens, the one always rode passively with its legs folded under it in the solid-floor bed of its car. The other kitten could walk, and in doing so it controlled its own movements around the track as well as the movements of the other kitten. This clever arrangement assured that the visual experience of the two kittens was very similar except in the important detail that one kitten, by its own movements, *caused* the apparent movement of the stripes past its eyes, while for the other kitten the apparent movement of the stripes was *not* under its control – it was the movements of the first kitten that caused the apparent movements of the stripes.

The visual development of kittens raised under these two conditions proved to be strikingly disparate. The visual spatial perceptual systems of the kittens which

316

controlled their own visual experience developed significantly faster, even in such primitive aspects as the protective blink reflex in response to objects thrust toward the eye.

In "Feline indolence: Cats prefer free to response-produced food," Koffer and Coulson (1971) reported that they first trained six cats to press an easily accessible panel to earn small portions of food, and then gave them sufficient days of practice to assure that the panel-press response had stabilized in all cats. In subsequent test trials, the same food was presented in an open dish in the test chamber for "free feeding" (FF) in competition with continuing opportunities to "earn" the food by panel-pressing. In all test trials, each cat was placed in the chamber oriented toward the "work" manipulandum, away from the "free" feeding dish. Nonetheless, it was found that on all test sessions after the first one, every cat went immediately, or after at most only a few panel presses, to consume the food provided in the free food dish. In short, these cats did not choose the opportunities for honest labor, but seemingly without shame or remorse, unhesitatingly freeloaded for all they were worth!

This finding is of special interest to us because cats may be the most complex or advanced species (except for children, who were involved in one project) to have been studied in the "contrafreeloading" research. The findings that relatively simpler species such as rats, gerbils, pigeons, and the like, often seem to prefer "earned" reinforcements, while cats prefer to get them free, runs directly contrary to our predictions that members of higher species will prefer to control their own experience more strongly and consistently, in comparison to members of lower species.

On the other hand, anyone who has observed the behavior of cats may object to Koffer and Coulson's characterization of cats as "indolent." *When a cat chooses* to be indolent, to be sure, there are few species which can rival *felis domesticus* in the total dedication, the soaring aptitude and well-practiced proficiency which the average individual of this species brings to the task of sleeping. It is true, moreover, that the average cat, among those we have known, seems to choose to spend a large proportion of its "waking hours" sleeping. If we can tear our reader away, however, from this persuasive evidence of the validity of Koffer and Coulson's charcterization of cats as "indolent," we find it instructive to observe this species on those admittedly rare occasions when it chooses *not* to be indolent. At such rare moments we are struck by the degree to which the behavior of this species approaches whatever might be the *penultimate opposite of indolence*. The domestic cats we have known have, without warning, demonstrated convincingly their genetic link to the cheetah, accelerating in split seconds from less than a standing start to blinding speed, twisting and turning with breathtaking agility, batting crumpled paper balls into the air and keeping them aloft for several seconds in deft, vigorous, beautifully coordinated virtuosity with the same paws which other members of their species were so reluctant to employ in the panel-pressing task – honest labor paid fair wages – as arranged through the benevolence of Koffer and Coulson.

With such contrasts in mind, we prefer to regard the contrafreeloading literature from the standpoint of personal control. Cats and members of other more advanced species will freeload and in other ways act indolently when it suits the purposes for which they are controlling their behavior at the moment. They will also behave with startling energy and vigor when that suits their motivational states.

From his review of the substantial literature then available on contrafreeloading, Osborne (1977) concluded that "responding for food in the presence of free food is importantly controlled by stimulus changes attendant upon response-dependent food presentation" (abstract, p. 221). He identifies this as "the stimulus-reinforcer effect on behavior," and he indicates that it and all of the other findings on this issue are consistent with the results of modern research on operant learning and reinforcement. We have no quarrel with that hypothesis concerning the contrafreeloading research – indeed, we will regard it as interesting and valuable if all of the phenomena which we are addressing can be subsumed under a few simple, time-tested principles. But, as stated above, we are not confident that that will be possible. Some separate motive to control events in the environment may prove necessary if all of the facts are to be accounted for.

Wolfe & Kaplon (1941) varied the size of feed pellets fed to chickens and found that learning was more strongly reinforced when four pecks were necessary (small pellets) than when one peck was sufficient (large pellets) for consumption of the same amount of feed. It was concluded that "chicks would rather peck than eat," a finding in line with the contrafreeloading hypothesis.

A possible problem in that study is that the larger pellets may have fit less comfortably in the mouths or esophagi of the chickens – pecking *per se* may not have been reinforcing, and may even have been aversive, but small pellets may have been less aversive than larger ones.

Sterritt and Smith (1965) used a different approach which avoids this problem hindering interpretation of the Wolfe & Kaplon study. A tube was implanted in the esophagus of each of a number of leghorn chicks within 24 hours of hatching. The implanted tube could be attached to a light spiral of the same tubing suspended from an overhead track, permitting the chick to move freely in a Y-maze. These chicks never ate or drank during their lifetimes, but received all food and water by injection through the implanted tube directly into the esophagus.

Chicks prepared in this way were (in experiment 3) randomly assigned to four groups given different reinforcements in the positive goalbox of the maze. A 2×2

318

factorial design was used, with opportunities to peck at a tactually and acoustically responsive target (which consistently elicited pecking) *vs.* no target as one dimension, and delivery of a food-water mixture *via* the tube directly into the esophagus *vs.* no tube-feeding in the goalbox as the other dimension of the design.

Access to the peck target by itself was not a significant reinforcer of maze learning. Tube-feeding in the absence of the pecking target was a significant reinforcer, but this event was significantly *less* reinforcing than was being tube-fed while pecking at the pecking target. Chicks with no prior experience in feeding themselves would *not* "rather peck than eat," but they *are* significantly more reinforced by opportunities to peck and receive food in the crop simultaneously, by contrast to the reinforcement provided by either event, pecking or being tube-fed, alone.

These findings are consistent with other indications, e.g., those of the contrafreeloading literature, suggesting that vigorous activities in the service of "controlling" interactions with the environment are often preferred over events yielding the same "need reductions" but requiring less vigorous responding, even in submammalian species. The Wolfe & Kapion and the Sterritt and Smith findings, as well as our observations concerning the occasional non-indolence of cats, lead us to disagree with Osborne's (1977) conclusion that animals generally obey a "law of least effort" in their dealings with the world. We believe that organisms, especially "higher" ones, usually behave in ways that maintain and enhance their control over the environment. Control efforts often will be directed toward minimizing effort, but in many well-documented instances, e.g., the examples cited above, effort will be strikingly greater than the minimum required to satisfy biological needs.

Assumption II can be restated with reference to the profoundly negative consequences of "learned helplessness." Originally formulated by Overmier and Seligman (1967) and Seligman and Maier (1967), the learned helplessness hypothesis was revised by Abramson, Seligman and Teasdale (1978). In each form this hypothesis has generated a wide range of important research, which was summarized recently by Peterson and Seligman (1984).

The learned helplessness model deals with the perception by individuals that certain "bad events" may happen to them which they are powerless to avoid or escape. Symptoms of learned helplessness include "passivity; cognitive deficits; emotional deficits including sadness, anxiety, and hostility; a lowering of aggression; a lowering of appetitive drives; a set of neurochemical deficits; and an increase in susceptibility to disease. In addition, the symptom of self-esteem loss is sometimes one of the symptoms of helplessness" (Peterson and Seligman, 1984, page 349).

These symptoms seem likely to decrease significantly the abilities of afflicted individuals to survive and reproduce. To the degree that predispositions toward learned helplessness may be determined in part genetically, there would seem to be selection pressure against individuals having that genotype. Thus, in the long run, we would expect that inborn tendencies toward learned helplessness, if indeed any such tendencies exist on a genetic basis, may be purged from the gene pool by negative selection pressure.

In any event, tendencies toward learned helplessness appear to be conceptually opposite to the positive striving to achieve personal control over outcomes, which we perceive to be the essence of intelligence. Thus tendencies toward learned helplessness acquired by an individual during his lifetime would seem to diminish his abilities to exercise intelligence, as we define intelligence, and conversely any training, therapy or other experience which diminishes tendencies toward learned helplessness should improve the ability of the individual to apply his intellectual powers. The increased optimism and expectations of personal effectiveness which should follow from reductions in personal learned helplessness might in some cases cause some individuals to "stick their necks out" in ways that increase their exposure to dangerous risks, reducing their chances for survival and reproduction of their own genes. In general, however, we expect the opposite – reductions in learned helplessness should enable individuals to live happier, more intelligent lives, and we expect that such individuals will be more likely to survive long enough to reproduce their own genes.

Assumption III

The reinforcement value of perceived *control* has at least two vectors: (i) The first is that it serves to maintain a quiescent or nonanxious state. Novelty in the environment can be an attraction; it can also be disconcerting, as if it often raises the question, What else might happen? To the degree that behaviors are generated to the end of gaining perceived (if not real) control, quiescence is achieve. (ii) The second vector is the conservation of energy. An organism needs to emit a given behavior only when the previously perceived consequences of that behavior are needed. With increasing ability to predict contingencies between one's behavior and the contingencies of the environment, there is a benefit in terms of energy saved. Selection for competence in "inferring contingencies produced by behaviors" would seem to be a potent force in the evolution of human intelligence. Such competence would be particularly advantageous for organisms struggling to adapt in rapidly changing contexts or environments.

We propose that the belief, the confidence that one understands why the world,

320

its elements, and its events are as they are, is primary. The validity of the belief is secondary – unless the belief might lead to behaviors that risk survival. Without survival, of course, there can be no reproduction. So survival is paramount.

At the time of the first hominids it seems probable that little was understood that went beyond the boundaries of the activities and elements necessary for life. Consequences in the natural course of events – e.g., what animals would do, what volcanoes might do, why storms occurred, what caused wild game to appear and disappear, what caused the seasons to change, what caused fruits and grains to be present only at certain times, and why some, but not all, fruits tasted good – were not known. The stage was set for superstitions which defined courses of action to gain control over the awesome, the overwhelming, and the powerful events of the world they sensed. To the degree that behaviors generated were irrelevant to true cause-effect mechanisms, survival and reproduction were at risk. To the degree that those behaviors were congruent with cause-effect mechanisms of the natural world, real control could be achieved. And to the degree that this occurred, there could be selection for what we now call intelligence – a constellation of operations which are aimed at perceptions of "what caused what" in the natural world.

Conclusion

The long and short of what is being asserted is that to the degree an organism is intelligent, and not just biologically smart, we can expect it to be interested in determining how things work and really are. More and more precise cause-effect relationships will be sought. And, when the search falsifies a perspective which it has spawned, the incentive to persevere is greater than when hypotheses are affirmed – at least in science such is the norm.

To the degree that behaviors are generated and selected on the basis of correctly perceived cause-effect relationships between behaviors and consequences, we have the possibility of rapid evolutionary advance in (i) learning through thought, not just stimulus-response associations, (ii) capacity to generate new behaviors, new options for solving problems, (iii) facile problem solving based in the extrapolation of principles, where such is not possible, on the basis of behaviors extrapolated on the basis of specifics, (iv) symbolization, (v) observational learning, which quite marvelously provides for the transduction of visual information, obtained through perceptions of others' behaviors, into the production of functionally equivalent motor patterns, (vi) referencing, (vii) use of arbitrary symbols to facilitate referencing between individuals in social interaction, (viii) creativity in many forms, and (ix) the emergence of teaching of the young from an ever-increasing base of cultural knowledge. (A related outline of possible levels of intelligence was presented by Campbell, 1974.)

This paper has attempted to explicate in terms of survival value why organisms are not necessarily satisfied with mere survival, and how human intelligence might have evolved. To the maximum degree possible, animals have come to try to do that which brings environmental contingencies under their control. Those species which have succeeded best in controlling their environments have been most successful in surviving, and the most intelligent species have been most outstanding in their capacities to control the parts of the world in which they have lived. Paradoxically, it is undoubtedly humankind's incredible success in controlling our environment that poses the greatest threats to the survival of ourselves and of all other living things on earth. Perhaps this is an inherent liability in the intimate association we are proposing between intelligence and powers to control the environment. As others before us have noted, the survival of our planet would seem to depend upon our capacity to develop, over a very short span of time, a more intelligent sense of responsibility in our use of our powers of control, one that must be commensurate with the awesome magnitude of those powers, or all is lost.

Acknowledgment

Preparation of this paper was supported by two grants: NICHD–06016 and RR–00165.

References

Abbott, B. B. (1985). Rats prefer signaled over unsignaled shock-free periods. *Journal of Experimental Psychology: Animal Behavior Processes,* **11**, 215–223.

Abramson, L. Y.; Seligman, M. E. P.; and Teasdale, J. D. (1978). Learned helplessness in humans: Critique and reformulation. *Journal of Abnormal Psychology, 87,* 49–74.

Campbell, D. T. (1974). Evolutionary epistemology. In P. A. Schilpp (Ed.) *The philosophy of Karl Popper. The Library of Living Philosophers,* Vol. 14–I, pp. 413–463. LaSalle, IL: Open Court Publishing Co.

Daly, H. B. (1985). Observing response acquisition: Preference for unpredictable appetitive rewards obtained under conditions predicted by DMOD. *Journal of Experimental Psychology: Animal Behavior Processes,* **11**, 294–316.

Daly, H. B. and Daly, J. T. (1984). A mathematical model of reward and aversive nonreward in appetitive learning situations: Program and instruction manual. *Behavior Research Methods, Instruments & Computers,* **16**, 38–52.

Goldstein, K. (1940). *Human nature in the light of psychopathology.* Cambridge, MA: Harvard University Press.

Gould, J. L. (1983, April). *The invertebrate mind*. Paper presented at the Second National Zoological Park Symposium, Animal Intelligence, Washington, D.C.

Groos, K. (1901). *The play of man* (Trans. by E. L. Baldwin). New York: D. Appleton.

Hebb, D. O. (1949). *The organization of behavior: A neuropsychological theory*. New York: Wiley.

Held, R. and Hein, A. (1963). Movement-produced stimulation in the development of visually-guided behavior. *Journal of Comparative & Physiological Psychology,* **56,** 872–876.

Koffer, K. and Coulson, G. (1971). Feline indolence: Cats prefer free to response-produced food. *Psychonomic Science,* **24,** 41–42.

Osborne, S. R. (1977). The free food (contrafreeloading) phenomenon: A review and analysis. *Animal Learning & Behavior,* **5,** 221–235.

Overmier, J. B. and Seligman, M. E. P. (1967). Effects of inescapable shock upon subsequent escape and avoidance learning. *Journal of Comparative & Physiological Psychology,* **63,** 28–33.

Peterson, C. and Seligman, M. E. P. (1934). Causal explanations as a risk factor for depression: Theory and evidence. *Psychological Review,* **91,** 347–374.

Piaget, J. (1952). *The origins of intelligence in children* (Trans. by M. Cook). New York: International University Press.

Popper, K. R. (1963). Conjectures and refutations. New York: Basic Books.

Rumbaugh, D. M. and Pate, J. L. (1984). The evolution of cognition in primates: A comparative perspective. In H. L. Roitblat, T. G. Bever, and H. S. Terrace (Eds.), *Animal cognition* (pp. 569–587). Hillsdale, NJ: Lawrence Erlbaum.

Seligman, M. E. P. and Maier, S. F. (1967). Failure to escape traumatic shock. *Journal of Experimental Psychology,* **74,** 1–9.

Skinner, B. F. (1953). *Science and Human Behavior*. New York: Macmillan.

Sterritt, G. M. and Smith, M. P. (1965). Reinforcement effects of specific components of feeding in young leghorn chicks. *Journal of Comparative & Physiological Psychology,* **59,** 171–175.

White, R. W. (1959). Motivation reconsidered: The concept of competence. *Psychological Review,* **66,** 297–333.

Wolfe, S. B. and Kaplon, M. D. (1941). Effect of amount of reward and consummative opportunity on learning in chickens. *Journal of Comparative Psychology,* **31,** 353–361.

Cognitive Explanations and Cognitive Ethology

RITA E. ANDERSON
*Department of Psychology, Memorial University of Newfoundland, St.
John's, Newfoundland, Canada A1B 3X9*

My growing belief that evolutionary and ecological considerations might be important to the understanding of human cognition, coupled with a long-standing personal interest in animal behavior led me to spend a sabbatical year (1983–1984) exploring the world of ethology. My rather vague goals were two-fold: to learn something about the art of evolutionary thinking and to find out what was known about how nonhuman animals code and act on the information in their various environments. What I soon discovered was that ethology itself is an interdisciplinary endeavor, crossing the boundaries of biology and psychology. Moreover, within this interdisciplinary mix, I discovered a multiplicity of subdisciplines and an emphasis on different levels of analysis. Thus, although "core" ethology, behavioral ecology, neuroethology, and comparative psychology are all concerned with animal behavior, each approaches the topic from a different perspective and at a different level of analysis. Because each area and perspective seemed to have something to offer, I found myself sampling from each, suffering from information overload, and almost losing my cognitive identity. Nonetheless, I persevered, and my comments today reflect the perspective of a cognitive psychologist who has gained some appreciation for the complexities involved in the study of animal behavior.

The modern day scientific study of human cognition has focused mainly on the characteristics of the mundane, nonconscious processes involved in the selection, encoding, organization, storage, and retrieval of information. People are assumed to be active and constructive participants in the processing of information, not passive recipients of environmental stimulation. Because active processing often depends upon prior knowledge, much of current cognitive research is focused on questions concerning how people structure and organize their knowledge of the world. Thus far, a number of theoretical knowledge structures, such as semantic networks, prototypes, schemas, and many others, have been proposed. In general, these knowledge structures represent our generic concepts of animte and inanimate objects, situations, events, and actions at various levels of detail and of abstraction.

Bechtel, W (ed), Integrating Scientific Disciplines. ISBN 90-247-3242-5.
© *1986, Martinus Nijhoff Publishers, Dordrecht. Printed in The Netherlands.*

They organize and represent our interpretations of our current experiences and guide our subsequent perceptions and actions. As such, they are not only crucial for comprehending our current reality, they provide the basis for our expectations and predictions.

Investigation into higher order cognitive problems, such as consciousness and will, has only recently begun (cf. Mandler, 1983; 1985); Norman and Shallice, 1980; Tulving, 1985). Although there is general agreement that consciousness is important, there is little agreement as to what it is or what function it serves in human cognition. In some accounts, consciousness is a sometimes effective executive of mental society, while in others, it is merely a public relations officer. In any case, current explanations rely heavily on our understanding of the knowledge structures and processes that support perception and action. For example, Mandler (1983, 1985) develops a constructivist account of consciousness. Conscious states consist of the combined information from a small number of currently relevant preconscious knowledge structures that have been activated in response to the current intentions and needs of the individual and the demands of the environment. In this framework, the schema is the building block of consciousness. Tulving (1985) links different states of consciousness to different types of memory systems. He relates self-knowing conscious states to personal remembering (the episodic memory system), introspective awareness of the world to impersonal knowing (the semantic memory system), and non-knowing reactivity to the world to adaptive action (the procedural memory system). Both Mandler and Tulving assume that our awareness is limited to the products, not the processes, of cognition. As such, although conscious states are important, they are only one aspect of our cognitive reality.

I was therefore surprised and somewhat dismayed to discover that much of the current cognitive ethology literature focused on the mental experiences of animals, raising questions about topics ranging from consciousness and self awareness to intentionality in action and communication (Griffin, 1981, 1984). How I wondered, as has Sober (1983), can anyone hope to understand the mental experiences of animals without tackling the more fundamental cognitive problems? Although Griffin's interest in the mental lives of nonhuman animals has been critical in rekindling interest in the animal mind and has raised many important issues, I fear his emphasis on conscious experience may have deflected interest from the types of cognitive events and processes (conscious or not) that are basic to the understanding of mental experiences.

Thus, I welcome the focus of the current papers. Although Rumbaugh and Sterritt address the question of the continuity of mental abilities, specifically the evolution of intelligence, they attempt to do so at a more tractable level. And, neither they nor Mason are concerned explicitly with the nature of conscious experience. Instead, both are interested in knowledge: how it is acquired, how it influences behavior, and how and when it is used for prediction and control. Questions about the types and structures of knowledge animals might have and how they might use it are amenable to scientific investigation given our current technology (see Dennett, 1983, and related commentary for potential additions to current approaches).

Although I assume that cognitive ethology is aligned more closely with animal behavior than with cognition, no one has satisfactorily explained how cognitive ethology relates to the study of animal behavior as a whole. While I agree with Mason that much of ethological research is ultimately about knowledge, it is not clear why this perspective should be of interest to other ethologists or scientists. In what follows, I will spend some time developing my vision of the role of cognitive ethology in animal behavior and will then attempt to deal with the question of the evolution of intelligence from that perspective.

As Mason so clearly stated, there is no objective fact, theory, or science. The questions we ask and the answers we receive are biased by our perspective. And so it is within biology and ethology. Mayr (1961) argued that most biological phenomenon require at least two different levels of explanations; those invoking ultimate (evolutionary) and those invoking proximate (individual) causation; Tinbergen (1963) further argued that a complete account of ethological phenomenon required at least four levels of explanation. Within his scheme, ultimate levels of analysis serve to explain behavior either on the basis of the phylogenetic history of the species or on the adaptive significance of the behavior in current ecological time. In contrast, proximate levels of explanation focus either on the developmental history of the individual or on the psychological, hormonal, and neural mechanisms involved in the control of the behavior of an individual. No single level of explanation is primary; each complements the other in the explanation of why and how a given behavior occurs. In practice, although most research programs focus only on one or two levels of analysis, research at one level is sensitive to what is known at the other levels. Thus, a neuroethologist assumes that the design features of the nervous system reflect the life style of the species and hence, ties his or her neurophysiological investigations to the ecological realities of the species.

Where do cognitive explanations fit into this scheme? Nowhere, for all intents and purposes. Although cognitive explanations may eventually be related to neurological mechanisms, they do not depend upon physical, mechanistic accounts. Obviously, cognitive explanations that are consistent with known neurological functioning are preferable to explanations that are neurologically implausible. Even so, cognitive explanations simply attempt to explain how mental states and knowledge are causally implicated in behavior, and to paraphrase Mason, have a legitamacy in their own right. The structures and processes discussed by most cognitive psychologists are usually abstract entities (e.g., schemas, activation processes) which are unlikely to correspond to specific neural mechanisms or processes. Hence, cognitive explanations are not mechanistic explanations in the true sense of the term.

To cope with a similar problem, Paul W. Sherman and Steven E. Glickmon (personal communication) proposed a fifth level of analysis, that of hypothetical constructs or the presumed cognitive and motivational processes involved in the production of behavior. The five levels of analysis in the Sherman/Glickmon expanded scheme are as follows: phylogeny, adaptive significance, hypothetical constructs, ontogeny, and mechanism. Mason's conclusion that the study of cognitive processes is a unifying theme in the endeavors of ethology and comparative psychology can be interpreted in this light to mean that a cognitive approach may form a pivotal link in the causal nexus. A cognitive level of analysis, incorporating motivational concerns, will focus on the types of knowledge an animal might need to carry out its life tasks and the cognitive structures and processes needed to explain how animals acquire, organize, and use that knowledge. The cognitive explanations offered by cognitive ethology should be sensitive to (and perhaps can help to mediate) the two axes of ultimate and proximate explanations. On the one hand, cognitive explanations should be responsive to the balance between the phylogenetic and ontogenetic influences on the ways of knowing. On the other hand, cognitive explanations should specific the types of knowing that mediate between the environmental pressures on the organism and the design features of the mechanisms that produce behavior.

A cognitive level of analysis requires two assumptions. First, to paraphrase Mason, we must assume that the world is orderly, structured, and intelligible to animals within the limits imposed by their sensory and motor capabilities. Second, we must assume that individuals act as agents with a unity of organization and purpose. Neither these assumptions nor the inference of knowledge structures or cognitive processes in animals are anthropocentric so long as we respect the world view of the animal. An animal's cognitive apparatus and ways of knowing evolved within the constraints of its world realities. If it is to survive, there must be some congruence between its knowledge of the world and the world itself. Furthermore, it must have the necessary cognitive apparatus to use this knowledge as required for survival. When environmental events are highly predictable in specific contexts over transgenerational time, knowing and behaving may take very fixed forms. In these cases, little is to be gained by extensive learning and much may be lost (e.g., one's life or progeny). Nonetheless, even easily canalized behaviors reflect responses to processed environmental input and hence, reflect behavior based on a form of knowledge. When salient environmental events are variable, flexibility of response may be beneficial. In this case, behaving and knowing may depend upon highly evolved cognitive processes which support the flexible acquisition of knowledge and allow for selective information processing in response to changing environmental conditions. The worlds of honey bees and dogs surely differ, as does their cognitive apparatus; nonetheless, both worlds are comprehensible to each species in its own terms.

To clarify the multiple ways that an animal acquires and acts on knowledge, cognitive ethology needs to consolidate the relevant behavioral and cognitive data according to species, rather than according to phenomenon. Thus, the "dumb" Herring Gull, mindlessly brooding a supranormal egg stimulus, may also be shown to have certain memory and assessment abilities which enable it to behave flexibly and intelligently in such matters as mate and nest site selection, foraging, and interactions with conspecifics. The behavior of most animals, human and nonhuman alike, will reflect the influence of both computationally-cheap and computationally-expensive ways of knowing [see Mayr (1974) for a related discussion on closed and open behavior programs]. Only when the full range of behaviors is assembled can we begin to distinguish between the relative influence of the different ways of knowing. As such, a lifetime cognitive profile for each species needs to be compiled which includes information about the characteristics of the elementary information processing capabilities available to the species, the categorical or functional structure of the species' knowledge of the world, and its biological predispositions.

Development of a cognitive profile necessarily begins in observation of the behaviors exhibited by the different species, and continues in experimentation. This is where anthropomorphism is likely to be a problem. Because I specialize in the study of human cognition, anthropomorphic thought has never been a problem for me, except when the occasional subject makes me question his or her human identity. Nonetheless, observation is subject to at least three classes of anthropocentric bias. First, as Mason noted, our initial observations of any behavior are necessarily anthropomorphic – we interpret behavior through the filter of our human experience and knowledge. Unless we have access to a genie that can put us into the head of our target animal so that we can experience and subsequently remember the world as that animal does, the interpretation of our observations will be colored by our human perspective. Since such genies do not yet exist, the only alternative is to develop a species-specific data base which includes information about all aspects of an animal's life: its sensory and motor capacities, its mating, parenting, and foraging activities, the nature of its conspecific interactions and communication skills, its developmental history, its interactions with the environment, and so on.

Second, observation is selective, subject to anthropocentric bias and interest. Unique, conspicuous, or species-typical behaviors have historically received more attention than the background behaviors which support daily life. For instance, the behavioral sequence described by Mason probably would have received less attention than a behavioral sequence involving intense competition between two conspecifics during the mating season. Yet, as is becoming apparent, day-to-day, routine behaviors may reveal as much about the nature of the information an animal abstracts from its environment as conspicuous displays. For example, although much research has been devoted to understanding the spectacular feats of bird migration and homing, we have only recently begun to appreciate the less spectacular, though equally important, spatial knowledge of other species. Chimpanzees, rats, birds, and even honey bees appear to have some form of spatial maps. I suspect that any organism which routinely forages away from a fixed home base will have some means of orienting in space, whether it be odor trails, as in the ant, or some higher order cognitive map. This may not be a novel insight to an ethologist, but it probably would be to the cognitive psychologist who has ignored the demands of daily life; after all, we have only recently begun to tackle the problem of spatial cognition in humans.

The third form of inadvertant anthropomorphism is to be found in experimentation. Nonhumans are no more likely to reveal the full range of their cognitive prowess in artificial, anthropocentrically defined tasks than humans are to excel in honhuman defined tasks. For instance, humans cannot "read" the information at scent posts anywhere nearly as well as dogs can. In fact, the discovery of many of the basic principles of human cognition was delayed for over half a century by the use of artificial stimuli, such as nonsense syllables. True progress in understanding the processes involved in the acquisition and organization of human knowledge began only when attention turned to the processing of meaningful materials, such as words, stories, or coherent scenes. By ignoring the practicalities of daily life, researchers failed to appreciate the importance of spatial cognition, as mentioned above, or, as another example, the control process involved in multitask situations, such as driving a car. Today, an increased concern with ecological validity (cf. Neisser, 1976) is leading to a more coherent understanding of human cognition.

A parallel development can be seen in the study of animal cognition. For example, current research on food hoarding birds has adopted an ecological approach to experimentation (Shettleworth, 1983). Both marsh titmice and Clark's nutcrackers regularly cache food supplies in widely dispersed locations and may not retrieve it for days or even months. Experiments in the field and in laboratory analogues of field situations have shown that these birds have a remarkably accurate, large capacity, long duration memory for the location of cached food. Yet, artificial tasks which might be thought to tap a similar memory system suggest that pigeons can barely remember the location of a target light for a few seconds. Although this comparison is confounded by species and task differences, it is quite possible that the poor performance by the pigeon reflects some aspect of the impoverished cues offered in the laboratory task. Lacking full insight into the cognitive realities of nonhumans, we might be well advised to focus initially on the knowledge and elementary cognitive processes used by an animal as it goes about its life. First, let the animal define the task. Then gradually strip the task of its complexities so as to determine the limits of the inferred cognitive processes.

The foundation for a cognitive profile consists of the molar behaviors an animal performs in the course of daily life, as viewed from a cognitive perspective. By observing how the animal samples its environment and noting the distinctions it makes, we can begin to determine something about the perceived correlational structure of the animal's world. For instance, the types of distinctions an animal makes between different prey and predator species, conspecifics, and environmental stimuli and events may suggest how the animal lumps and splits its world (von Uexkull, 1957; Marler, 1982). The types of knowledge and the cognitive processes that operate on that knowledge may be suggested by the degree of flexibility and choice an animal demonstrates in a variety of life problems. In principle, the behavior patterns across a number of life problems should provide converging evidence to suggest the extent of an animal's knowledge and cognitive capabilities. Does the animal only reveal evidence of spatial knowledge of its environment or can it also recognize individuals and their states, independent of the spatial context? Are most of its exploratory and foraging activities dependent upon taxes and kineses, or does it demonstrate flexibility and choice in its activities? The overall pattern should reveal the extent to which its cognitive functions reflect the operation of relatively inflexible control mechanisms, specialized cognitive adaptations which allow for behavioral flexibility in specified domains, or more generalized cognitive adaptations. By knowing how these abilities are used by an animal in solving its life problems, we are less likely to fall into the anthropomorphic trap of assuming that an animal's knowledge and cognitive apparatus function in the same way as a human's. A full cognitive profile, of course, would go beyond a behavioral profile to examine the characteristics of the structure of knowledge available to the animal and of the proposed cognitive abilities in detail.

By anchoring itself firmly within ethology as a whole, with a clear identity and focus, cognitive ethology has the potential to complement, amplify, and constrain research at the other levels of analysis. By providing information about the knowledge animals have and how they use it, a cognitive analysis provides a necessary link between mechanism and behavior. By providing a different perspective on behavior, it may raise new questions for the other levels of analysis to tackle, as for example, the work on the ontogeny of alarm calls in vervet monkeys (Marler, 1982). Finally, the information gained by a cognitive analysis may constrain the types of models offered to account for behaviors, especially at the functional level. For example, many current models in behavioral ecology assume that animals have considerable knowledge about their environments and the ability to make sophisticated decisions and cost/benefit analyses. The types of models that might be considered would be nicely constrained by an understanding of the knowledge and cognitive apparatus that are available to an animal. Significant beginnings in the study of foraging behavior are already evident (Kamil and Sargent, 1981). And, of course, a focus on the cognitive lives of animals should facilitate the search for the roots of intelligence, the next topic of discussion.

Unfortunately, cognitive psyhologists have generally ignored both the question of individual differences and of intelligence, preferring instead to focus on the normative properties of the human information processing system. Over the past ten years, however, interest in both questions has increased. In line with the perspective outlined by Rumbaugh and Sterritt, intelligence is not viewed as a unitary concept, nor should it be viewed only as a "thing-in-the-head". If anything, human intelligence appears to be an emergent property of a constellation of highly coordinated and finely-tuned component processes and knowledge structures that operate within a given person-environment context. To the extent that any component functions less than optimally in a given context, the apparent level of intelligence decreases, following the principle of graceful degradation (Norman and Bobrow, 1975). That is, performance (intelligence) does not crash; it simply deteriorates gradually. Note that this cognitive principle says nothing about the underlying neural mechanism. For instance, some form of behavioral compensation may mask the complete failure of a neurological processor.

Within the cognitive tradition, two main lines of research on intelligence have developed in recent times. Differences in task performance have been related either (i) to differences in the operation of automatic information processes or (ii) to differences in the amount and organization of task-specific knowledge. The goal of the cognitive-correlates approach is to identify the cognitive components that best account for the differences observed in the performance of complex skills. For example, Hunt (1978) found that the speed of activation of highly overlearned information in long-term memory and the functioning of active memory both correlate fairly well with measured verbal ability (SAT scores) over a wide range. In contrast, the goal of the knowledge-based approach is to determine how differences in the amount and organization of domain-specific knowledge affects task performance. As Chi, Glaser, and Rees (1982) argue, a major component of intelligent behavior in a particular domain is knowledge. They have found substantial differences between experts and novices not only in the amount of task-specific knowledge at their disposal, but in how that information is represented and structured in memory. These differences profoundly affect the way experts and novices approach and solve problems.

In this critical reiew of current cognitive research on human intelligence, Keating (1984) argued that if more emphasis were placed on the context (e.g., goals, activation, culture) in which intellectual functioning occurs, many of the apparent inconsistencies in research on human intelligence would be eliminated. This argument appears equally appropriate to research on humans and nonhumans alike. As Hodos (1982) noted in his comments on the nature of animal intelligence,

> ... behavior is seen as intelligent or not according to the context in which it occurs. The same pattern of responses that in one situation is regarded as highly intelligent and adaptive might be regarded as very unintelligent and self-defeating in another situation. (p. 34).

The admonition to pay attention to the context in which behavior occurs is reminiscent of the call for greater ecological validity of experimentation. As Rumbaugh and Sterritt argue, cats, and presumably members of a number of other species, will act when it suits their purposes. Their purposes will be a function of the fit between their perceived needs and desires and their perceptions of the current environmental opportunities and demands. In other words, the context in which behavior occurs is vitally important to the interpretation of that behavior.

What about the evolution of intelligence? I have only begun to appreciate the extent of the difficulties associated with comparative research. Since we haven't mastered the art of comparing human intelligence or even the component cognitive processes across ages, races, or cultures, I shudder to think of the barriers to cross-species comparisons. Yet, I agree with Mason and with Rumbaugh and Sterritt that questions about the origins of mental abilities should be addressed. As might be gathered from my previous comments, my preference would be to trace the development of basic ways of knowing across phylogenetic lines and habitats. For example, one might attempt to trace the evolution of "memory" from largely innate ways of knowing, to memory systems specialized for certain types of information, to general purpose memory systems. The memory systems of insects, birds, and mammals surely differ in important respects. Within taxa, the memory systems of

332

different species are likely to reflect the demands of their ecological niche and social organization. Thus, although the knowledge and memory systems of birds must reflect something about life as a bird in general, there may be considerable differences in sophistication across bird species. For example, the complex social structure and cooperative breeding and foraging activities of the white-fronted bee-eaters of Kenya (Hegner, Emlen, and Demong, 1982) suggests that their knowledge structures and, perhaps, their memory system may differ considerably from other less social species.

Causal inference is dependent upon memory and comparison abilities. As such, tests of causal inference share the interpretational difficulties of traditional tests of intelligence. If an animal performs poorly, you don't know why. Is it because the animal has the requisite cognitive abilities, but cannot make causal inferences or because the requisite abilities simply don't exist? Consider one of the examples alluded to by Rumbaugh and Sterritt. In order for a rat (or honey bee) to react to diminished rewards or the failure of the present to live up to expectations from the past, the rat must remember the past event. Furthermore, that memory must either be actively retrieved or be reinstated somehow by the current situation. Finally, the rat must be able to compare the salient properties of the memory event to the current situation and be able to detect a difference between the two events. Failure to react to diminished rewards could reflect several possibilities. Knowledge of the past event may not exist. Or it may exist, but not be accessible. This could occur if the knowledge was encoded as part of a larger event, such that the relevant knowledge could not be abstracted and matched to the current situation. Alternatively, the retrieval or reactivation processes may fail. Even if the knowledge is accessible, inference might fail because the ability to compare past knowledge with present realities or the ability to detect a difference is lacking. Finally, the task may be outside of the animal's domain of expertise. If its cognitive apparatus is specialized for the processing of certain types of information, it may be able to do causal inference in some, but not other tasks.

In spite of these misgivings, I find the idea of Control Theory and causal inference appealing. Simply being able to remember is not the same as being able to use that information for purposes of prediction and control. As such, the pattern of causal inference abilities across phylogenetic lines, habitats, and social structures has the potential of revealing something about the evolution of mental abilities. Cognitive profiles might be useful for determining which species tasks are likely targets for which tests of causal inference. In addition, some leverage may be gained by adopting the cognitive-correlates approach to individual differences on causal inference tasks within species. Is there a relationship between the extent of causal inference abilities and some aspect of the operation of the memory system?

Although Rumbaugh and Sterritt's distinction between biologically smart behaviors (e.g., those which allow for migration, nesting, etc.) and cognitively smart behaviors (e.g., those which allow for causal inference) seemed intuitively obvious at first, I became confused when I tried to pin it down. At the extreme, if the criterion for a biologically smart behavior is the biological preparedness or predisposition for a given behavior, human language would have to be classified as a biologically smart behavior! Because even responses to "releasing stimuli" can be modified by experience, the potential for learning cannot be the sole criterion for cognitively smart behaviors. Since many behaviors reflect a combination of cognitive and not so cognitive influences, I suspect that they wish to distinguish between different ways of knowing, rather than different ways of behaving. For instance, successful bird migration appears to depend upon a number of factors. Some factors, such as the conjunction of appropriate hormonal state and light-dark cycle that set the stage for the triggering of migration, are relatively immune to cognitive control. Other factors, such as those that facilitate the return to the same locale year after year, appear to depend upon the use of acquired knowledge. Since familiarity with the local environment confers a number of advantages to residents, returning to the same locale may not only make for a biologically successful bird, but may also affect the quality of life.

I am not sufficiently familiar with motivational accounts to know how Control Theory fits with past models. However, it is clear that cognitive variables must be related to motivational concerns and Control Theory appears to be one step in the right direction. In line with my previous comments, I feel the need to note that the cognitive induction of a nonquiescent state is dependent upon more fundamental cognitive processes. Knowledge of the temporal and spatial regularities of the salient objects and events in an animal's life and an ability to detect a discrepancy between that knowledge and external reality are likely to be prerequisits for the development of cognitively-based anxiety. To the extent that an animal's cognitive apparatus is tied to specific domains, cognitively-based anxiety may develop only within those restricted domains. I also have difficulty making a clear separation between homeostatic and cognitive drives. If I am hungry, my perceptions and cognitions often reflect that biological state. And knowledge may affect biological states, as suggested by the social control of reproduction observed in a number of species (see Snowdon, 1983, for a brief review).

The major difficulty with Control Theory, as applied to nonhumans, is that it appears to be a run-away model of the evolution of intelligence. There are no brakes on the system. Once an animal acquires the necessary cognitive apparatus to move beyond biologically-driven homeostasis, it should rapidly become addicted to prediction and control and all that entails. Perhaps that is why animals seem to be

334

getting smarter with every year! More seriously, the costs, as well as the benefits, of inference and the role of knowledge need to be considered, as outlined in the following example. Experienced song sparrow females are more agitated by the presence of a female brown-headed cowbird mount near the nest than naive yearling females; the experienced females are also more likely to be parasitized (Smith, Arcese, and McLean, 1984). These data could be interpreted to suggest that experienced females can make the causal inference between the presence of cowbirds and being parasitized. However, this knowledge backfires in that their agitation provides the cowbird with information regarding the location of the nest sites of experienced, high quality female hosts. Song sparrows may not have evolved either the ability to perform the next level of causal inference (i.e., to infer the consequences of their agitation) or other cognitive abilities needed to act appropriately on their knowledge. Alternatively, they may have the abilities, but simply don't live long enough to acquire all the relevant information. Given the seasonal nature of the cowbird problem, they may need one breeding season to learn about cowbirds and another to learn about the consequences of their agitation. By the third season, they may not be around to use this knowledge. Regardless of the interpretation of these data, song sparrows appear "guilty" of knowing too much for their own good. Although the ability to draw causal inferences may lead to rapid evolution, it will only do so when knowledge acquisition and other cognitive capabilities develop accordingly.

The excesses of the introspective approach toward human mentality and an uncritical anthropomorphic stance toward nonhuman mentality in the early 1900s led to the rise of radical behaviorism and the virtual elimination of the scientific study of mental processes. By the 1960s, it became clear to many psychologists that a full account of human behavior had to include the mind, and thus, cognitive psychology was born. At that time, the task of cognitive psychology appeared simple: to flesh out the operating characteristics of the information processing systems involved in human perception, attention, memory, language, reasoning, and problem solving. Initially, it appeared that these systems could be studied in the abstract, in isolation from the contexts in which they normally occurred and the functions that they normally served. These beliefs have been challenged and today a number of cognitive psychologists are beginning to consider the ramifications for human cognitive theory of such issues as the evolutionary history of humans (Lachman and Lachman, 1979), the ecological validity or invalidity of our experimental procedures (Neisser, 1976), and the fact that survival, not cognition, is our task in life (Norman, 1981). It is as though we are overcoming our aversion to "animal" research that grew out of the revolution from the tyranny of radical behaviorism and are reclaiming our place in the natural order of the world.

Regardless of the motivation, the interests of cognitive psychology and of ethology are beginning to overlap. As such, I have great hopes for a viable cognitive ethology, one that will enrich the understanding of human and nonhuman cognition alike.

Acknowledgments

I wish to thank the members of the Department of Psychology, Cornell University, for hosting me while I was on sabbatical leave during the 1983–1984 academic year, and the members of the Neurobiology and Behavior Section, especially Elizabeth Adkins-Regan Steven Emlen, Bob Johnston, and Paul Sherman for providing many opportunities to learn about the study of animal behavior. I am especially indebted to Bob Johnston for many fruitful discussions on a wide range of topics and his critical reading of the draft of this paper that was presented at the conference. I am also grateful to Bill Bechtel, Carolyn Harley, Bill Montevecchi, and Anne Storey for providing a range of perspectives and associated commentary on the final versions of this paper. All shortcomings of the present paper should be attributed to my failure to fully attend to the excellant tutelage I received at Cornell and at the hands of my colleagues at Memorial.

References

Chi, M. T. H.; Glaser, R.; and Rees, E. (1982). Expertise in problem solving. In R. J. Sternberg (Ed.), *Advance in the psychology of human intelligence, Vol. 1* (pp. 7–75). Hillsdale, NJ: Erlbaum.

Dennett, D. C. (1983). Intentional systems in cognitive ethology: The "Panglossian paradigm" defended. *The Behavioral and Brain Sciences, 6,* 343–390.

Griffin, D. R. (1981). *The question of animal awareness,* (2nd ed.). New York: Rockefeller University Press.

Griffin, D. R. (1984). *Animal thinking.* Cambridge, Mass.: Harvard University Press.

Hegner, R. E.; Emlen, S. T.; and Demong, N. J. (1982). Spatial organization of the white-fronted bee-eater. *Nature, 298,* 264–266.

Hodos, W. (1982). Some perspectives on the evolution of intelligence and the brain. In D. R. Griffin (Ed.). *Animal mind – human mind* (pp. 33–56). New York: Springer-Verlag.

Hunt, E. (1978). Mechanisms of verbal ability. *Psychological Review, 85,* 109–130.

Kamil, A. C. and Sargent, T. D. (eds.). (1981). *Foraging behavior: Ecological, ethological, and psychological approaches.* New York: Garland Press.

Keating, D. P. (1984). The emperor's new clothes: The "new look" in intelligence research. In R. J. Sternberg (Ed.), *Advances in the psychology of human intelligence, Vol. 2* (pp. 1–45). Hillsdale, NJ: Erlbaum.

Lachman, J. L. and Lachman, R. L. (1979). Theories of memory organization and human evolution. In C. R. Puff (ed.), *Memory organization and structure* (pp. 133–193). New York: Academic Press.

Mandler, G. (1983, June). Consciousness: Its function and construction. CHIP Technical Report No. 117, Center for Human Information Processing, University of California, San Diego.

Mandler, G. (1985). *Cognitive psychology: An essay in cognitive science.* Hillsdale, NJ: Erlbaum.

Marler, P. R. (1982). Avian and primate communication: The problem of natural categories. *Neuroscience & Behavioral Reviews, 6,* 87–94.

Mayr, E. (1961). Cause and effect in biology. *Science, 134,* 1501–1506.

Mayr, E. (1974). Behavior programs and evolutionary strategies. *American Scientist, 62,* 650–659.

Neisser, U. (1976). *Cognition and reality: Principles and implications of cognitive psychology.* San Francisco: Freeman.

Norman, D. A. (1981). Twelve issues for cognitive science. In D. A. Norman (Ed.), *Perspectives on cognitive science,* (pp. 265–295). Hillsdale, NJ: Erlbaum.

Norman, D. A. and Bobrow, D. G. (1975). On data-limited and resource-limited processes. *Cognitive Psychology, 7,* 44–64.

Norman, D. A. and Shallice, T. (1980, December). Attention to action: Willed and automatic control of behavior. CHIP Technical Report No. 99, Center for Human Information Processing, University of California, San Diego.

Shettleworth, S. J. (1983). Memory in food-hoarding birds. *Scientific American, 243,* 102–110.

Smith, J. N. M.; Arcese, P.; and McLean, I. G. (1984). Age, experience, and enemy recognition by wild song sparrows. *Behavioral Ecology and Sociobiology, 1984, 14,* 101–106.

Snowden, C. T. (1983). Ethology, comparative psychology, and animal behavior. *Annual Review of Psychology, 34,* 63–94.

Sober, E. (1983). Mentalism and behaviorism in comparative psychology. In D. W. Rajecki (Ed.). *Comparing Behavior: Studying Man Studying Animals.* Hillsdale, NJ: Erlbaum.

Tinbergen, N. (1963). On aims and methods of ethology. *Zeitschrift fur Tierpsychologie, 20,* 410–429.

Tulving, E. (1985). How many memory systems are there? *American Psychologist, 40,* 385–398.

vonUexkull, J. (1957). A stroll through the world of animals and man (1934). In C. H. Schiller (Ed.), *Instinctive behavior: The development of a modern concept.* NY: International Universities Press.

Editor's Commentary

Compared to the other cross-disciplinary endeavors within the life sciences that have been considered in earlier parts of this volume, cognitive animal ethology is at a very early stage of development. Ethology has a long history within this century but one that has been marked, as noted by Wimsatt's paper earlier in this volume, by a conflict between the European tradition, whose allegiances were more with evolutionary theory, and an American tradition, strongly influenced by behavioral psychology. As this division suggests, the roots of ethology lie both within biology and psychology so that, as Anderson notes in the introduction to her paper, ethology itself is an interdisciplinary endeavor, involving numerous perspectives and levels of analysis. The introduction of a cognitive framework into ethology thus has the appearance of an expansion of an already existing interdisciplinary research cluster. This might suggest it will be comparable to the cases, discussed in the two previous parts of the volume, of possible extensions to the evolutionary synthesis and to cognitive science. However, the effect of bringing in a cognitive framework to this interdisciplinary research cluster may be quite different than the introduction of additional disciplines to those existing clusters. As the papers in this part indicate, the introduction of a cognitive perspective may serve to unify those disciplines already incorporated into animal ethology, not to add another discipline.

In particular, the introduction of a cognitive perspective may serve to reconcile the opposition between the nativistic emphasis of the European tradition and the behaviorist emphasis of the American tradition. There is something of an irony in this, since the theoretical frameworks that have been developed for human cognitive psychology have, until recently, had a strongly nativistic element. However, as I discussed in my commentary to the unit on extending cognitive science, the gap between nativist approaches in cognitive psychology and empiricist approaches in learning theory currently seems to be shrinking. As the results of studies done on nonhuman animals develop, they may suggest new perspectives that can be applied to the study of human cognition that will further reduce the opposition between nativism and empiricism.

Bechtel, W (ed), Integrating Scientific Disciplines. ISBN 90-247-3242-5.
© *1986, Martinus Nijhoff Publishers, Dordrecht. Printed in The Netherlands.*

For the most part, cognitive animal ethology consists in introducing a perspective (and, to a lesser degree, explanatory models) developed for describing and explaining human cognition to the pursuits already followed in animal ethology. Thus, a major force leading toward the development of cognitive animal ethology has been the success of the "cognitive revolution" in the study of human behavior. Many of the topics investigators are now exploring in animal cognition, such as representation, memory, perception, and reasoning, are direct cognates of topics studies in human cognitive psychology, and some of the explanatory concepts and models show a similar origin. However, what has actually been carried over from the one domain to the other has been less than one might initially have expected. At present it is mostly the language and concepts of human cognition that are being extended and not the research techniques. The reason has to do with the significant species differences between humans and other animals. Many of the techniques for studying cognition in the human realm depend on verbal reports which, with the possible exception of a few primates, will never be available for studies of other animals.

Before looking at the specific proposals developed in the preceding papers, it is worth considering why a cognitive framework is only being vigorously developed in animal ethology. There are a variety of reasons for this. The first involves the long domination of a behaviorist methodology throughout psychology, especially radical behaviorism which prohibited the use of any mentalistic terms in developing psychological explanations. Behaviorism held as firm a grasp on animal psychology as on human psychology. The second reason has to do with the particular factors that led away from radical behaviorism and towards more cognitive approaches in human psychology: the development of transformational models in Chomskian psycholinguistics and of artificial intelligence models. These had less impact on animal psychology. The reason psycholinguistics had little impact on animal studies has already been alluded to. As for artificial intelligence, computer simulation of animal cognition is, in principle, as reasonable if not more so than simulation of human cognition. However, much of the interest in artificial intelligence was to develop programs to simulate humans or perform tasks better than humans; hence, little attention was paid to simulating animals (but see Boden, 1983). A third reason has to do with the fear of anthropomorphizing. As Anderson notes, the fear of anthropomorphism has not been a concern in human psychology (although one of the arguments for behaviorism – the claim that mental language is too much based on our pre-scientific view of human agents and so is not explanatory – seems to be comparable to a charge of anthropomorphic to mentalistic explanation in the human domain).

Mason's paper is, in large part, directed to the concern about how to use cognitive language in discussing animal behavior without being merely anthropomorphic. Mason rejects the tradition of radical behaviorism that repudiated the use of mentalistic language on the grounds that we cannot even describe the behavior of animals unless we attribute needs, perceptions, memories, intentions, and so forth. To distinguish the phenomena to be explained, actions, from mere physical motions, he claims we must construe them as based on information and need. He presents such mentalistic characterizations as providing the unifying theme to comparative psychology and ethology. However, Mason also wants to carefully limit this use of mentalistic discourse and not to assume that it has "substantial explanatory value."

What Mason seems concerned to avoid is treating mentalistic terms as referring to modularized operations in the mind (see the earlier paper by Richardson for some discussion of the role of modularity assumptions in current endeavors at cognitive modelling). Rather, he wants to maintain a focus on the behavior of the whole integrated organism. In many respects this approach of Mason parallels Dennett's (1983) use of the intentional stance. Dennett explicitly presents his intentional characterizations of animal behavior as instrumentalistic, denying that they describe processes occurring within the organism. (I have argued that Dennett should not endorse instrumentalism. I have argued that he can still maintain that intentional attributions apply to the whole system but yet treat them as real characterizations of the way the system interacts with its environment. (See Bechtel, in press.) Yet Dennett takes a further step into an area where it is unclear that Mason would follow. Dennett allows for a design stance account which explores the internal processing of the system in functional, not neurological terms. It is at this design level that most modelling in human cognitive psychology is described. Such modelling postulates procedures through which information is processed.

It is in making the move to functional characterizations inside the system that Anderson seems to differ with Mason. Insofar as they can provide an ecologically valid account of the knowledge organisms have and the procedures they have available for processing that knowledge, they view functional cognitive accounts as providing an intermediate account between that which focuses on the behavior of the organism and that which specifies the physiological mechanisms generating that behavior. She construes an abstract account of those mechanisms as providing a useful guide in the search for those mechanisms. As well, she suggests that such cognitive accounts can provide a useful constraint on the development of ecological and evolutionary scenarios that describe the organism's relationship to its environment. Thus, Anderson sees the cognitive framework, when moved into animal studies, as providing the same kind of intermediary account that some practitioners present it as playing in the human domain.

In her paper, moreover, Anderson suggests a further way of dealing with the dangers of anthropomorphism. In developing a cognitive account of an organism, she proposes that researchers should not try to interpret its behavior in human terms, but in terms of the cognitive demands placed upon it by its environmental. For this, she proposes developing a cognitive profile for each species that would capture the cognitive requirements imposed on the animal by its environment as well as its information processing capabilities. One might well object that we do not have independent access either to the internal processing capacities of a species or to the cognitive demands imposed by its environment. This objection, however, misses the point of Anderson's proposal which is that, although we do not have independent access to a full account on either dimension, we can use what we can ascertain from field and experimental studies to mutually constrain accounts on both dimensions, thus bootstrapping our way to a more adequate theory.

One thing that should be noted about Anderson's approach is that she is not just proposing a cross-disciplinary move to employ cognitive models in animal studies, but a multi-disciplinary conception of cognitive animal ethology itself. That is, she envisions several approaches to theorizing (an ethological approach, a cognitive approach, a neurological approach, as well as two others that cut across this hierarchy – an ontogenetic approach and a phylogenetic approach) that each serve to constrain the others. This multi-faceted conception of cognitive animal ethology may partly reflect the motivation that Anderson presents as leading her to turn to this field. She turned to animal ethology from her work on adult human cognition as a result of deciding that something was lacking in the study of human cognition, namely, consideration of evolutionary and ecological factors. One way to gain appreciation of evolutionary and ecological factors is to acquire a comparative framework, and so she turned to animal ethology. Since her interest was to integrate evolutionary and ecological factors with the study of human cognition, it is natural that she conceives of cognitive animal ethology as an integrative exercise itself.

The direction of Rumbaugh and Sterritt's interest parallels Anderson's in the sense that they, too, are interested in issues in cognitive animal ethology in part because of the light it can potentially cast on human studies. The exploration of this question, however, initially seems to be blocked by an incongruity between theoretical approaches that have been adopted within the human and animal domains. In the human domain, intelligence has been introduced as a property to explain individual differences in ability to succeed in a variety of tasks. This notion of "intelligence," though, is anthropocentric, insofar as tests of intelligence tend to test particular human cognitive skills and, in particular, to require the use of language.

To overcome this anthropocentrism in the concept of intelligence, Rumbaugh and

Sterritt have been led to develop an alternative perspective on intelligence through their notion of Control Theory. The basic idea is that intelligence involves recognition of cause-effect relationships that allow an organism to control its environment. At first it might not seem that the notion of control provides a way out of anthropomorphism, in that we impute knowledge of causal relationships to animals on the basis of how we recognize them. But Rumbaugh and Sterritt are aware of that problem. They are concerned to differentiate intelligence (control through perception of causal relations) from biological smartness (genetically controlled behavior under stimulus control). (See Anderson, though, for some cautions about this distinction.) They also suggest a way of applying the distinction. One tests whether animals prefer an environment where they enjoy control to one where they lack it. If they differentiate the environments and choose one over the other, that is evidence that control is important *to them*. Although they note that organisms may choose the particular context where they want to exercise control, and not express the desire for control in all aspects of their existence, Rumbaugh and Sterritt interpret the contrafreeloading data as evidence for the claim that the animals showing this behavior prefer to exercise control. If Rumbaugh and Sterritt's Control Theory succeeds, it provides a way for us to discover the point in phylogeny where a preference for control begins, and in what domains animals are able to perceive and use causal relationships.

There is, however, a further twist to Rumbaugh and Sterritt's control theory, for they view the desire to control one's environment not simply as something that evolves, but as a force for further evolution. Insofar as an organism is questing for control, it will be dissatisfied when its attempts at controlling its environment are thwarted. Hence, the organism will have a motivation to develop more adequate representations of the causal connections in its environment. What control theory projects is not quite a search for the truth, but something quite similar, a quest for representations of causal connections that are not confuted by the environment. Moreover, it projects this quest as rooted deep in our phylogenetic tree.

What Rumbaugh's cross-disciplinary theorizing has done is to address a question concerning the evolution of intelligence, which has its roots in human psychology and human cognition, through a study in a related discipline that explores animal behavior. In the course of that theorizing, though, they are forced to reshape that question so that it can be addressed in that domain. There is another feature of their interfield theorizing, though. If the reformulated question is taken seriously, it will have a transforming impact on animal psychology, for it invites reversing the behaviorist approach of trying to bring animal responses under stimulus control through varying contingencies. Instead, it directs psychologists to investigate how animals bring contingencies of the environment under their control as they acquire

342

knowledge about the cause-effect relations in their environment. This potential of transporting a concept across disciplinary boundaries and resulting in the transformation of another discipline has been illustrated several times in this volume.

As Anderson notes, the approach Rumbaugh and Sterritt take to intelligence is quite different from that which has appeared in recent cognitive literature. What the information processing tradition suggests is that there are different ways in which an organism can fail to exercise control over its environment. It may, for example, lack appropriate knowledge or not have the knowledge in an accessible form, or it may not have the cognitive procedures for applying that knowledge to current circumstances. What Anderson's comments indicate is a way of further developing the phylogenetic study of intelligence by considering how these different features themselves arise. This development might turn out to be informative both as a framework in ethology and for gaining a further perspective on human cognition.

The three papers thus offer a variety of perspectives on the way in which cognition may fit into animal ethology. As I suggested at the outset, for the most part of the introduction of a cognitive perspective, in any of the three forms it is proposed here, will serve primarily to unify the interdisciplinary endeavor within animal ethology, and not build a new interdisciplinary complex. While there may be implications from this research for human psychology, it seems unlikely that major new bridges between the two are in the offering. Most of the theoretical proposals made in these papers are not of the order to result in a new interfield theory. Rather, if pursued they will serve either to characterize the phenomena constituting the domain of the interfield research cluster in animal ethology (Mason) and perhaps provide a framework for integrating the various activities already occurring in the cluster (Anderson). Rumbaugh and Sterritt's perspective on intelligence, especially if elaborated in the ways Anderson suggests, has more potential for building a theoretical framework linking human psychology and animal ethology through a phylogenetic perspective. That would, however, require more interaction between those interested in human psychology and those working in animal ethology than appears to be happening at the moment.

Regardless of whether more connections are drawn between human psychology and the domain of animal ethology, the cognitive perspective seems to be gaining rapid attention. Although to my knowledge there are no specific academic units for the study of cognitive animal ethology (which would not be expected in light of the fact that animal ethology is already a well-developed area) or established professional organizations, there have been at least three major conferences whose proceedings have been published (Hulse, Fowler, and Honig, 1978; Griffin, 1983; and Roitblatt, Bever, and Terrace, 1984). Additionally, several articles and commentaries in *The Behavioral and Brain Sciences* have focused on animal cognition and cognitive ethology (most recently, Dennett, 1983, and accompanying commentaries). The fruits of these continuing endeavors will undoubtedly provide further material for analysis by those interested in how conceptual frameworks borrowed from one domain are altered in the course of deployment in a closely related domain.

References

Bechtel, William (in press). Realism, instrumentalism, and the design stance. *Cognitive Science.*

Boden, Margaret A. (1983). Artificial intelligence and animal psychology. *New Ideas in Psychology,* **1,** 11–33.

Dennett, Daniel C. (1983). Intentional systems in cognitive ethology: The "Panglossian paradigm defended. *The Behavioral and Brain Sciences,* **6,** 343–390.

Griffin, Donald R. (1983). *Animal mind-human mind.* Berlin: Springer-Verlag.

Hulse, S. H.; Fowler, H.; and Honig, W. K. (eds.) (1978). *Cognitive processes in animal behavior.* Hillsdale, NJ: Erlbaum.

Roitblatt, H. L.; Bever, T. G.; and Terrace, H. S. (1983). *Animal cognition.* Hillsdale, NJ: Erlbaum.

Index of Names

Subject Index